D1161323

Electron Spectroscopy of Crystals

PHYSICS OF SOLIDS AND LIQUIDS

SUPERIONIC CONDUCTORS
Edited by Gerald D. Mahan and Walter L. Roth

HIGHLY CONDUCTING ONE-DIMENSIONAL SOLIDS
Edited by Jozef T. Devreese, Roger P. Evrard, and Victor E. van Doren

ELECTRON SPECTROSCOPY OF CRYSTALS
V. V. Nemoshkalenko and V. G. Aleshin

Electron Spectroscopy of Crystals

V. V. Nemoshkalenko *and*
V. G. Aleshin
Ukrainian Academy of Sciences
USSR

Translated from Russian by
Irina Curelaru
Chalmers University of Technology
Göteborg, Sweden

Translation Editors
Stig Lundqvist *and*
Per-Olof Nilsson
Chalmers University of Technology
Göteborg, Sweden

PLENUM PRESS · NEW YORK AND LONDON

Library of Congress Cataloging in Publication Data

Nemoshkalenko, Vladimir Vladimirovich.
Electron spectroscopy of crystals.

(Physics of solids and liquids)
Translation of Elektronnaia spektroskopiia kristallov.
Includes index.
1. Electron spectroscopy. 2. Crystals–Spectra. I. Aleshin, Valentin Grigor'evich, joint author. II. Title.
QC454.E4N4513 548'.8 78-12954
ISBN 0-306-40109-6

The original Russian text, published by Naukova Dumka Press in Kiev in 1976, has been corrected by the authors for the present edition. This translation is published under an agreement with the Copyright Agency of the USSR (VAAP).

ЭЛЕКТРОННАЯ СПЕКТРОСКОПИЯ КРИСТАЛЛОВ
В. В. Немошкаленко, В. Г. Алешин

ELEKTRONNAYA SPEKTROSKOPIYA KRISTALLOV
V. V. Nemoshkalenko, V. G. Aleshin

Printed in the United States of America

Foreword

This book is conceived as a monograph, and represents an up-to-date collection of information concerning the use of the method of X-ray photoelectron spectroscopy in the study of the electron structure of crystals, as well as a personal interpretation of the subject by the authors.

In a natural way, the book starts in Chapter 1 with a recapitulation of the fundamentals of the method, basic relations, principles of operation, and a comparative presentation of the characteristics and performances of the most commonly used ESCA instruments (from the classical ones—Varian, McPherson, Hewlett Packard, and IEEE—up to the latest model developed by Professor Siegbahn in Uppsala), and continues with a discussion of some of the difficult problems the experimentalist must face such as calibration of spectra, preparation of samples, and evaluation of the escape depth of electrons.

The second chapter is devoted to the theory of photoemission from crystalline solids. A discussion of the methods of Hartree–Fock and Hartree–Fock–Slater for the calculation of bonding energy levels in multielectronic systems is presented, and the necessity of including in the theory both relativistic and relaxation effects is argued. A review of the methods of calculation of energy bands and wave functions in crystals (the OPW and APW methods, the method of Green's functions, the method of the pseudopotential) serves as a basis for the comparison of experimental data with the theoretical calculations, which follow in the next chapters. The increased interest, in recent years, in the study of disordered systems (the alloys of transition and noble metals) justifies the inclusion of the coherent-potential method in the same chapter. Further, the information given by the study of angular distribution of photoelectrons is summarized.

Chapter 2 ends with a discussion of the possibilities, limitations, and perspectives of the present efforts to correlate the results of photoelectron spectroscopy and X-ray emission spectroscopy to the data obtained by using other experimental methods such as Mössbauer spectroscopy and nuclear magnetic resonance.

The next four chapters discuss the presently available experimental data in the study of the electron and X-ray spectroscopy of crystalline elements, alloys, and compounds, grouped as follows:

Chapter 3: Metals and alloys
Chapter 4: Sphalerite-type crystals
Chapter 5: Halides of alkali and alkaline-earth metals
Chapter 6: Compounds of the transition metals

In Chapter 3, in which metals and alloys are discussed, the implications of the finite resolution of presently available instruments are emphasized, as they affect various regions in the periodic table of elements. The latest achievements of the coherent-potential method in the interpretation of the highly complex spectra of alloys are mentioned.

In Chapter 4, the crystals of the zinc blende type are discussed. For this simple tetrahedral lattice, theoretical models are developed up to an advanced level. A detailed discussion is given about the possibilities of the pseudopotential model to describe the band structure of several large groups of crystals such as diamond, graphite, silicon, germanium, A_3B_5, A_2B_6. Special attention is paid to the possibilities and limitations of far-ultraviolet spectroscopy in the study of valence and conduction bands of semiconductors. The importance of correlation of photoelectron spectroscopy, X-ray emission, and optical data is shown. The necessity of considering nonlocal pseudopotentials is commented on, and the possibility of improving in this way the agreement with experimental data in the upper part of the valence band is shown. The open problems in the further development of the theory are presented, especially concerning the interpretation of the lower part of the valence band.

In Chapter 5, the class of alkali halide crystals is discussed. Old data obtained by X-ray emission and absorption and by absorption and emission in the far ultraviolet are reviewed in light of the new possibilities opened up by the methods of X-ray photoelectron spectroscopy, in which spectra belonging to different energy regions can be placed on the same scale.

Chapter 6 is devoted to the problem of local states in compounds of transition metals. Long-range electron–electron correlation effects on the physical properties of compounds of transition metals are considered. Typical transitions from the metallic to an insulating state, by varying the temperature, are discussed in terms of d-electron long-range interactions in partially filled bands of these compounds.

In Chapter 7, the study of solid surfaces using a group of experimental methods (Auger electron spectroscopy and X-ray photoelectron spectroscopy) is presented from the point of view of the scientist interested more in the practical technological applications than in theoretical research. The topics of adsorption and oxidation processes, quantitative and qualitative surface analysis, and catalysis are covered.

The last chapter of the book (Chapter 8) contains an analysis of the structure of photoelectron spectra. Multiplet splitting, shake-up and shake-off processes, plasmon excitations, and asymmetry of photoelectron lines in the case of unfilled d bands are explained, and some guiding recommendations for their interpretation are given.

This monograph collects a large amount of information—mainly experimental and phenomenological—about the electron band structure of crystals that at present is widely spread in the literature. The book contains more than 500 references and gives a fairly complete survey of the experimental situation up to 1975. It is very complete, indeed, in the description of work done in the USSR, where the authors themselves have been major contributors. The emphasis is primarily on the discussion of experimental data and phenomenological methods of their interpretation. Since the theoretical discussion is limited to basic concepts and standard theoretical methods, physicists who are more interested in recent theoretical developments in photoelectron spectroscopy will have to find other texts to supplement this volume.

We feel that this volume will serve as a useful introduction to the field as well as a collection of important experimental data.

Stig Lundqvist
Per-Olof Nilsson
Irina Curelaru

Preface

Our knowledge of the electron structure of atoms, molecules, and solid materials is based mainly on the study of the interaction of photons with electrons in bound atomic states. As a result of such interactions, secondary photons as well as electrons are emitted. Investigation of the energy spectra of these electrons provides information about the electron structure of the sample material. The incident and the emitted photons may have energies within a wide range of values, from the optical region to the X-ray region. The experimental method of investigation is similar for the whole energy range.

The present monograph is devoted mainly to problems related to the study of the electron structure of crystals by the method of X-ray photoelectron spectroscopy. The basic aim of these studies is to extract information about the electron structure of gaseous, liquid, and solid samples from the energy and angular distributions of the photoelectrons emitted under bombardment with an incident beam of photons of given energy.

Prior to the large-scale development of X-ray photoelectron spectroscopy, studies of the electron structure of materials in various physical states were carried out using the method of ultraviolet photoelectron spectroscopy. This method was developed largely through the work of F. I. Vilesov at Leningrad University. In ultraviolet photoelectron spectroscopy, the energy of the incident photons falls in the range of 10–15 eV.

At present, He (21.22 eV) or He^+ (40.81 eV) resonance lines are used as radiation sources. A number of papers have been published describing the use of synchrotron radiation in the study of photoelectron spectra in the energy range of 10 to 100 eV. The synchrotron radiation is characterized by a high intensity and a high degree of polarization. By using the synchrotron radiation in the energy region of about 100 eV, it is possible to study not only the valence and conduction bands, but also some of the core levels. Since the chemical bonding in molecules and solid materials is brought about by the valence electrons, the interest and efforts of scientists were for a long time focused on the study of valence states. It was considered that the study of core-level states cannot provide useful information about the nature of chemical bonding, since core electrons do not participate directly in the formation of chemical bonds.

However, the work of K. Siegbahn and his school at Uppsala University has shown that the study of core states can give a great deal of valuable information about the type of chemical bonds in molecules and solids. It has been shown that transitions from one chemical compound to another are accompanied by a shift of the deeply lying electron levels. The magnitude of this shift usually amounts to 2–3 eV, and in some cases it can reach values of up to 10 eV. Since the resolution of most commonly used electron spectrometers is typically between 0.6 and 1.1 eV, and the accuracy in the determination of the position of sufficiently narrow core-level lines is of the order of 0.1 eV, it is evident that, for the majority of chemical compounds, X-ray photoelectron spectroscopy represents an efficient experimental method for the investigation of the nature of chemical bonding. For this reason the method has been named ESCA (Electron Spectroscopy for Chemical Analysis). The principal results obtained by Siegbahn and his co-workers in the field of chemistry were included in a monograph [1] published in Uppsala in 1967. A second monograph published later by Siegbahn and his colleagues [2] was devoted to the study of molecules in the gaseous state. Following the development and construction of commercial electron spectrometers, the number of experimental studies in this field has increased rapidly. X-ray photoelectron spectroscopy has at present become a powerful analytical method which gives information about the electronic and structural constitution of molecules and about the qualitative and quantitative elemental composition of the investigated specimens.

It has been found that this method may also be successfully used in the study of the electron structure of solid materials (metals, insulators, and semiconductors). X-ray photoelectron spectroscopy can yield important and reliable data about the energy-band structure of crystals. Studies performed on chemical compounds and crystals with unfilled *d* and *f* shells have revealed multiplet splitting effects resulting from the interaction of the spin and orbital momenta of deeply lying electrons with the spin and orbital momenta of electrons in unfilled shells.

Investigation of multiplet splitting effects can lead to a better understanding of the physical mechanisms in the process of photoelectron emission. Study of the change in magnitude of the multiplet splitting for different compounds provides information about the space localization of electron wave functions. Also important are processes of the "shake-up" and "shake-off" type in which, simultaneously with the emission of one photoelectron, excitation of one or several electrons in bound states as well as emission of another photoelectron may take place. Such processes, involving participation of several electrons, are among a class of multielectron effects that occur relatively frequently in X-ray photoelectron spectroscopy. When photoelectrons travel through the sample, they can lose part of their energy both continuously, through inelastic collisions with other electrons, and in discrete quantities. The spectra of these plasma losses may also contain peaks related to the excitation of surface plasmons.

Investigation of the angular distribution of photoelectrons shows that different shells make different contributions to the total angular distribution of photoelectrons. This effect allows the determination of the symmetry type of the spatial distribution of the electron density.

A field of research in which X-ray photoelectron spectroscopy may be efficiently applied is the study of surface phenomena. It makes possible the investigation of the properties of layers adsorbed on the specimen surface, and consequently of adsorption and catalytic processes.

Thus the method of X-ray photoelectron spectroscopy offers the means of solving many interesting problems in physics and chemistry. It should not be assumed, however, that this method will provide solutions to all problems related to the electron structure of crystals and to the electron energy spectra of atoms and molecules. Other complementary methods are also necessary. In X-ray photoelectron spectroscopy, the excited electron may be considered as being free, and therefore the conduction band states provide a small contribution to the observed structure of photoelectron spectra. Consequently, the method of X-ray photoelectron spectroscopy is useful for the study of valence band electron states. Since ultraviolet photoelectron spectroscopy is characterized by a higher resolution than X-ray photoelectron spectroscopy, it is suitable for the study of structures related to rotations and oscillations of molecules in gases. X-ray emission spectroscopy may, in its turn, offer the means to determine the symmetry type of electron states localized in the valence band of crystals, while X-ray absorption spectroscopy gives information about the states situated at the bottom of the conduction band. It is also useful to combine information about chemical shifts obtained by X-ray photoelectron spectroscopy with that obtained by γ-resonance spectroscopy, since the latter is more sensitive to changes of s-electron density in the vicinity of the atomic nucleus. It has also been observed that there is a correlation between the shifts observed in X-ray photoelectron spectroscopy and those determined by nuclear magnetic resonance.

At present the method of X-ray photoelectron spectroscopy is being applied on a very large scale to the study of crystals. The book of Siegbahn and co-workers [1] represents only an introduction to the topic of X-ray photoelectron spectroscopy and its applications in chemistry.

In the Soviet Union, a review article by Nefedov [3] was published in the series *Advances in Science and Technology*. The article gives a general survey of the results obtained by the method of X-ray photoelectron spectroscopy as applied in the study of different classes of chemical compounds. At the time of writing, there exists no similar review article or monograph that is devoted to solid state materials.

The present work represents an attempt to fill this gap. It is devoted mainly to problems related to the application of X-ray photoelectron spectroscopy to the study of the electron structure of crystals. Theoretical and experimental studies of crystal band structures by the method of X-ray emission spectroscopy are treated in two other monographs by Nemoshkalenko [4] and by Nemoshkalenko and Aleshin [5]. The latter is devoted to a detailed treatment of the general theoretical methods and to some specific examples of band structure calculations for various classes of solid state materials. In the present work, therefore, the basic concepts concerning the structure of energy spectra and the theoretical methods of their study are considered only briefly.

Although the present monograph is concerned mainly with results obtained in the study of the electron structure of solids by the method of X-ray photoelectron spectroscopy, we have entitled it *Electron Spectroscopy of Crystals*. The term "electron spectroscopy" as used in the present context seems to be appropriate since it extends the limits of the method by also including the spectroscopy of Auger electrons emitted in the processes of interaction of X-ray quanta with electrons bound in core levels.

The monograph does not claim to be exhaustive in the field of electron spectroscopy of solids. Detailed discussion is restricted to the most significant results obtained under the best experimental conditions. Less attention is paid to the problems related to the use of X-ray photoelectron spectroscopy in the study of the electron structure of crystal surfaces and of physicochemical surface processes such as chemisorption and catalysis. Similarly, little space is allotted to the influence of many-electron effects on the shape and energy position of X-ray photoelectron lines. In these cases, as well as in the case of problems related to chemistry, the authors discuss only those basic results that contribute to an understanding of effects in X-ray photoelectron spectra, those that demonstrate the physical characteristics of photon-initiated electron excitation processes in crystals, or those that illustrate the practical applications of the method. Nevertheless, the authors have attempted to select the data in order to offer the reader a comprehensive background against which can be presented the results of original research obtained in the last four years at the Institute of Metal Physics of the Ukrainian Academy of Sciences. The authors realize that such a composition of the monograph is not ideal. However, the volume of experimental results obtained in the last few years is so tremendous that specialists involved in the study of the electron structure of solids by other methods can hardly cope with them, and therefore the time has now become appropriate to undertake their generalization and systematization. To what extent the authors have succeeded in accomplishing this task is for the reader to judge. Any comments and proposals that might contribute to improvement of the book will be met with gratitude.

V. V. Nemoshkalenko
V. G. Aleshin

Contents

1

Fundamentals of the Method of Photoelectron Spectroscopy

Materials bombarded with X-ray quanta emit electrons. If the electron is emitted from a deeply lying level, the vacancy created on this level is filled by an electron from other, less tightly bound shells, including the valence band. As a result of such a transition X-ray quanta may be emitted. Moreover, there is a finite probability that another competitive, radiationless process will occur. In this alternative process, the result of the vacancy filling is that one of the electrons from the same level or from another, less tightly bound shell is emitted from the crystal. The emitted electron is called an Auger electron. The study of both of the above-mentioned processes is in itself of particular interest. In X-ray emission spectroscopy, the energy distribution of the emitted photons is investigated. If the photoelectron does not leave the sample, but is transferred into the states of the conduction band, absorption of incident X-ray quanta takes place. The study of this process is the object of X-ray absorption spectroscopy. In X-ray absorption spectroscopy, the intensity distribution of absorbed photons is measured. In X-ray photoelectron spectroscopy, the energy distribution of the emitted photoelectrons and of the Auger electrons is studied.

The rapid progress of electron spectroscopy is mainly the result of the importance of the information that is obtained about binding energies of electrons in atoms, molecules, and solid state materials. Determination of binding energies is based on the measurement of the kinetic energy of the electrons emitted from the system. This kinetic energy can vary over a wide range up to values approximately equal to the energy of the incident photons, as in the case of valence-electron ionization processes. In the case of ionization of core electrons, the kinetic energy of the emitted photoelectrons is considerably lower than the energy of the incident photons. For the study of different regions of the energy spectrum, various types of radiation sources may be employed. Valence states are usually studied by ultraviolet photoelectron spectroscopy, whereas core-level states are studied by X-ray photoelectron spectroscopy. At present, however, the resolving power of X-ray photoelectron spectrometers has been significantly increased so that both core-level states and valence states can be successfully studied by the method of X-ray photoelectron spectroscopy. The progress in the construction of X-ray photoelectron spectrometers has been made possible by the development of high-intensity sources of monochromatic radiation, by the increase of the resolving power of electron analyzers, by the availability of high-performance detecting systems, and by the development of powerful mathematical methods of processing experimental data.

Though the experimental technique has reached a high level of development, some of the principal problems of photoelectron spectroscopy have not yet been completely resolved. Such problems are, for example, the determination of the charge built up on nonconductive samples, the determination of the photoelectron escape depth, and the occurrence in the photoelectron spectra of the surface and bulk properties of the samples under investigation. Therefore, before going into the discussion of the basic results obtained by X-ray photoelectron spectroscopy of crystals, it seems appropriate to consider first a series of problems related to the determination of the binding energies of electrons in metals, insulators, and semiconductors, as well as to the determination of the effective escape depth of electrons from the sample, the calibration of electron spectra for nonconducting samples, and the basic characteristics of electron spectrometers currently available.

Basic Equation of X-Ray Photoelectron Spectroscopy

The basic relation that describes the process of photoelectron emission is the Einstein equation:

$$h\nu = E'^{f}(k) - E^{i} + E'_{\text{kin}}, \tag{1}$$

where $h\nu$ represents the incoming photon energy; E^i is the total initial energy of the atom, molecule, or solid state sample; $E'^f(k)$ is the total final-state energy of the system after emission of the electron from the state characterized by the quantum number k; and E'_{kin} is the kinetic energy of the emitted electron. The energies E^i and $E'^f(k)$ include the contributions from the electrons themselves and, for the case of molecules, from the vibrational, rotational, and translational movements.

$E'^f(k)$ in equation (1) may be written as

$$E'^{f}(k) = E^{f}(k) + E^{r},$$

where E^r represents the recoil energy. Siegbahn *et al.* [1] have shown that the magnitude of the recoil energy corresponding to photon energies of the order of 1500 eV is equal to 0.1 eV for the valence electrons of lithium, and to 0.04 eV for the valence electrons of sodium (in the case of valence states, the recoil energy is maximal).

Thus E^r can be neglected, since the resolving power of currently available spectrometers does not allow for detection of effects associated with the exisistence of the recoil energy. Consequently, from equation (1) one obtains

$$h\nu = E'_{\text{kin}} + E^f(k) - E^i.$$

Since the magnitude of the rotational and vibrational excitations falls outside the limits of sensitivity of currently available photoelectron spectrometers, the study of $E^f(k)$ can be restricted to consideration of the pure electronic states. If the binding energy $E_B(k)$ of an electron in the k state is defined as the energy necessary for the emission of this electron to infinity, with a kinetic energy equal to zero, then

$$E_B(k) = E^f(k) - E^i$$

and therefore

$$h\nu = E'_{\text{kin}} + E_B(k).$$

It follows from equation (1) that in order to describe the process of interaction of photons with matter, it is necessary to measure the kinetic energy of the emitted photoelectrons, E'_{kin}. The main components of the instruments used in photoelectron spectroscopy to determine E'_{kin} are the X-ray source, the electron analyzer (which allows the measurement of the kinetic energy of emitted electrons), the detector (which gives information about the number of electrons having a given kinetic energy), and the system commanding the operation of the spectrometer. Any X-ray photoelectron spectrometer should include all of these basic components. The construction of electron spectrometers is discussed in more detail in a later part of this work. We only mention here that in all electron spectrometers, the binding energy of the electrons in solid state samples is determined relative to the Fermi level of the material from which the spectrometer's electron analyzer itself is constructed.

In order to determine the electron binding energy, the photoelectron kinetic energy E_{kin} is measured by the spectrometer analyzer. The magnitude of E_{kin} is different from the value E'_{kin}, the latter corresponding to electrons that have left the sample but have not yet reached the spectrometer. This difference is caused by the existence of an accelerating or decelerating field between the sample and the entrance slit of the analyzer, created by the difference between the work functions of the sample (φ_S) and of the spectrometer (φ_{sp}).

Figures 1 and 2 show schematically how binding energies are determined for metals and for insulators. In the case of metal samples, owing to the fact that a good electric contact exists between the sample and the spectrometer, the Fermi levels of the sample and of the spectrometer coincide. Consequently, for metal samples, the binding energy measured with respect to the Fermi level of the spectrometer E_B^F is given by the expression

$$E_B^F = h\nu - E_{\text{kin}} - \varphi_{sp}, \tag{2}$$

where $h\nu$ is the photon energy, and φ_{sp} is the work function of the spectrometer material. In some metals, the position of the Fermi level can be distinguished as as steep onset of the photoelectron spectrum. Therefore, in such cases, the binding energies of the sample, as related to the Fermi level E_B^F, can be determined directly, without any need to use relation (2). In order to determine the binding energies relative to the vacuum level of the sample the following relation is used:

$$E_B^V = E_B^F + \varphi_s. \tag{3}$$

Since the value of the work function of the sample φ_S cannot be determined by using the method of photoelectron spectroscopy, it has to be determined by some other, independent measurement.

In the case of semiconductors and insulators, the determination of binding

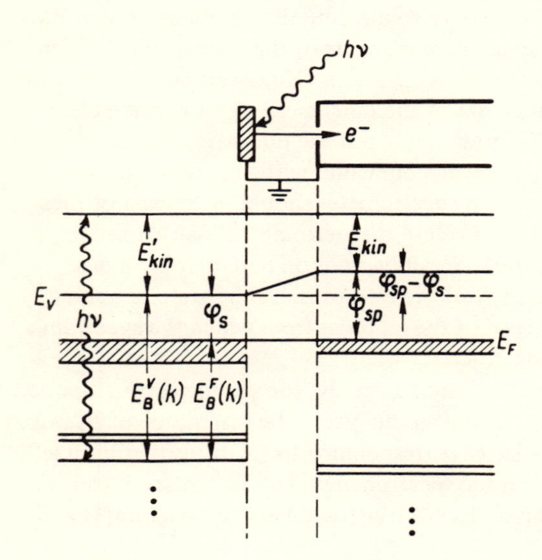

Figure 1. Determination of binding energy in metals.

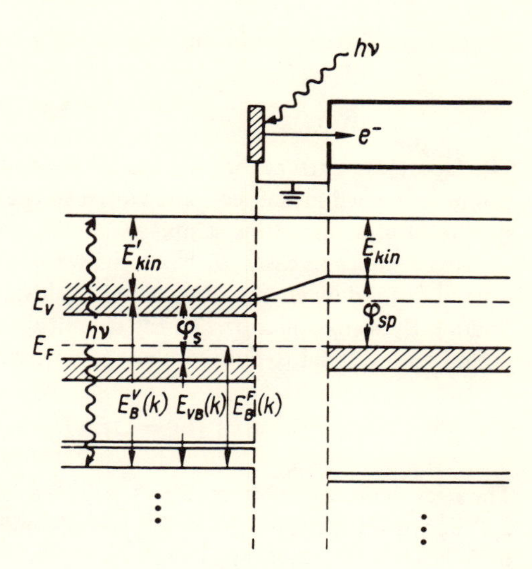

Figure 2. Determination of binding energy in dielectrics and semiconductors.

energies is even more difficult. As a result of the loss of electrons through emission, the samples investigated by photoelectron spectroscopy become charged, and the effective Fermi energy of nonconductive crystal samples undergoes a shift with respect to the Fermi level of the spectrometer. In this case, the value of E_B^F cannot be determined simply by measuring the photoelectron kinetic energy. Moreover, in the spectra of insulators and semiconductors there exists no steep onset corresponding to the Fermi level. In insulators, it is possible to determine the position of the Fermi level, but its actual physical significance is at present not yet definitely understood. Therefore, it is convenient to make the measurement of $E_B(k)$ values relative to the energy of the top of the valence band. In the following, the electron binding energy as determined with respect to the top of the valence band will be denoted by E_{VB}.

In the process of photoelectron emission, X-ray quanta generate photoelectrons that, in their turn, can excite secondary electrons. Many of these secondary electrons have a relatively high energy, sufficient to enable them to leave the sample surface. Consequently, the surface gradually becomes positively charged until the current of photoelectrons together with the current of secondary emitted electrons I_e are balanced by the neutralizing current I_n generated as a result of the space charge formed in the vicinity of the sample surface. The magnitude of the currents I_e and I_n depends on the particular type of spectrometer, on the sample properties, and on the sample mounting system. The magnitude of sample charging φ is determined for the stationary state corresponding to $I_e = I_n$. Since the sample becomes positively charged, the whole photoelectron spectrum is displaced toward lower kinetic energies. The values of binding energies are determined in this case by the relation

$$E_B^F = h\nu - E_{\text{kin}} - \varphi_{sp} + \varphi. \tag{4}$$

The factors that affect the magnitude of φ and the methods of its determination will be discussed in the section devoted to the calibration of photoelectron spectra of nonconducting samples.

In a number of cases, if the atoms are bound in various chemical compounds, instead of binding energies the chemical shifts of the k atomic core level are considered. By using equation (3), the chemical shift characteristic for a given compound with respect to another may be written as follows:

$$\Delta E_B^V(k) = \Delta E_B^F(k) + \varphi_{S_2} - \varphi_{S_1}.$$

The occurrence of the term $\Delta\varphi_S = \varphi_{S_2} - \varphi_{S_1}$ in this formula, where φ_{S_1} and φ_{S_2} are the work functions of the two considered samples, represents a source of difficulties when theoretical and experimental results are to be compared, because in theoretical calculations the energies are referred to the vacuum level. As a consequence, in the interpretation of experimental data, the difference between electron work functions is in most cases neglected. The work of Siegbahn *et al.* [1, 2] indicates that the magnitude of $\Delta\varphi$ does not have a significant influence on the interpretation of experimental results. However, it has been observed that in the diagrams representing the correlation between experimentally measured chemical shifts and theoretical calculations, a larger scatter of values occurs for solids than for gases. This is consistent with the fact that in gases the chemical shifts are determined with respect to the vacuum level of gas molecules.

X-Ray Sources and the Principle of X-Ray Monochromatization

The first X-ray photoelectron spectrometer was constructed by K. Siegbahn [1]. As a source of X rays it used the aluminum $K\alpha_{1,2}$ emission line, having an energy of 1486.6 eV.

Figure 3 shows the shape of the $K\alpha_{1,2}$ emission line of aluminum, resulting from superposition of the $K\alpha_1$ and $K\alpha_2$ lines. Since the energy separation between these two lines is small, this doublet may be treated as a single line having a full width at half-maximum (FWHM) of approximately 1 eV. The magnesium $K\alpha_{1,2}$ line is also frequently used in X-ray photoelectron spectroscopy. This has an energy of 1253.6 eV and a full width at half-maximum somewhat lower than that of the aluminum line. Both the $K\alpha_{1,2}$ line of aluminum and the $K\alpha_{1,2}$ line of magnesium have satellites generated by $2p$–$1s$ electron transitions in doubly

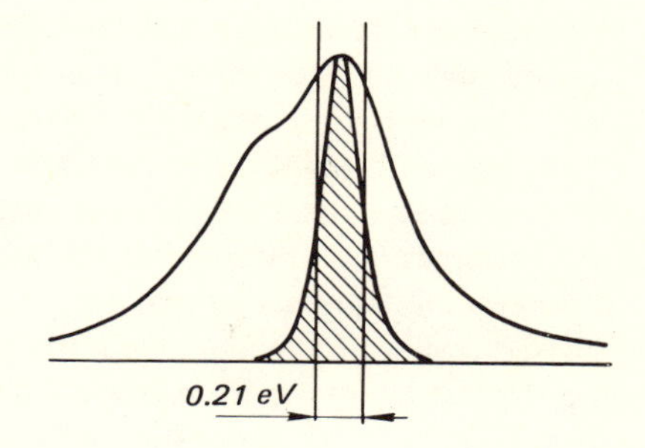

Figure 3. $K\alpha_{1,2}$ *emission line of aluminum before and after monochromatization.*

or even triply ionized atoms. In Figure 4, the 1s photoelectron line of carbon is shown, together with its satellite lines. The most intensive satellite that accompanies the $K\alpha_{1,2}$ lines of both aluminum and magnesium is the $K\alpha_{3,4}$ line, having an energy some 10 eV higher than the principal $K\alpha_{1,2}$ line. The intensity of these satellites is approximately 15% of the intensity of the $K\alpha_{1,2}$ line. All the other satellites have a much lower intensity (of the order of 1%). Also seen in Figure 4 is the contribution of the magnesium $K\beta$ line to the photoelectron spectrum. The satellite generated by the $K\beta$ line is situated at a distance of approximately 50 eV from the $K\alpha_{1,2}$ line, and has a relative intensity of the order of 1%. In order to improve the signal-to-noise ratio in electron spectrometers, one makes use of filters made of aluminum foils approximately 6 to 9 nm thick. The function of the filter is to lower the background of bremsstrahlung radiation. The intensity of this radiation depends on the angle θ (measured with respect to the incidence direction of the electron beam on the anode) as $\sin^2\theta$, for electron energies up to 10 keV, which is just the range of energies characteristic of X-ray photoelectron spectrometers. If the electron beam that excites the characteristic radiation is normal to the anode surface, then the intensity of bremsstrahlung radiation emitted in this direction is a minimum. However, in practice, an oblique electron incidence angle is used, since in

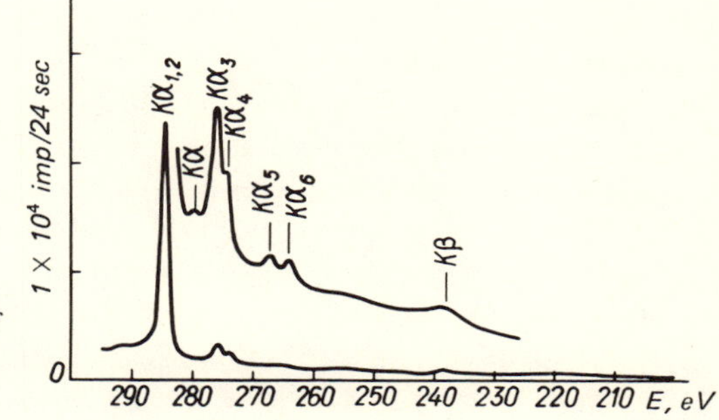

Figure 4. Photoelectron spectrum of 1s electrons of carbon in graphite. The satellite lines are also shown.

this case the intensity of the characteristic radiation is higher. By proper choice of the angle between the anode and the sample to be investigated, it is possible to realize an optimum intensity ratio of the characteristic-to-bremsstrahlung radiation, corresponding to the given direction.

At present, the $K\alpha_{1,2}$ lines of magnesium and aluminum are still the standard sources of X rays. In fact, excitation energies of the order of 1200 eV and 1500 eV allow the study of a great number of energy levels of interest. The choice of magnesium and aluminum is determined by the fact that the elements up to magnesium cannot be used as anode materials since, in the elements up to neon, the $2p$ levels form a broad valence band; in addition, neon is a gas and sodium is not a stable anode material. The elements situated after aluminum ($Z = 13$), as can be seen from Figure 5, are characterized by broad $K\alpha_1$ and $K\alpha_2$ lines, and by a large separation between them. Use of elements after aluminum as anode materials, without a preliminary monochromatization, is meaningless.

For the study of the valence bands of crystals, it is, in principle, possible to use the same radiation sources as in ultraviolet photoelectron spectroscopy, namely, the resonance lines of He and He^+. They are characterized by high intensity and small linewidth. However, in the region of energies of the order of 40 eV, the structure of photoelectron spectra is strongly influenced by the structure of the conduction band, and this results in a more difficult interpretation of the experimental data. Yttrium has a narrow M_ζ line ($4p_{3/2}$–$3d_{5/2}$ transition, FWHM equal to 0.45 eV). However, it has a high surface sensitivity for impurities and a high oxidation reactivity, and the energy of the emitted radiation, equal to 132.3 eV, is insufficient even for the study of carbon $1s$ states.

Monochromatization of X rays for the purposes of X-ray photoelectron spectroscopy was again first realized by K. Siegbahn [2], who, in the last few

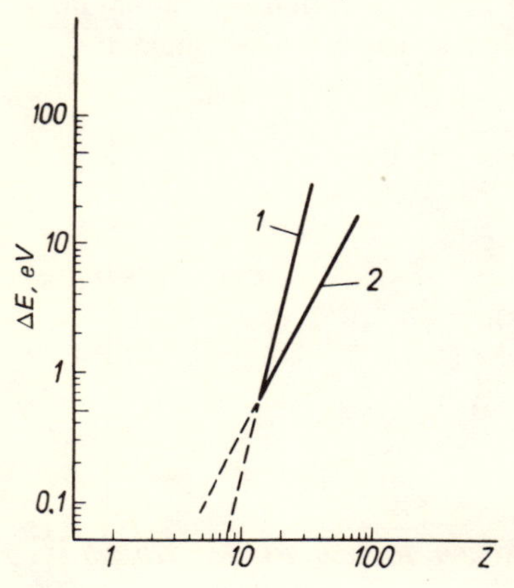

Figure 5. Energy difference between $K\alpha_1$ and $K\alpha_2$ lines (1) and width of $K\alpha_1$ line (2) as a function of atomic number.

years, has significantly improved the experimental technique [6–8]. In the new electron spectrometers constructed in Uppsala, a series of new devices has been developed. In particular, mention can be made of the new X-ray source, comprising a high-power electron gun, a rotating anode, and a monochromator that includes a spherically bent quartz crystal.

The X-ray monochromators used in electron spectrometers are based on the monochromatization principle proposed by Rowland. The diffracting crystal, with radius of curvature $2R$, is mounted tangentially to a circle of radius R, on which both the X-ray source and the sample are placed. If the directions of the incident and diffracted X rays make equal angles θ with the crystal surface, then the diffracted X rays that satisfy the Bragg law will be focused on the sample. The Bragg law relates the X-ray wavelength λ to the characteristic crystal interplanar distance d through the relation $n\lambda = 2d \sin \theta$, where n is an arbitrary integer. Those X rays that do not satisfy this relation will not fall on the sample surface.

Monochromatization of the $K\alpha_{1,2}$ line of aluminum by using reflection from the (010) plane of a spherically bent quartz crystal results in an improvement of the resolution of up to 0.16 eV as compared to the $K\alpha$ linewidth of 0.83 eV. Moreover, it removes the background caused by bremsstrahlung radiation and the $K\alpha_{3,4}$ satellites, which results in a considerable increase of the signal-to-noise ratio. In practice, however, it is difficult to realize a resolution of 0.16 eV because of monochromator aberrations and because of radiation absorption in the crystal itself. Absorption in the crystal also causes a decrease of the $K\alpha$-peak intensity down to 45%. Gelius *et al.* [8] have calculated the shape of the aluminum $K\alpha_{1,2}$ line reflected on the (010) plane of an α-quartz crystal. The calculated line shape is shown in Figure 6 together with experimental points measured with a point X-ray source obtained by using a spherically bent crystal. For shorter wavelengths the resolution increases and the intensity of the reflected radiation approaches 10%. It is important to mention that crystal imperfections represent another factor that contributes to a limitation of the resolving power.

In practice, efforts directed toward achieving a high resolving power en-

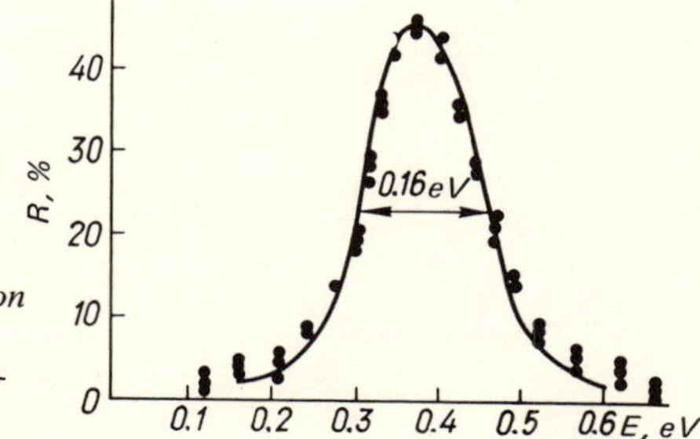

Figure 6. Bragg reflection of the Kα radiation of aluminum on the (010) plane of a spherically bent α-quartz crystal. (· · ·), experimental values; (——), theoretical calculation.

counter a number of difficulties. Only a small part of the entire $K\alpha_{1,2}$ line is reflected by the crystal. If higher-energy photon lines are used as the source of incident radiation, the emitted photoelectrons will also have higher energies. However, retardation of electrons in the electron lens results in a decrease of the intensity by a factor of $(E_i/E_0)^{1/2}$, where E_i is the energy of the photoelectrons entering the lens, and E_0 is their energy after retardation. Other factors that contribute to resolution limitation are aberrations associated with the finite dimensions of the dispersing crystal in the plane of the Rowland circle, possible asymmetric mounting of the crystal on the Rowland circle, a shift of the crystal from the Rowland circle, and the finite dimensions of the X-ray source.

In earlier monochromator constructions, cylindrically bent quartz crystals were used. With this type of monochromator it was possible to eliminate efficiently the bremsstrahlung background and the satellite contribution. Since the linewidth at the sample site is practically unaffected, improvement of the monochromatization can be achieved by use of a slit placed in front of the sample. This method of monochromatization is called *slit filtration* (Figure 7). Its greatest drawback is that it causes a drastic decrease of the X-ray intensity falling on the sample.

A more elaborate method of monochromatization is the method of dispersion compensation (Figure 8). Since the energy of photons incident on the sample varies approximately as much as 1 eV over the sample surface, the emitted photoelectrons originating from the same energy level will have different energies for different points of emission on the sample surface. Before entering the electron analyzer, the electrons are retarded and focused by an electron lens. The dispersion of the electron analyzer is chosen in such a way that electrons with different energies are focused at the exit along a narrow

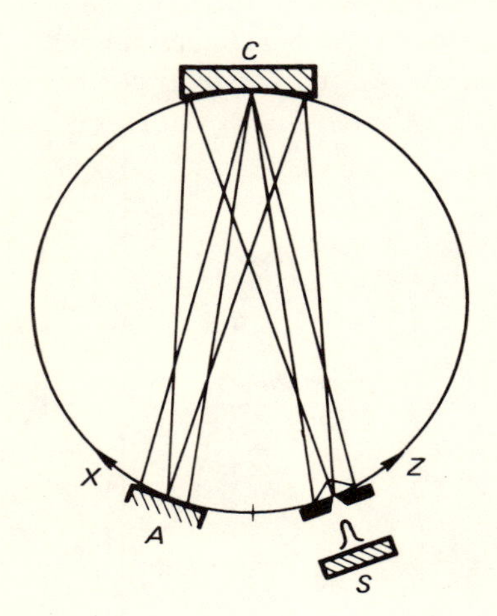

Figure 7. Principle of monochromatization by slit filtration: A–anode; C–crystal monochromator; S–sample.

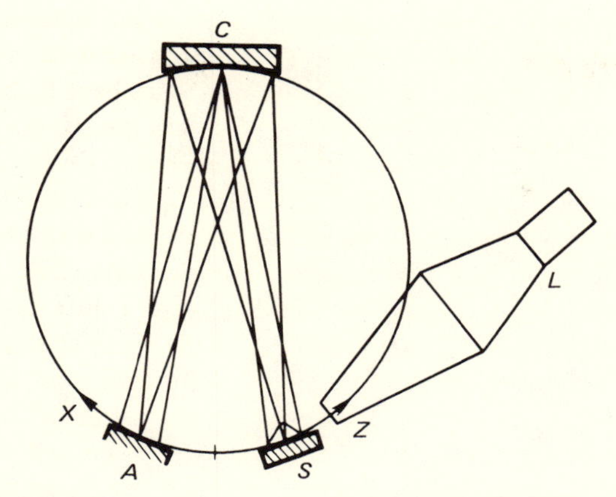

Figure 8. Principle of monochromatization by dispersion compensation: A–anode; C–crystal monochromator; L–electron lens; S–sample.

line. The electrons emitted from another energy level are characterized by another mean energy and are consequently focused at another position at the exit of the electron analyzer. This type of monochromatizing system allows the use of X-ray sources having a relatively large area. Care should be exercised in this case to minimize the surface roughness of the source.

The most advanced monochromatization method is at present the method of fine focusing. In this method, as can be seen from Figure 9, a rotating anode is employed, which allows the attainment of sources that are practically ideal point sources of X rays. As the diffracting component, it is most convenient to use spherically bent crystals. With point sources, the Bragg-law condition is satisfied most closely. Since the rotating anode offers the possibility to extract high-

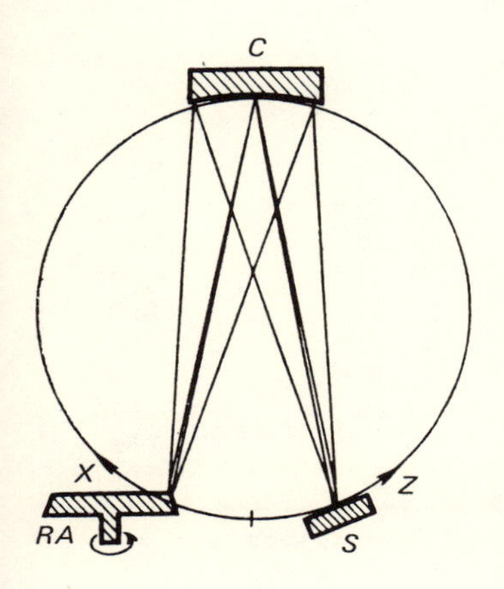

Figure 9. Principle of monochromatization by fine focusing: RA–rotating anode; C–crystal monochromator; S–sample.

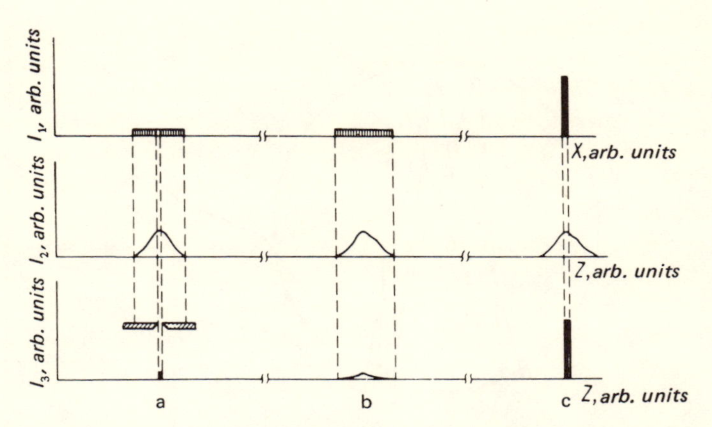

Figure 10. Efficiency of various methods of monochromatization: a–slit filtration; b–dispersion compensation; c–fine focusing; I_1–radiation intensity at the anode; I_2–X-ray line profile; I_3–radiation intensity after monochromatization.

intensity primary radiation, an additional slit can be mounted in front of the sample. Moreover, the method of fine focusing obviates the need for specific sample shapes.

The efficiencies of various monochromatization methods are schematically illustrated in Figure 10. Characteristic data for radiation sources used in the X-ray and ultraviolet regions of the radiation spectrum are given in Table 1. As can be seen from Table 1, titanium, chromium, and copper may be used as

TABLE 1. Sources of Radiation in the Far Ultraviolet and X-Ray Region

Source	Type of radiation source[a]	Energy, eV	Typical value of the intensity, photons/sec	Linewidth, eV
He		21.2	1×10^{12}	0.003
He^+		40.8	1×10^{11}	0.017
Y M_ζ		132.3	3×10^{11}	0.450
Mg $K\alpha_{1,2}$		1254	1×10^{12}	0.680
Al $K\alpha_{1,2}$		1486	1×10^{12}	0.830
Al $K\alpha_{1,2}$	M	1486	1×10^{11}	0.165
Al $K\alpha_{1,2}$	M, RA	1486	3×10^{12}	0.165
Ti $K\alpha_{1,2}$		4510	5×10^{11}	2.000
Ti $K\alpha_{1,2}$	M, RA	4510	3×10^{10}	0.027
Cr $K\alpha_{1,2}$		5417	1×10^{12}	2.100
Cr $K\alpha_{1,2}$	M, RA	5417	1×10^{10}	0.016[b]
Cu $K\alpha_{1,2}$		8055	2×10^{12}	2.550
Cu $K\alpha_{1,2}$	M, RA	8055	3×10^{9}	0.004[b]

[a]M = monochromatic radiation; RA = X rays obtained with a rotating anode.
[b]Because of crystal imperfections the linewidth will probably be larger than 0.020 eV.

anode materials. In the case of titanium, for example, the energy of the $K\alpha_{1,2}$ radiation is approximately three times greater than that of the corresponding line of aluminum. This means that for its monochromatization the same diffracting crystal plane may be used, but this time in the third order, whereas the Bragg reflection angle will be only slightly different. Titanium, however, is characterized by a very low thermal conductivity, and according to the data from Table 1, moreover, the intensity obtained will be two orders of magnitude lower.

For chromium and copper, after monochromatization (which is absolutely necessary for these metals), relatively low radiation intensity is obtained (see Table 1). This is perhaps the reason why at present no experimental work has been reported in which monochromatized radiation generated from these elements is used.

Construction Features of Modern Electron Spectrometers

Let us consider in more detail some of the main features of the new electron spectrometers developed in the past few years at Uppsala. We have already mentioned that these instruments include high-power electron guns, electrostatic electron analyzers, and a more advanced monochromatizing system using the method of fine focusing. In monochromators based on the principle of fine focusing and using rotating anodes, it is necessary to use high-power electron guns.

In spite of its high electron density, the electron beam should exhibit a fine focus. The position and shape of the focal spot should be rigorously fixed. Focusing is realized by using an electrostatic field. The electron gun is equipped with an indirectly heated cathode. A small electron gun, using a tungsten spiral filament, is used to heat a disk-shaped tungsten cathode. Since during operation the electron gun reaches temperatures up to, e.g., 2400°C, the isolators are made of beryllium oxide. The electron gun developed by Gelius *et al.* [6] provides emission currents of the order of 500 mA, at accelerating voltages from 6 to 15–20 kV. The latest electron gun developed at Uppsala [8] allows the use of accelerating voltages of 20 kV and currents of 1 A. An overall view of this gun is shown in Figure 11.

The electron beam falls obliquely on the anode surface so that the shape of the focal spot is an ellipse of 4 mm length and 2 mm width. The power developed by the electron gun over this spot may be even greater than 100 kW/cm^2. Such power intensity cannot be tolerated by any material at rest. Therefore it is necessary to use rotating anodes. Such an anode was manufactured of an aluminum alloy (97.5% Al, 1% Si, 0.8% Mg, 0.7% Mn) and was provided with an efficient water cooling system. For a power of 6 kW in the X-ray source and an anode rotation speed of 5000 rpm, the useful anode surface is heated up to a temperature of 575°C. This temperature is low enough to preclude evaporation

Figure 11. Outer view of the electron gun.

of the aluminum. The anode has a diameter of 11 cm. Its cooling is achieved by a water flow of 5 liters/min. An overall view of the rotating anode is shown in Figure 12.

Monochromatization of X rays by the method of fine focusing is achieved by using a spherically bent quartz crystal (Figure 13). The crystal has a diameter of 60 mm, and the radius of the Rowland circle is 389 mm. To allow bending, the crystal is very thin, having a thickness of 0.1 mm. As the diffracting plane, the (010) plane is used. The Bragg diffraction angle for the $K\alpha_{1,2}$ line of aluminum is equal to 78.5°. The shape of this line after monochromatization is shown in Figure 3.

The first electron analyzers used in X-ray photoelectron spectroscopy were of magnetic type. A detailed description of these analyzers may be found in the literature [1]. The main drawback of magnetic analyzers is their high sensitivity to accidental magnetic disturbances. In order to obtain energy resolutions of the order of 0.01%, it is necessary to reduce the extraneous magnetic fields to 10^{-4} G, which is realized by using large compensating coils. These coils caused the large dimensions of the magnetic spectrometers. Another drawback is the fact that in magnetic spectrometers it is difficult to use the technique of pre-

Figure 12. Outer view of the rotating anode.

Figure 13. The spherically bent crystal monochromator.

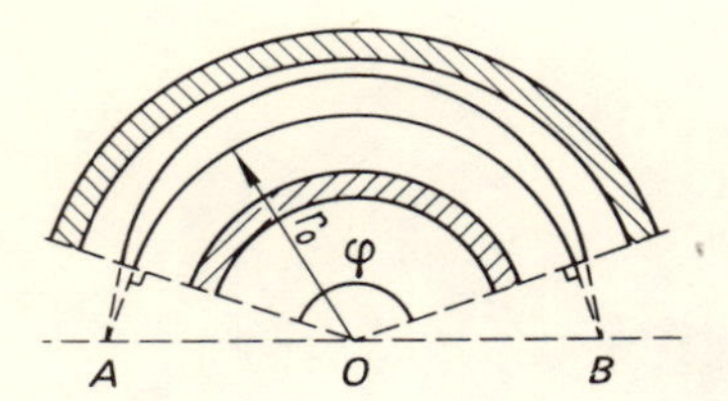

Figure 14. Principle of operation of the spherical condenser electron analyzer.

liminary electron retardation. For this reason, electrostatic-type analyzers are being used in X-ray photoelectron spectroscopy.

The new electron spectrometer developed in Uppsala includes an electrostatic analyzer of the spherical condenser type shown schematically in Figure 14. The central electron trajectory is situated in the plane of the figure. Its radius of curvature r_0 is the same as the radius of the corresponding equipotential surface in this plane. Since the field has spherical symmetry, this equipotential surface also has the same radius of curvature r_0 in the orthogonal direction. The electron source is placed at the point A. The angle φ is chosen as 157.5° and the radius of the central trajectory as 360 mm. The spherical electrodes are placed 80 mm apart. In the azimuthal direction, the spherical sector is confined inside an opening angle of 60°.

In order to give an idea of the progress of experimental techniques, Figure 15 shows the spectrum of the gold $4f$ level, as recorded (1) by the first [1] and (2) by the latest, most advanced [6] electron spectrometers. The resolving power has in the second case become sufficient to reveal that the actual shape of the line is described by a Lorentz distribution. A great difference is also observed in the spectra of trifluoracetate. As can be seen from Figure 16, the new photoelectron spectrum exhibits different ratios of the line intensities, which is due to the absence of overlapping effects.

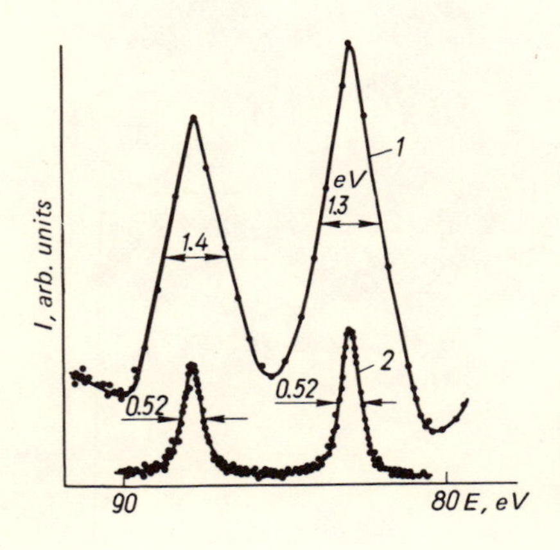

Figure 15. Photoelectron spectra of $4f_{5/2}$ and $4f_{7/2}$ electrons of gold as obtained (1) without and (2) with monochromatization of the exciting X rays.

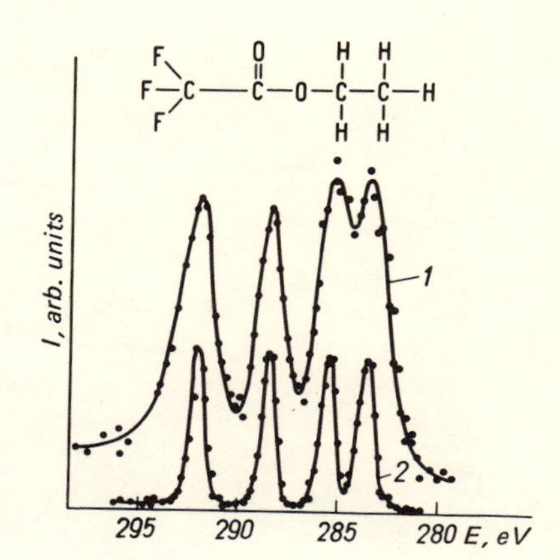

Figure 16. Photoelectron spectrum of carbon 1s electrons in trifluoracetate as obtained (1) without and (2) with monochromatization of the exciting X rays.

Construction Features of Commercial Electron Spectrometers

The new systems developed in the last few years in Uppsala and incorporated in new photoelectron spectrometers have not yet been exploited in currently available commercial electron spectrometers. For example, these spectrometers do not make use of rotating anodes and do not apply the principle of fine focusing for the monochromatization of X rays.

The first in the series of commercial electron spectrometers was the IEE-15 spectrometer manufactured by the firm Varian in 1969. At present, it is one of the most widely used instruments and is to be found in a great number of laboratories all over the world. An overall view of the IEE–15 spectrometer is shown in Figure 17, and its principle of operation in Figure 18. The IEE-15 spectrometer is a highly automated instrument. Its principal components are the source of X rays, the vacuum system with two pumps (a titanium sublimation pump and a turbomolecular pump), the electrostatic electron analyzer, an oscilloscope, a teletypewriter, an X–Y recorder, and a computer of VARIAN 620/*i* type. The spectrometer analyzer is protected against accidental magnetic field perturbations (in particular the earth's magnetic field) by a shield made of μ-metal. The vacuum level in the spectrometer is of the order of 10^{-7} torr. The spectrometer is equipped with a special preparation chamber inside which the samples can be cleaned, cooled down to $-180°$C, or heated up to $+250°$C. As X-ray source, the spectrometer uses the $K\alpha_{1,2}$ line of aluminum or magnesium. An aluminum filter is placed between the sample and the anode. It improves the signal-to-background ratio and also damps the satellite $K\alpha_{3,4}$ radiation.

By varying the voltage on the electrostatic electron analyzer, the width of the analyzed line can be changed. For example, the width of the $4f_{7/2}$ line of gold is 1.8 eV for 100 V on the analyzer, and 1.1 eV for 30 V on the analyzer, when an aluminum anode is used. One drawback of the IEE-15 spectrometer is

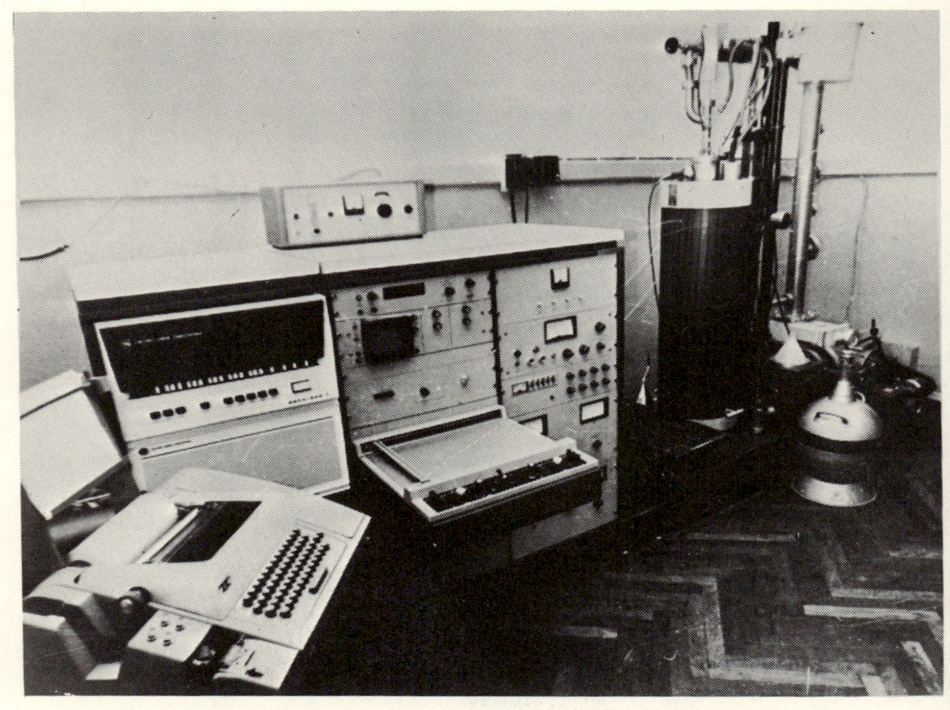

Figure 17. Outer view of the IEE-15 electron spectrometer.

its relatively low resolving power. This can only partially be compensated for by mathematical processing of the experimental data with the computer connected on-line to the spectrometer.

The computer is used not only as a multichannel analyzer but can also perform some of the operations required in the mathematical processing of the experimental data. Thus, for example, special programs have been developed for the calculation of mean values and even for improving the resolving power of the spectrometer by eliminating the broadening introduced by the analyzer. These programs make use of the technique of Fourier analysis. Figure 19 shows as an example the results of the mathematical processing of the photoelectron spectrum of the chlorine $2p$ doublet in KCl. This doublet (curve 1) was recorded in 10^3 sec with a voltage of 40 V on the analyzer. The photoelectron spectrum of this doublet, after Fourier analysis, is shown as curve 2 in Figure 19. Convolution of this spectrum with the "filter" 3 gives the Fourier spectrum 4, which by reversed Fourier analysis leads to the spectrum 5. Comparison between the initial spectrum 1 and the final spectrum 5 shows a substantial improvement in resolution, which is, however, accompanied by some

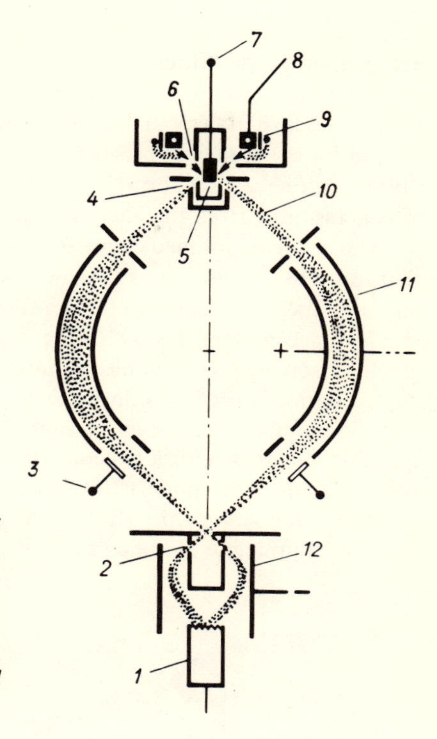

Figure 18. Principle of operation of the IEE-15 electron spectrometer: (1) electron multiplier; (2) ring slit; (3) focusing control; (4) ring slit; (5) sample; (6) X-ray beam; (7) scanning voltage; (8) anode of the X-ray source; (9) cathode of the X-ray source; (10) photoelectron trajectories; (11) spherical condenser; (12) cylindrical condenser.

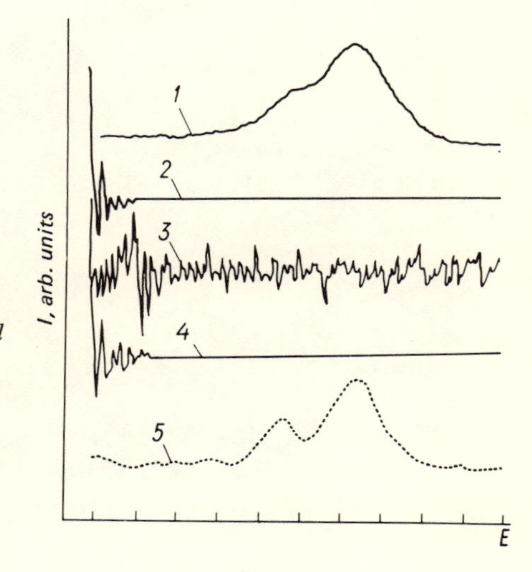

Figure 19. Successive steps in the mathematical processing of the photoelectron spectrum of the chlorine 2p doublet in KCl by the method of Fourier transformation: (1) photoelectron spectrum of chlorine 2p doublet in KCl; (2) Fourier image of the photoelectron spectrum; (3) filter; (4) convolution of the Fourier image and the filter; (5) finally processed photoelectron spectrum.

decrease in sensitivity. In order to perform this transformation, it is necessary to use an appropriate "filter." In the above example, the filter was constructed by using the Fourier-type spectrum of the 1*s* line of carbon in graphite, produced with energies of 100 eV and 50 eV on the analyzer. Such a filter can correct the spectra for experimental broadening introduced by the analyzer. Other types of "filter" can be devised, however, that provide correction of the spectra for effects resulting from the X-ray source (line asymmetry, satellites).

The above-mentioned example demonstrates that since experimental effects that result in broadening of the lines may be corrected for by choosing appropriate filters, well-resolved spectra can finally be obtained within reasonable times.

The spectrometer HP-5950A developed by the firm Hewlett–Packard has a greater resolving power. A monochromator based on the principle of dispersion compensation is used in order to remove the satellites accompanying the $K\alpha_{1,2}$ radiation of aluminum or magnesium, as well as the bremsstrahlung radiation that introduces an additional contribution to the structure of photoelectron spectra resulting from inelastically scattered electrons.

An overall view of the HP-5950A electron spectrometer is shown in Figure 20, and its principle of operation in Figure 21. The monochromator incorporates

Figure 20. Outer view of the HP-5950A electron spectrometer.

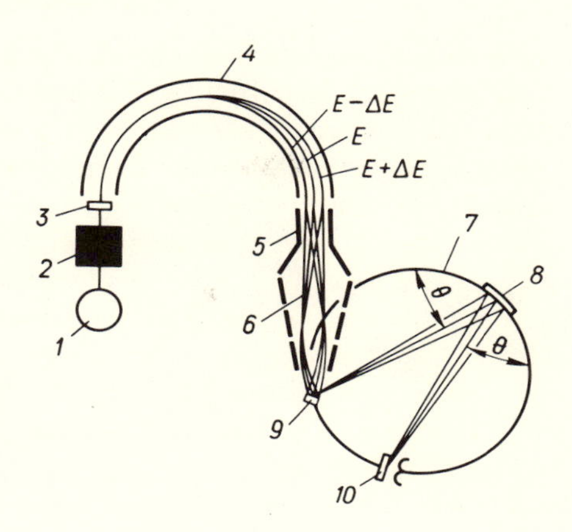

Figure 21. Principle of operation of the HP-5950A electron spectrometer: (1) display; (2) multichannel analyzer; (3) multichannel detector; (4) electron analyzer; (5) electron lens; (6) photoelectron trajectories; (7) Rowland circle; (8) crystal monochromator; (9) sample; (10) anode of the X-ray source.

three bent crystals, one of which is shown in Figure 21. The other two crystals are situated symmetrically with respect to that shown, in two planes that also pass through the X-ray source and the sample. In this way, each of the crystals is situated on its own Rowland circle. This special construction was adopted in order to increase the X-ray intensity on the sample. This monochromator design and the small sample size necessitate the use of advanced electron detectors. A single electron falling on the active surface of such detectors causes the emission of approximately 10^8 electrons. These electrons irradiate a phosphorescent screen, and the generated light signals are recorded by a television tube. The signals are then directed toward a multichannel analyzer. The most noteworthy characteristic of the analyzer in this system is the absence of the slit. The spectrometer has a high resolution—0.6 eV.

Besides the IEE-15 and HP-5950A instruments, there exist several other types of spectrometers. The firm McPherson (USA), for example, has developed the spectrometer ESCA-36, and Vacuum Generators (United Kingdom) the spectrometer ESCA-2 and subsequently the more elaborate ESCA-3 spectrometers. Since these spectrometers have performances similar to those of the IEE-15 instrument we shall discuss here only those characteristics that make them different from the IEE-15 spectrometer. In the ESCA-36 it is possible to change the anode rapidly without breaking the vacuum. This allows the study of binding energies over a wide energy interval 0–4000 eV. It is also possible to change samples without breaking the vacuum. The vacuum level in the spectrometer is approximately 10^{-9} torr.

The ESCA-3 spectrometer is operated at an even better vacuum, of the order of 10^{-9}–10^{-10} torr. An advantage of the spectrometer is the possibility of investigating samples of small dimensions (approximately 1 cm^2), and also the capability of data handling and processing with the help of a dedicated programmer. Among the facilities available on the data system are: spectral averaging, smoothing and deconvolution of peaks, background subtraction, and peak position determination.

Let us now consider the parameters that may be used in comparing various types of electron spectrometers:

$$\frac{S}{B} = \frac{\text{signal}}{\text{background}} = \frac{I_P - I_B}{I_B},$$

$$\frac{S}{N} = \frac{\text{signal}}{\text{noise}},$$

$$\text{sensitivity} = \frac{I_P - I_B}{C_A},$$

where I_P is the intensity (number of counts) of the peak of the photoelectron line, I_B is the corresponding value of the noise level, and C_A is the concentration of A-type atoms on the surface of the sample under investigation. The spectrometer resolution is characterized by the quantity ΔE determined by the relation

$$(\Delta E)^2 \approx (\Delta E_x)^2 + (\Delta E_a)^2 + (\Delta E_s)^2.$$

TABLE 2. Characteristics of the Principal Types of Electron Spectrometers[a]

Electron spectrometer type	X-ray photo-electron line	I_p, imp/sec	FWHM	$\frac{S}{B}$	$\frac{S}{N}$	Sensitivity	ΔE, eV	Precision of the peak position determination	Reproducibility
Varian IEE-15	Au 4*f*	1 000 000	1.6	7	—	8800	—	—	—
	Ag 3*d*	24 000	0.88	13	—	220	0.9	—	—
	C 1*s*	11 700	1.0	17	—	110	—	0.1	0.03
Hewlett-Packard HP-5950A	Au 4*f*	120 000	0.8	120	—	1200	—	—	—
	Ag 3*d*	26 000	0.87	33	—	250	—	—	—
	C 1*s*	12 000	0.8	200	1500	120	0.5	0.1	0.03
Mc Pherson ESCA-36	Au 4*f*	75 000	1.35	14	350	700	—	—	—
	Ag 3*d*	18 700	0.88	19	270	180	0.88	0.04	0.15
	C 1*s*	14 500	1.00	53	470	140	—	—	—
Vacuum Generators, Ltd. ESCA-2	Au 4*f*	30 000	2.36	—	—	—	—	—	—
	Ag 3*d*	—	—	—	—	—	—	—	—
	C 1*s*	10 000	1.83	—	—	—	1.0	0.1	0.05
Vacuum Generators, Ltd. ESCA-3	Au 4*f*	35 000	1.05	10	—	—	—	—	—
	Ag 3*d*	25 000	0.95	—	—	—	—	—	—

[a]For the HP-5950A and ESCA-2 spectrometers, the data refer to the aluminum anode. Before the measurements of the characteristics of the HP-5950A spectrometer, the surface of the gold sample was cleaned by argon ion bombardment.

Here, ΔE_x is the width of the X-ray line incident on the sample, ΔE_a the effective width of the probed core level, and ΔE_s the magnitude of the spectrometer aberration. Some basic parameters of the most widely used spectrometers are shown in Table 2 using data from Lucches [9, 10].

At present, many other different types of spectrometers are being constructed. We have considered here only those instruments that have been most widely used.

Calibration of Electron Spectra

When nonconducting samples are investigated, in order to obtain accurate values of binding energies, as given by equation (4), it is essential to determine the magnitude of sample charging φ under the action of X radiation. This can often be determined by measuring the charge of hydrocarbon layers formed on the surface of the sample under investigation, or the charge of a thin gold layer evaporated onto the sample surface. To ensure that these methods are sufficiently reliable, it should be demonstrated that the magnitude of the charges developed on the sample surface layer is equal to that of the hydrocarbon or gold overlayer. It would be desirable to perform the measurements under such experimental conditions that sample charging is minimal. The problem of sample charging is important and difficult. In the following, we shall discuss briefly the results obtained in this field.

The simplest case is the study of metals covered by an oxide layer. Johansson *et al.* [11] have shown, using the photoelectron spectrum of a polished aluminum foil as an example, that the incident X radiation causes charging of the surface oxide layer but does not alter the position of the photoelectron line corresponding to the pure metal. The difference between the metal and oxide $2p$ lines is 2.6 eV. If an external positive potential of +5 V is applied to the sample, the magnitude of this difference is reduced to 2.4 eV, whereas if a negative potential of −5 V is applied the difference is increased to 2.8 eV. In both cases, the metal lines are displaced in proportion to the magnitude of the applied potential.

A far more difficult situation is encountered in the case of nonconducting samples. Khatowich *et al.* [12] have reported results obtained for such samples deposited on gold or palladium substrates and coated with a sputtered layer of gold or palladium. The positions of the photoelectron lines corresponding to the surface layer and to the substrate were compared, and the energy difference was attributed to the charging effect. By applying an external potential to the substrate, the lines corresponding to the sample itself and to the noble metal layer suffered a shift of the same magnitude. For the $4d_{5/2}$ level of barium the same binding-energy value was obtained regardless of whether gold or palladium was used as the sputtered layer. Therefore, it can be concluded that the thin noble-metal layer has the same potential as the surface of the nonconducting sample.

Consequently, the electron lines corresponding to such sputtered overlayers may be used for calibration purposes.

A more difficult experiment has been performed by Johansson *et al.* [11]. On the sample—a NaCl monocrystal with a 0.5-cm^2 surface area mounted on a platinum substrate—a small (1-mm diameter) spot of gold was deposited. The observed shift of the carbon 1*s* line from the NaCl crystal surface and from the platinum foil was determined as 3 eV. The measured energy value of the gold $4f_{7/2}$ line, as compared to the $4f_{4/2}$ line of platinum, corresponded to a binding energy of 86.8 eV; the correct value is 83.8 eV. This result also shows that the magnitude of the charge is equal to 3 V. When a voltage of +5 V or −5 V was applied to the specimen holder, the lines corresponding to the substrate, i.e., the $4f_{7/2}$ line of platinum and the 1*s* line of carbon, were also shifted by −5 eV or +5 eV on the kinetic energy scale, but the $4f_{7/2}$ line of the gold, the Auger line of sodium in NaCl ($KL_2L_3(^1D_2)$), and the 1*s* line of carbon from the crystal surface were shifted by only −2.5 eV and +1 eV, respectively. The photoelectron lines of gold were also shifted as much as the lines corresponding to the holder, if electric contact was established between the gold sheet and the sample holder. Experimental measurements showed that the lines of carbon, gold, and sodium are shifted equally when the voltage on the sample holder is changed. It follows that, with this arrangement, the lines of gold and carbon may be used for the determination of sample charging and therefore for the calibration of spectra.

The magnitude of sample-charging effects depends on the sample thickness. Johansson *et al.* [11] deposited a thin NaCl layer of approximately 20 Å thickness on a gold substrate. It was observed that the shift of the carbon 1*s* line arising from the hydrocarbon layer on the sample surface was 0.2 eV greater than the shift of the carbon 1*s* line arising from the hydrocarbon layer on the substrate, toward greater binding energies. For thicker layers (of the order of 2000 Å), shifts as great as 0.6–0.8 eV were observed. For monocrystals, the charging effects are even greater, reaching values of the order of 2–3 eV.

The magnitude of sample charging depends on the substrate material. This was found experimentally in the case of a thin NaCl layer deposited simultaneously on beryllium, aluminum and gold foils. Although the magnitude of sample charging for the sample deposited on aluminum was found to be twice as great as that for the sample deposited on gold, the charge for the beryllium substrate was intermediate between those for the aluminum and gold substrates. This experiment shows that, in order to minimize the magnitude of sample charging, gold is a good substrate material.

Calibration of photoelectron spectra may also be achieved by mixing the sample to be investigated with a suitable reference material. It has been shown experimentally that graphite is suitable as a calibration material. In this case, the mixture should be pressed and mounted on a conducting substrate. However, not all substances can be used for calibration purposes, and particular caution should be exercised when this method of calibration is considered. Bremser and Linneman [13], for example, have suggested the use of LiF for calibration of photoelectron spectra. However, as is demonstrated by the work of Johansson *et al.* [11], the magnitude of LiF charging depends on the duration of the spectrum recording. Furthermore, as a result of nonuniform charging in samples mixed with LiF, spurious lines may appear. For example, in the case of BN mixed with LiF, two 1*s* lines of nitrogen were observed.

It may be concluded from these studies that interaction of X radiation with nonconducting samples leads to the emission of photoelectrons, Auger electrons, and secondary electrons from the sample surface, leaving it electrically charged.

There exist a number of factors that contribute to a reduction of the charging effect. These include the surface conductivity of the sample stimulated by the X radiation and the existence of a photoelectron current from the window of the X-ray tube. The studies of Ebel and Ebel [14] performed on the electron spectrometer ESCA-36 have shown that, at a constant current, the magnitude of sample charging increases when the voltage applied to the anode is increased. In the region of low anode voltages (of the order of 2.5 kV), the sample may become negatively charged. In order to elucidate the effect on sample charging resulting from the electron current arising from the aluminum foil (0.008 mm thick) that covers the window of the X-ray tube, the window was covered with a gold layer on the side facing the sample. Under identical excitation conditions, the number of electrons emitted by gold was about five times greater than the number of electrons emitted by aluminum. In this case, the magnitude of the specimen charge was found to be lower. A variation of the anode potential from 2 to 6 kV resulted in a variation of the charge from approximately 1 to 1.5 eV. This experimental fact shows that the electrons arising from the aluminum foil lower the sample charging.

Ebel and Ebel [14] also studied the electron spectrum excited by the X radiation and by the electrons scattered from the anode. For this purpose, the experimental arrangement was modified as follows. The window of the X-ray tube was left open and the foil was placed in the spectrometer so that the electrons emitted by it could be collected. In this case, the electrons scattered by the anode had a considerable effect, since they caused secondary electron emission from the aluminum foil. In the region of anode potentials of the order of 3–6 kV, the number of secondary electrons was proportional to the anode current i_a and was nearly independent of the anode voltage. However, the flux of electrons arising from the anode, and the current i_1 corresponding to it, is determined mainly by the anode voltage and only to a much lower extent by the anode current.

Consequently, one can understand why in some cases the sample may get negatively charged. The flux of electrons emitted by the sample decreased quickly when the anode potential decreased, but at a constant anode current (i_a) the flux of electrons emitted by the aluminum foil (i_2) was independent of the anode potential and remained constant. Therefore, at low voltages (of the order of 2.5 kV) the number of electrons hitting the sample was greater than the number of electrons leaving it. Since both the electron emission from the sample and the flux of electrons coming to the sample are proportional to i_a, it follows that the magnitude of sample charging, being the difference between two linear variations, also has a linear dependence on i_a. In fact, at low currents a linear behavior is observed, whereas at high currents a saturation does occur. For different samples, the first region of the curve of φ versus i_a may be different, but in all cases φ becomes independent of i_a at a current of $i_a \sim 20$ mA. The empirically determined dependence of φ on i_a can be represented by the simple relation

$$\varphi = A \frac{Bi_a}{1 + Bi_a}, \tag{5}$$

where A and B are constants. This dependence suggests that a further mechanism may exist, which has not yet been considered and which causes a decrease of sample charging. It is presumably caused by the X radiation and results in a flux of electrons directed toward the charged surface layer of the sample. It can be supposed that such a mechanism arises from photoconductivity.

Let us consider the current i through the circuit formed by the sample surface and the sample holder. It has been found experimentally that this current depends linearly on i_a. By using the Kirchhoff law, one obtains for the current i the following expression (see Figure 22):

$$i = i_1 - i_2.$$

Since $i_1 = K_1 i_a$ and $i_2 = K_2 i_a$, it follows that

$$i = \frac{\varphi}{R} = (K_1 - K_2)\, i_a.$$

By making use of the empirical formula (5), one may write

$$\sigma = \frac{1}{R} = \frac{K_1 - K_2}{AB} + \frac{K_1 - K_2}{A}\, i_a.$$

Since the flux n of photons hitting the sample is proportional to the current i_a, this expression agrees with the formula giving the dependence of the photoconductivity σ on n, namely,

$$\sigma = \sigma_0 + En.$$

This formula, which is characteristic for photoconductivity, is similar to formula (5), and this provides evidence of the significant influence of photoconductivity

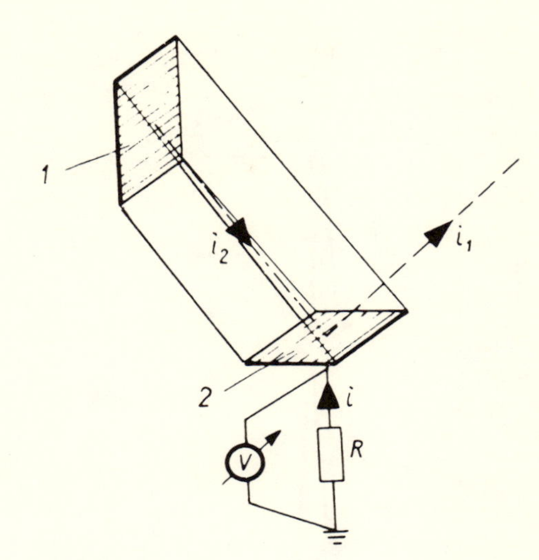

Figure 22. Schematic diagram of the circuit for determination of the current passing through the sample and the sample holder: (1) window of the X-ray tube; (2) sample surface.

on the process of sample charging. The experimental data reported by Ebel and Ebel [14] were obtained for glass covered by gold, for Teflon, and for Teflon covered by a silver foil.

The magnitude of sample charging depends on the state of the surface. If the sample surface is covered by oxide particles, the magnitude of the charge on each particle will depend on its dimensions. It will also vary over the surface of each particle, and will be different for particles of different shapes. These problems, however, have not yet been studied experimentally. One fact is clear, namely, that the sample to be investigated should have a smooth surface.

Before proceeding with a consideration of methods of calibration of photoelectron spectra of insulators and semiconductors, let us discuss the experiment of Hedman *et al.* [15], in which *n*- and *p*-type silicon samples were studied. The samples were prepared by alloying silicon with a concentration of 2×10^{19} atoms/cm^3 phosphorous and with 4×10^{19} atoms/cm^3 boron. At such concentrations, the mixed levels form a band, and the conductivity has in both cases a metallic character. Consequently, the samples do not become charged under the action of X radiation. It follows that the Fermi level is situated at the top of the valence band in the case of *p*-type silicon, and at the bottom of the conductivity band in the case of *n*-type silicon. To avoid changes of the characteristics of the alloyed samples, their surfaces were not exposed to any chemical poisoning.

The measurements were performed under a vacuum of 10^{-6} torr. In the photoelectron spectrum, a low-intensity line was observed on the high-energy side of the silicon 2*p* binding-energy peak. The occurrence of this line was due to the presence of an oxide layer on the sample surface. As a result of the different position of the Fermi level in the samples, it was to be expected that a

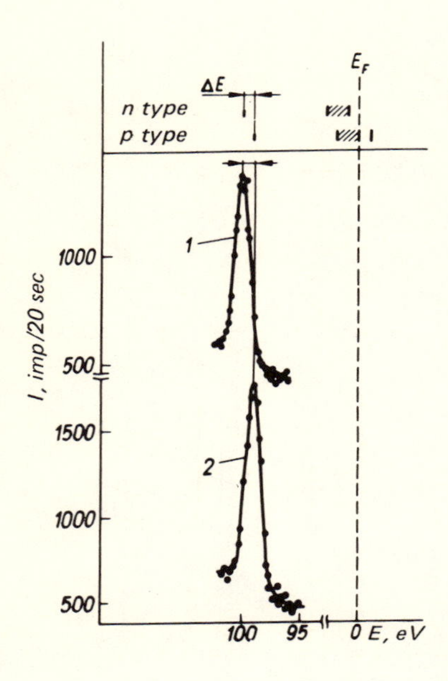

Figure 23. Photoelectron spectra of 2p electrons in highly alloyed silicon: (1) n type; (2) p type.

shift of the silicon $2p$ core level would appear. Such a shift was, in fact, detected experimentally, and its magnitude, 1.0 ± 0.2 eV, was close to the value of the forbidden bandwidth (1.1 eV). For the two particular samples considered here, the following binding energy values were obtained: 100.0 eV for n-type silicon and 99.0 eV for p-type silicon. The photoelectron spectrum for silicon $2p$ electrons is shown in Figure 23. When the specimens were subjected to chemical poisoning, a binding energy of 99.5 ± 0.2 eV was obtained. This was the same value as that obtained for silicon with a low concentration of the second component.

Ley *et al.* [16] have studied the problem of the Fermi-level position in insulators and semiconductors. The magnitude of φ was determined as follows. After recording the photoelectron spectra of the sample, its surface was covered with a thin layer of gold, and the binding energy of the gold $4f_{7/2}$ electrons was measured, together with one of the intense core-level lines. Since in gold the position of the $4f_{7/2}$ line relative to the Fermi level is well known ($E_B^F = 84.0$ eV), it follows that the binding-energy value of the core level under investigation could be obtained with respect to the Fermi level of the deposited gold layer. This procedure makes possible the determination of the Fermi-level position with respect to the top of the crystal valence band, and also the determination of the magnitude of sample charging φ. Very careful experiments were performed in order to determine the charging effect for 26 semiconductors and insulators. Figure 24 shows the measured φ values of these crystals, as a function of the width of their

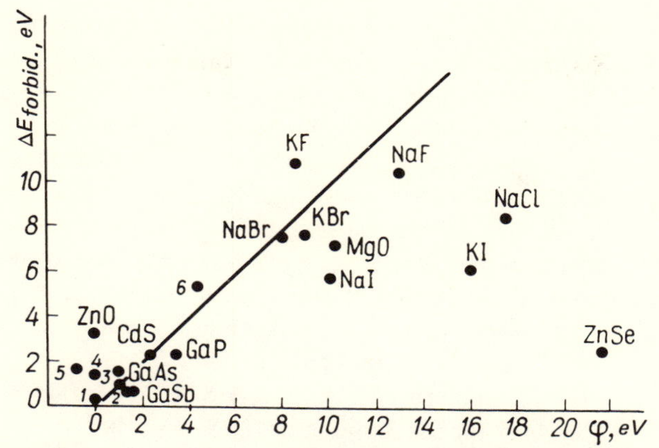

Figure 24. Dependence of the charge of dielectrics and semiconductors on the magnitude of the forbidden bandgap of the crystal: (1) PbS, PbSe, PbTe, InSb; (2) Ge; (3) Si (film); (4) CdTe; (5) CdSe; (6) diamond.

forbidden band. This figure shows clearly that some correlation exists between the charge φ and the width of the forbidden band of semiconductors and insulators. Materials with narrow forbidden bands become charged up to values close to or somewhat lower than the width of the forbidden band itself. Crystals characterized by wide forbidden bands become charged up to values close to or somewhat greater than the width of the forbidden band.

Unexpectedly, high charging was observed for the compound ZnSe (22 eV), which may result from the influence of the photovoltaic effect. The existence of the above correlation indicates that, at high values of the current that neutralizes the specimen charging, a mechanism of the Zener breakdown switches becomes active. This mechanism is particularly probable for materials with high carrier concentration. It is also necessary to take into account the surface conductivity, which can introduce a significant contribution to the neutralizing current in those cases in which the magnitude of charging is lower than the width of the forbidden band. Thus, the procedure described above allows a quite accurate determination of the Fermi energy position with respect to the top of the valence band. However, the significance of this E_F value remains as yet incompletely understood.

A similar situation arises when special techniques are used with the aim of canceling out the charging effect. This can be achieved by irradiating the sample under investigation with low-energy electrons or with ultraviolet radiation.

The position of the Fermi level, as determined from ultraviolet photoemission experiments, depends on the stoichiometry, on alloying, and on the surface states of the sample. Deformation of the electron energy bands caused by sample surface charging should be taken into account. Wagner and Spicer [17] have shown that, in order to determine the position of the Fermi level, each of the aforementioned parameters should be carefully examined. Consequently, the experiments should be performed on sample surfaces of atomic-scale purity, and it is also absolutely necessary to know the precise stoichiometric composition and the degree of alloying. These factors are usually more easily controlled in silicon and germanium than in binary semiconductors. Unfortunately, a great deal of experimental work has been performed on specimens for which at least one, and very often even several of the aforementioned parameters have not been determined. Exceptions to this are several photoemission measurements performed in the vacuum ultraviolet region [17, 18]. As is seen in Table 3, the Fermi level is preferentially localized near the center of the forbidden band.

Vesely *et al.* [19] have determined the position of E_F for CdS, CdSe, CdTe, ZnO, ZnSe, ZnTe, and for the semiconductors HgSe and HgTe with a small forbidden band, by comparing E_B^V obtained from measurements in the ultraviolet region of the spectrum with E_B^V obtained from X-ray photoelectron spectra. The different values of E_B^V explain the discrepancy between the results obtained by Ley *et al.* [16], on the one hand, and by Vesely *et al.* [19], on the other. Use of monochromatized X radiation is of great importance in the determination of the magnitude of E_{VB}, which in this case can be obtained by extrapolating the steep side of the low-energy valence band peak toward the background line.

In the spectra of the semiconductors and insulators investigated by Ley *et al.* [16], sharp and narrow core-level lines were observed, even when the magnitude of the charging φ was 10 eV or even more. Therefore, it can be concluded that in spite of the fact that the sample surface is charged throughout the whole depth

TABLE 3. Position of the Fermi Level inside the Forbidden Bandgap of Dielectrics and Semiconductors

Compound	E_F-E_{VB}	Magnitude of the forbidden bandgap	Approximate location of E_F, eV[a]
ZnTe	0.17	1.19	B
GaAs	0.0	1.40	B
GaP	0.15	2.26	B
InSb	0.12	0.18	C
ZnSe	1.13	2.80	C
ZnS	1.08	3.6	C
ZnO	1.63	3.3	C
CdTe	0.47	1.40	C
CdS	1.27	2.58	C
InAs	0.3	0.35	U
CdSe	1.88	1.84	U

[a] B = bottom; C = center; U = upper part of the forbidden bandgap.

of X-ray penetration (approximately 10^4–10^5 Å), the charge potential has a stepwise variation at the surface.

It should be noted that although the calibration of the photoelectron spectra of nonconducting samples by covering them with a thin layer of gold gave consistent and noncontradictory results for the semiconductors and insulators studied by Ley *et al.* [16], it is not completely certain that this calibration method is universally valid. This is indicated, for example, by the results of other experiments [20, 21], in which it has been found that deposition of gold on the sample surface may result in chemical reactions with the sample material, and consequently in a chemical shift of the gold $4f_{7/2}$ line toward greater or lower values of the binding energy. Another factor that affects the shape and the position of both the gold and the sample lines is the thickness of the deposited layer.

Effective Escape Depth of Electrons

The effective escape depth of electrons is an important parameter that determines whether the experimental information obtained characterizes the bulk or the surface properties of the sample under investigation. In order to elucidate this problem, it is necessary to be able not only to measure the effective depth of electron escape from the sample under the action of X radiation, but also to determine the values of the effective depths that correspond to the bulk or to the surface properties of the sample, respectively.

By ultraviolet photoemission studies the effective escape depth of electrons has been determined for a number of crystals. These measurements have been mainly made in the region of incident photon energies up to 10 eV. It has been found that the escape depth depends strongly on the kinetic energy of the electron. For electron kinetic energies that are greater than the energy of the Fermi level by some tenths of an electron volt, the effective escape depth can reach values of the order of 10^4 Å [22]. It decreases rapidly to values of the order of 20–30 Å when the electron kinetic energy E_{kin} increases.

A number of experiments have been performed for the determination of this parameter by X-ray photoelectron spectroscopy, for electron energies up to 3500 eV. It has been shown that, in the X-ray region of photoelectron excitation, typical values of the effective electron escape depth are of the order of 20–30 Å, corresponding to electron energies in the range 1–2 keV. If the electron energy is further increased, an increase of the effective escape depth is observed.

Two methods are employed for the determination of the effective escape depth in X-ray photoelectron spectroscopy. The first method consists in measuring the thickness of a layer deposited onto a substrate made from a different material; the intensity of a given line from the deposited layer sample is compared with the intensity of the same line from a bulk sample of the same material. In the second method, the substance under investigation is also deposited on a substrate, but in this case the intensity of a given characteristic line from the covered substrate is compared with the intensity of the same line from the substrate material.

When analyzing the intensities in photoelectron spectra, it may be assumed that the intensity of the X-ray photon beam does not suffer significant modification in the vicinity of the sample surface. Therefore, for the intensity of photoelectrons with given energy, one may write the following expression:

$$dI = F\sigma N k e^{-x/\Lambda} dx, \tag{6}$$

where F is the intensity of the X-ray photon flux, σ is the photoionization cross section for the given photon energy and given atomic level, N is the number of atoms of a given type inside unit volume of the sample, k is a factor determined by the geometry of the spectrometer and by the counting efficiency of the photoelectron detector, x is the distance traveled by the photoelectron inside the sample material, and Λ is the effective electron escape depth.

From formula (6), the following expression is obtained for the intensity of the photoelectron flux emitted from a surface layer of the specimen, of thickness d:

$$I = F\sigma_1 N_1 k \Lambda_1 (1 - e^{-d/\Lambda_1}) = I_\infty (1 - e^{-d/\Lambda_1}). \tag{7}$$

Here, I_∞ is the intensity of the photoelectron flux emitted by the bulk sample. The subscript 1 refers to the sample under investigation. From formula (7), by using the first of the above-mentioned methods, one can determine the value of Λ_1. By using the second method, the intensity of photoelectrons traversing a sample layer of thickness d can be obtained from

$$I = I_0 e^{-d/\Lambda_1},$$

where I_0 is the intensity of the photoelectron flux emitted from the uncoated substrate material, i.e., $I_0 = F\sigma_2 N_2 k \Lambda_2$. Here σ_2, N_2, and Λ_2 are the corresponding characteristic parameters of the given sample.

Particularly stringent requirements are imposed on the quality of the sample layer deposited onto the substrate. It should be homogeneous since the presence of lower-density regions or of uncovered regions would result in an increased value of the escape depth Λ. There are several methods for the deposition of uniform layers onto substrates. One of these methods, proposed by Steinhardt *et al.* [23], consists in depositing the layer through a narrow slit onto a rotating cylinder. It is desirable to perform the deposition inside the spectrometer chamber itself, in order to avoid atmospheric exposure of the sample even for relatively short times. If, for practical reasons, this requirement cannot be met, it is necessary to heat the specimen after mounting it in the spectrometer, until the 1*s*-peak signals of oxygen and carbon decrease and the characteristic signal of the deposited substance increases. Deposition of the desired layer on the substrate is usually achieved by electron bombardment of a separate piece of material. Carlson and McGuire [24] deposited thin layers of tungsten trioxide on tungsten substrates by anodization.

Let us consider the results of some experiments that have been carried out to determine the magnitude of Λ. In one of the first experiments, Klasson *et al.* [25] determined Λ for gold and for aluminum oxide (Al_2O_3). The magnitude of $1-e^{-d/\Lambda}$ for three energy values of the electrons emitted from gold is shown in Figure 25. The full lines were drawn through the experimental points by using the least-squares method. In experiments devoted to the determination of Λ, the intensity ratio I/I_∞ is usually measured, since it is less sensitive to surface impurity contamination of the bulk sample and of the deposited layer.

For gold, the magnitude of Λ was also determined at a photoelectron energy of 3208 eV. Its value was found to be 37 Å. With available experimental data it was possible to establish the energy dependence of Λ in the investigated energy region as $\Lambda(E) = CE^n$, where the value of n was found to be 0.5 ± 0.1 by the least-squares method. It should be mentioned that the accuracy in the determination of Λ depends to a great extent on the accuracy in the determination of the sample thickness. In the work of Klasson *et al.* [25], the latter was determined by the standard interferometric method [26] to within ±10%. For gold, Λ was determined with an accuracy of ±15%. For Al_2O_3, at a photoelectron energy of 1389 eV, Λ was determined as 13 Å, and at 3856 eV it was found to be 22 Å.

Let us now consider the experiment of Carlson and McGuire [24], in which Λ was determined for tungsten trioxide WO_3. The photoelectron spectra of the

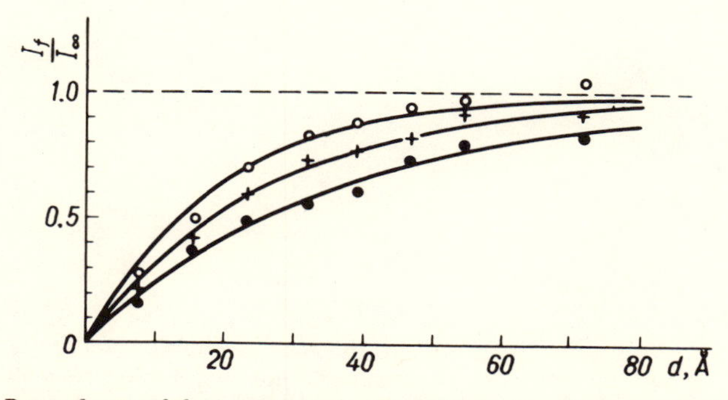

Figure 25. Dependence of the intensity of gold photoelectron lines on sample thickness, at three values of the kinetic energy of electrons: ○–*940 eV* ($4p_{3/2}$, *Al-*$K\alpha_{1,2}$); +–*1403 eV* ($4f_{7/2}$, *Al-*$K\alpha_{1,2}$); ●–*2671 eV* ($3p_{3/2}$, *Cr-*$K\alpha_1$).

$4f_{7/2,5/2}$ levels of pure tungsten and of its oxide WO_3 are different because of the chemical shift of these tungsten levels in the oxide (Figure 26). For an oxide layer of thickness d', the intensity ratio may be expressed as follows:

$$R = \frac{I_W}{I_{WO_3}} = \frac{N_2\Lambda_2}{N_1\Lambda_1} \frac{e^{-d'/\Lambda_1}}{(1 - e^{-d'/\Lambda_1})} \tag{8}$$

The ratio N_2/N_1 for tungsten and for WO_3 is known from the ratio of densities. The ratio Λ_2/Λ_1 may be obtained experimentally by measuring the intensity of $4f$ lines of pure tungsten and of pure WO_3, under identical conditions. Since the intensity of the photoelectron flux emitted from a bulk sample may be expressed by the formula

$$I = F\sigma Nk\Lambda,$$

it follows that

$$R^* = \frac{I_W}{I_{WO_3}} = \frac{N_2\Lambda_2}{N_1\Lambda_1}. \tag{9}$$

The ratio R^*, as determined in a separate experiment, is equal to 1.74. Equation (9) shows that if Λ is known for a particular energy and for a particular compound, it is possible to determine its value for any other compound containing the given element. As can be seen from Figure 26, when the thickness of the oxide layer increases, the signal corresponding to pure tungsten decreases, while the signal corresponding to WO_3 increases. It should be noted that the energy

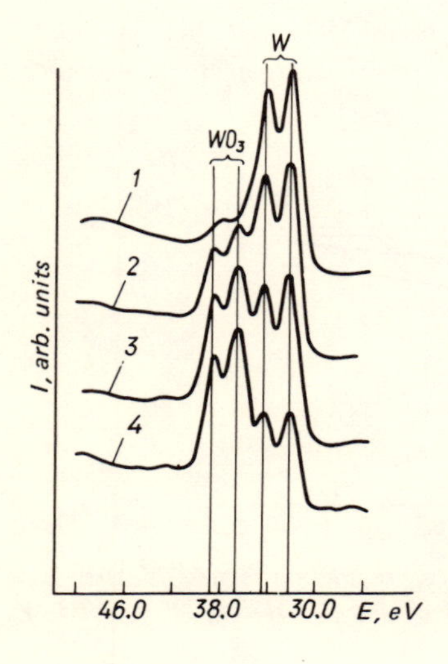

Figure 26. Photoelectron spectra of $4f_{7/2}$ and $4f_{5/2}$ electrons of tungsten and of tungsten oxide, for various thickness of the oxide layer: (1) d = 2.3 Å, (2) 14.9 Å, (3) 21.3 Å, (4) 30.6 Å.

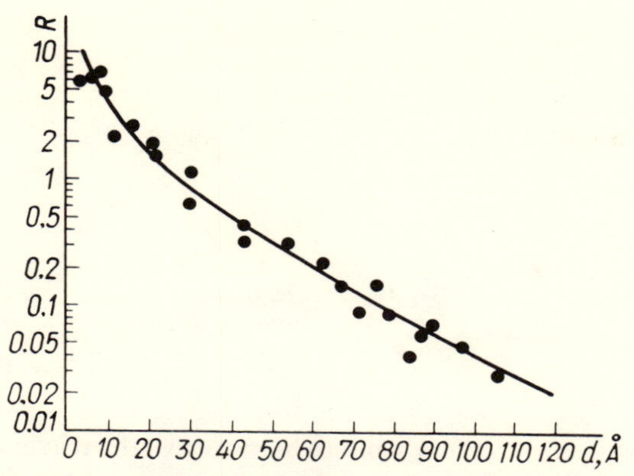

Figure 27. Intensity ratio of photoelectron lines of tungsten and of tungsten oxide as a function of the thickness of the oxide layer.

distance between the characteristic lines of pure tungsten and its oxide remains constant at all oxide thicknesses, except when the thickness of the oxide layer is approximately equal to one monolayer. The experimental values of the intensity ratio are shown in Figure 27. The curve in this figure was derived by using formula (8) with the experimental value $R^* = 1.74$ and by choosing the optimum value of Λ_1. This curve, giving the dependence of R on d', agrees well with the experimental data, in spite of the fact that the experimental values of R vary over three orders of magnitude. The best agreement with experiment was obtained for $\Lambda = 26.3$ Å. The value of Λ_2 in this case was 12.8 Å. The thickness of the oxide layer was determined by the method proposed by McCargo *et al.* [27], who determined a relation between the anodization potential and the thickness of the oxide layer. An additional check was achieved by irradiation of the tungsten sample with neutrons before anodization, followed by dissolution of the oxide layer in a solution of KOH and determination of its radioactivity. The authors estimated the error in the determination of Λ_1 to be ±20%. This error value includes the uncertainties in the determination of the layer thickness d' and of the ratios R and R^*.

Klasson *et al.* [28] have determined the value of Λ for silicon. A silicon film was deposited on copper and chromium substrates and the intensity of the photoelectron lines of silicon, copper, and chromium were studied as a function of the thickness of the deposited layer. The magnitude of the exponent n in the equation $\Lambda(E) = CE^n$ was evaluated for the energy range from 321 to 3574 eV. It was found to be 0.7 ± 0.2, which is fairly close to the value $n = 0.5 \pm 0.2$, obtained earlier for gold. Figure 28 shows the energy dependence of Λ for silicon and gold, based on the data from Klasson *et al.* [25, 28].

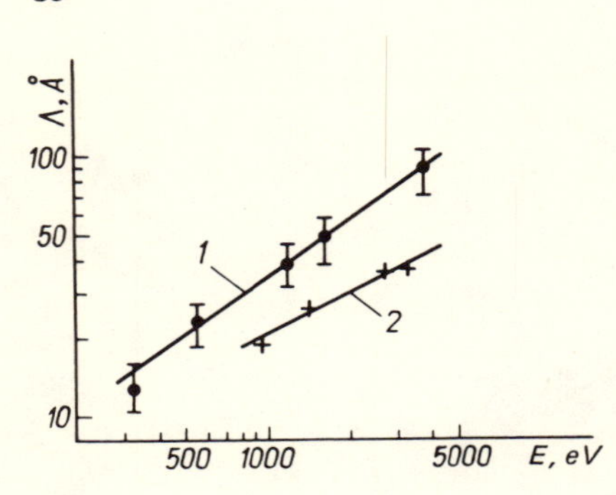

Figure 28. Effective escape depth of electrons from (1) silicon and (2) gold as a function of electron kinetic energy.

In the above discussion, our aim has been to give the reader a description of the experimental approach to the problem. Therefore, we have considered only a few, but characteristic experiments for the determination of the effective escape depth of electrons. All currently available data on the values of Λ, measured over a large energy range, for pure metals and for compounds, are shown in Figure 29.

We shall not discuss the methods used for the determination of Λ in ultravio-

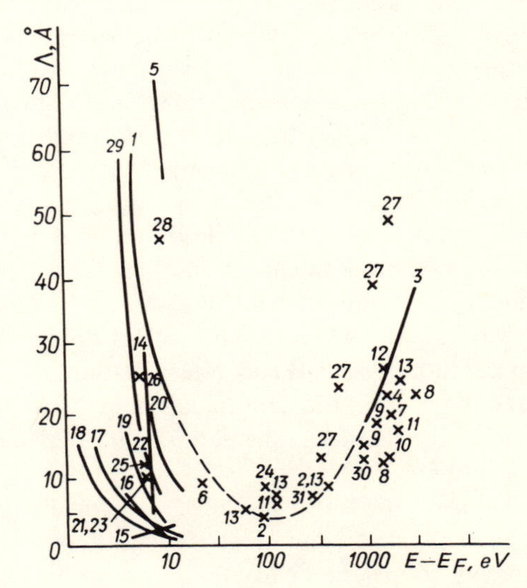

Figure 29. Effective escape depth of electrons as a function of the electron kinetic energy. The curves and the points are taken from the following papers: (1) Cu, Ag, Au [29–31]; (2) Ag [32]; (3) Au [25]; (4) Au [33]; (5) Al [34]; (6) Al [35]; (7) Al [36]; (8) Al_2O_3 [25]; (9) C [37]; (10) W [24]; (11) W [38]; (12) WO_3 [24]; (13) Mo [38]; (14) K [39]; (15) Cs [40]; (16) Sr [22]; (17) Ba [41]; (18) Yb [41]; (19) Ce [22]; (20) Cl [41]; (21) Gd [42]; (22) Y [42]; (23) Ni [42]; (24) Fe [43]; (25) Si [44]; (26) Si [45]; (27) Si [28]; (28) GdTe [46]; (29) NaKSb [47]; (30) Be [48]; (31) GeO_2 [49].

let photoelectron spectroscopy. Even there, however, the method used is based on the deposition of a layer of one material onto a substrate made from a different material, followed by a study of the contribution of these two substances to the photoelectron spectrum as a function of the thickness of the deposited layer.

In the energy region down to approximately 10^{-2} eV, there exist great differences between the Λ values of different materials. For cesium, strontium, and barium, at energies of the order of 10 eV, small values of Λ have been found, i.e., $\Lambda \sim$ 1-2 Å. However, the physical significance of Λ remains unclear if its magnitude is lower than the lattice constant of the elementary cell. In such cases, particular care should be exercised in the interpretation of the experimental energy distribution of electrons. In the majority of cases, however, at photoelectron energies lower than 10 eV and greater than 1 keV, the electrons are usually emitted from a depth of approximately 8–10 monolayers. In the intermediate energy range, a tendency has been observed toward a decrease of Λ. In the future, the most promising development will probably come from the study of photoelectron spectra by using synchrontron radiation, since this enables measurements to be made over a wide range of excitation energies. Since within this energy region the magnitude of Λ decreases, the study of the photoelectron spectra will allow a comparison of the bulk and surface properties of solid materials.

The effective escape depth of electrons is, in fact, a measure of their mean free path. It is defined by the expression

$$\Lambda = \frac{\int_0^\infty x\,dI}{I}$$

and represents the thickness $d = \Lambda$ of the layer that causes a decrease of e times in the photoelectron line intensity. Sometimes, instead of Λ, the value $\Lambda_{1/2}$ is used, defined by the condition that the intensity decreases to half of its initial value. The values of Λ and $\Lambda_{1/2}$ are related to each other through the expression $\Lambda_{1/2} = \Lambda \ln 2 = 0.693\Lambda$.

The theoretical treatment of the problem of electron scattering processes in the energy range from ultraviolet light to ultrasoft X rays is at present insufficiently developed. The photoelectrons traveling toward the surface lose their energy as a result of electron–electron and electron–phonon scattering, whence we can write

$$\frac{1}{\Lambda} = \frac{1}{\Lambda_e} + \frac{1}{\Lambda_p}.$$

The electron–phonon scattering is characterized by a relatively weak energy dependence and by small energy losses, of the order of 10–50 meV. Energy losses caused by electron–phonon scattering processes are important only for metals, at energies lower than 5–6 eV. Therefore, at higher energies, the predominant effect is that of electron–electron scattering, which is characterized by a strong energy dependence and by large energy losses resulting from energy transfer in the collision processes. In the X-ray energy range, account should be taken of the contribution of core-electron ionization processes. However, this contribution is not great, amounting to approximately 4% at energies of 100 eV and approximately 14% at energies of 1500 eV [50]. The main difficulty is related to the evaluation of the contribution of valence electrons. In conclusion, in spite of the fact that the photoelectrons are emitted from a depth of material much larger than Λ, the effective depth of their escape from the sample is, as a result of inelastic collisions, relatively low, being of the order of 20 Å. At present, there are strong reasons to believe that, for such values of Λ, X-ray photoelectron spectroscopy gives information about the bulk rather than surface properties of the samples under investigation. This is confirmed by the good agreement between the shape of the experimental energy distribution of valence electrons and the density-of-states curves calculated without taking into account the surface states or the theoretical energy distributions of valence electrons. Further on, we shall return to a discussion of these problems, showing that by studying the angular distribution of photoelectrons emitted from solid materials it is possible to distinguish at least semiquantitatively between the contributions to the photoelectron spectra of the bulk and the surface atoms.

Sample Preparation

The studies of effective electron escape depth have established that, in the process of X-ray photoelectron emission, only thin surface layers of material (20–30 Å) are involved. Consequently, the surface condition of the sample may have a marked influence on the results obtained. This is particularly true of metal samples on which oxide layers can be formed by contact with the atmosphere. Metal samples are usually in the form of foils, plates, or cylinders. In order to remove surface oxide layers, various methods may be used.

Fadley [51] and Baer *et al.* [52] have studied the effect of heating the sample to high temperatures ($T > 500°C$) in a hydrogen atmosphere at 10^{-2}–10^{-3} torr. The method of X-ray photoelectron spectroscopy has a great advantage over other methods of studying the electron structure of solid state materials, since it allows measurement of the degree of sample surface contamination during the entire duration of the experiment. For this purpose, it is sufficient to monitor the core-level photoelectron spectra of oxygen, carbon, and other elements present on the sample surface. The intensity of the observed peaks will serve as a measure of the degree of cleanliness of the sample surface. Figure 30 shows the results obtained by Fadley [51] after several successive steps in the cleaning of an iron sample surface.

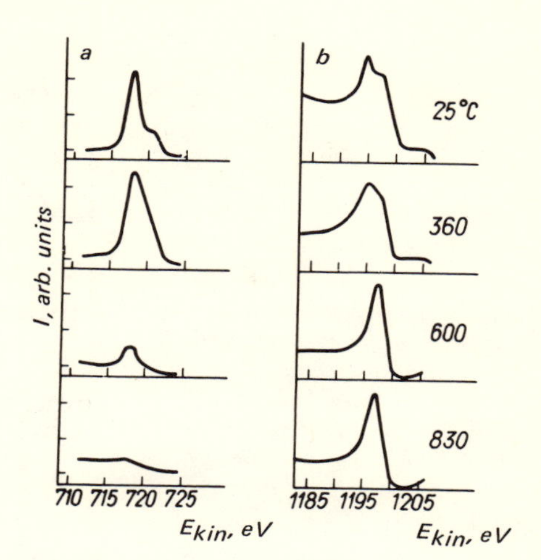

Figure 30. Photoelectron lines of oxygen 1s (a) and iron 3p (b) of a metallic iron sample, at various stages of surface cleaning in a hydrogen atmosphere.

As this figure shows, the oxygen 1*s* line at room temperature has a considerable intensity and contains two components. As the temperature of the sample was increased, a significant modification of the line intensity was observed. First, the component on the high-energy side, on the electron kinetic-energy scale (which corresponds to the low-energy side of the binding-energy scale), gradually disappeared. It seems reasonable to assume that this component originated from the oxygen molecules adsorbed on the sample surface, while the more intensive component originated from the iron oxides. It should be noted that after specimen cleaning the oxygen line did not disappear completely, but only decreased markedly in intensity. For the iron 3*p* photoelectrons a significant change in shape of the lines was observed. As the sample temperature was increased, the line component situated on the high-energy side, on the binding-energy scale, gradually decreased and finally disappeared completely.

The above method of cleaning sample surfaces is, however, not universally valid, since it causes removal of surface oxides, but not of carbides and nitrides. Moreover, hydrogen is an explosive gas and its handling in the experiment requires extreme care. A more efficient method consists in bombarding the surface to be investigated with argon ions. The method of surface cleaning by argon ion bombardment was first used by Hüfner *et al.* [53], and has subsequently been used in a great number of experiments. This method generally allows the removal of surface layers of sample material together with the impurities contained in them. Therefore, it can be sufficiently used for cleaning not only noble metals, but also metals characterized by a high reactivity, as well as insulators and semiconductors.

The method of surface cleaning by argon ion bombardment has also been used by the present authors. The cleaning was carried out in a special chamber,

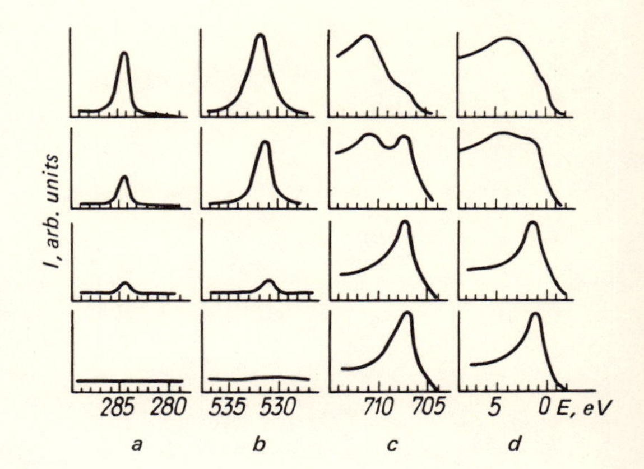

Figure 31. Dependence of the structure of the $2p_{3/2}$ and 3d photoelectron spectra of iron on the degree of cleanliness of the sample surface: (a) C-1s; (b) O-1s; (c) Fe-$2p_{3/2}$; (d) Fe-3D.

separated by a valve from the electron spectrometer. The argon pressure in the chamber was 0.1 torr, the voltage between the chamber walls and the sample holder 1 kV, and the current 15 mA. The cleaning was more efficient if a continuous flow of fresh argon toward the specimen was maintained. The duration of the cleaning process by argon discharge was 3–5 min.

Figure 31 shows how the spectra of iron, oxygen, and carbon change after progressive cleaning of an iron surface by argon ion bombardment [54]. Particularly pronounced changes are apparent in the valence band, which means that the study of the photoelectron spectra of pure iron is possible only on cleaned surfaces. As can be seen from Figure 31, this method of cleaning allows an efficient removal of oxygen and carbon from the surface layer of the sample. However, since the vacuum level in the spectrometer was only 10^{-7} torr, and the evacuation was performed by a turbomolecular pump, a layer of hydrocarbons was continuously built up on the sample surface, and consequently the 1*s* line of carbon appears in the photoelectron spectrum. The intensity of this line remained unchanged, however, for several hours. This is somewhat surprising and could perhaps be explained in terms of the formation of a surface hydrocarbon layer that protects the cleaned iron from oxidation. Of course, some argon ions are implanted into the sample, but the intensity of argon photoelectron lines is so low that this effect may be neglected.

The values of binding energies for a number of metals that have an even higher chemical reactivity than iron can be determined after mechanical cleaning of the sample surface. Let us consider, for example, determination of the binding energies of niobium $3d_{3/2}$ and $3d_{5/2}$ electrons. Figure 32 shows the niobium photoelectron spectrum over the energy range where the $3d_{3/2}$ and $3d_{5/2}$ lines are situated [55].

The spectra were recorded for foils of 0.2 mm thickness, made from thin sheets of metallic niobium. A magnetic electron spectrometer was used with the

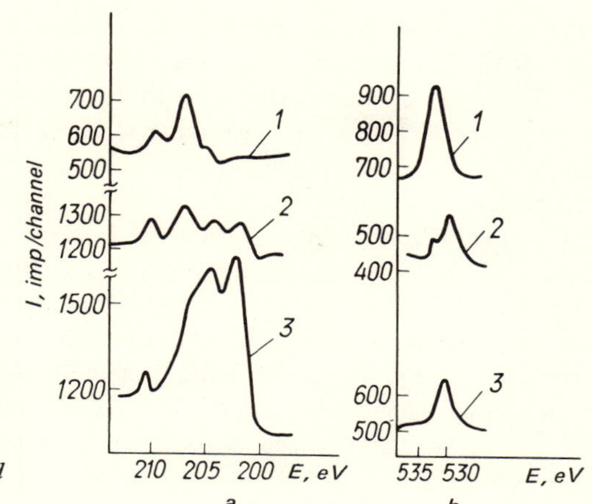

Figure 32. Photoelectron spectra of (a) niobium; (b) oxygen; (1) before cleaning, (2) after mechanical cleaning, (3) after mechanical cleaning and heating up to 400°C.

magnesium $K\alpha_{1,2}$ line as the source of photons. The experiment was performed in a rather poor vacuum of only 10^{-4} torr. As can be seen from Figure 32, the spectrum of the uncleaned specimen exhibits two well-defined peaks, situated at energy values corresponding to the $3d_{3/2}$ and $3d_{5/2}$ electronic states of metallic niobium. After mechanical cleaning and subsequent heating of the sample up to 400°C, the lines corresponding to the metallic state are increased in intensity. After mechanical cleaning, the intensity of $3d_{3/2}$ and $3d_{5/2}$ lines corresponding to the oxide state is also increased, as compared to the uncleaned foils. This demonstrates that on the sample surface, besides niobium oxides, there also exist adsorbed gases such as oxygen and hydrocarbons.

The presence of adsorbed oxygen on the sample surface is confirmed by the study of the oxygen $1s$ lines. For the uncleaned sample this line is rather broad (3.8 eV). The energy position of its maximum (532.4 eV) indicates the preponderance of adsorbed oxygen on the sample surface. After mechanical cleaning of the sample, a second component appears in the structure of the $1s$ oxygen line. The high-energy component is situated at approximately the same binding-energy value as the peak in the previous case, but the second component is shifted by 2.6 eV toward lower binding-energy values. This last line corresponds to niobium oxide. After heating up to 400°C, there remains only one peak in the photoelectron spectrum. This corresponds to oxygen in niobium oxide, with the adsorbed oxygen driven from the surface. Thus, in this case, even a simple operation of cleaning the sample surface makes it possible to determine the values of core-level binding energies for pure niobium as 204.3 eV for $3d_{3/2}$ electrons and 202.1 eV for $3d_{5/2}$ electrons. A drawback of this method is that in the spectrum obtained after mechanical cleaning the lines corresponding to metallic niobium are still rather broad. This introduces errors, albeit small (0.3–0.4 eV), in the determination of binding-energy values.

The study of the valence band photoelectron spectra of metals requires more stringent conditions, as far as sample surface cleanliness is concerned, than the study of core-electron spectra. As we have seen in the case of iron, for example, the structure of valence band photoelectron spectra may be strongly affected by the actual composition of the sample surface. For this reason, ultrahigh vacuum in the spectrometer is necessary and should be maintained at a level of about 10^{-10} torr during recording of the photoelectron spectra, especially in the case of samples with high oxidation affinity. One way to study oxidation-sensitive metal samples is to prepare them by evaporation onto a substrate under high-vacuum conditions. This method was used by Baer and Busch in the study of valence band photoelectron spectra of aluminum [56]. A thin aluminum layer was evaporated onto the optically flat surface of a quartz crystal substrate. During 1 min of rapid evaporation, the pressure increased to 1×10^{-9} torr, but 10 min after the evaporation it had decreased again down to 2×10^{-11} torr. However, even in this experiment, in spite of the very careful sample preparation, weak 1*s* lines of oxygen and carbon were observed. Their intensity was lower than 1% of the intensity of the aluminum 2*s* lines. The valence band photoelectron spectrum of aluminum obtained in this experiment is shown in Figure 33. The weak structure observed at 10-eV binding energy can be attributed to the 2*p* electrons of oxygen, since its intensity is directly correlated to the intensity of the oxygen 1*s* lines. This was shown by using sample layers evaporated under a vacuum of 10^{-7} torr.

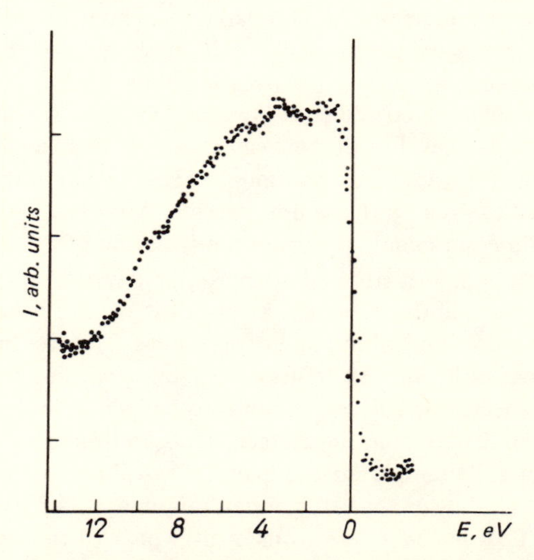

Figure 33. Valence band photoelectron spectrum of aluminum.

Fuggle *et al.* [57] have studied the photoelectron spectra of some alloys of noble metals with aluminum. The samples were obtained by successive evaporation of layers of both materials, one over the other, followed by diffusion under appropriate conditions. The total thickness of the evaporated layers was approximately equal to 1000 Å. The evaporation was performed twice, under a vacuum of 10^{-8} torr, and was followed by immediate monitoring of the presence of oxygen and carbon on the sample surface. Shortly after evaporation, the pressure level in the reaction chamber dropped down to 10^{-10} torr, and the sample was transferred to the analyzing chamber, which was kept constantly under a pressure of 5×10^{-10} torr. The observed intensity of oxygen and carbon 1*s* lines after each of the evaporation operations was insignificant. The sample with the two layers evaporated one over the other was heated up to a temperature greater than 180°C. The diffusion process was allowed to take place for about 10 min, and was considered to be complete when the photoelectron spectrum ceased to change with time. Various stages in the diffusion process are shown in Figure 34.

The cleaning of semiconductors and insulators is often performed by using an argon ion gun operated at a voltage of 1 kV and a current of 10 mA. Moreover, the sample is heated from behind by electron bombardment with an elec-

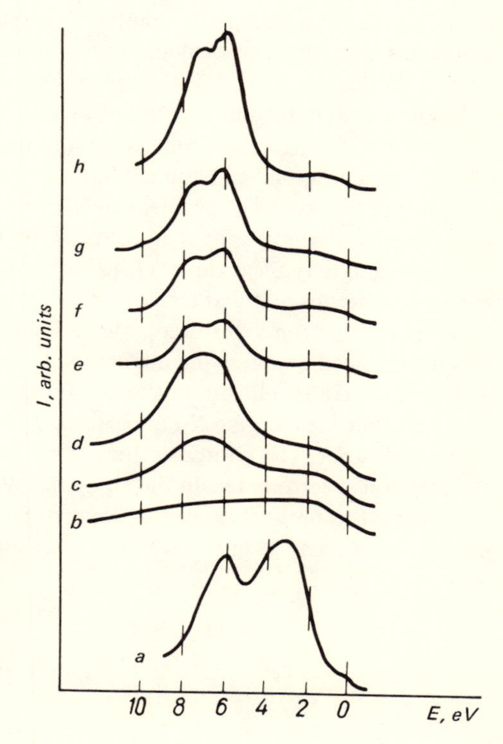

Figure 34. Valence band photoelectron spectra of aluminum and gold in Al–Au films at various stages of the diffusion process: (a) gold; (b, c, d, e, f, g) intermediate stages of the diffusion process; (h) final product of the diffusion process, Al_2Au compound.

tron gun. In order to increase the efficiency of photoelectron counting, the sample surface is first mechanically polished and then electropolished. Good sample flatness is also necessary for calibration by deposition of a gold layer on the sample surface.

For the preparation of clean sample surfaces, evaporation of the desired substance on a substrate under high vacuum conditions gives better results than bombardment with argon ions. Hüfner *et al.* [58] have shown that the cleaning of palladium samples by argon ion bombardment for a rather long time (up to 30 min) results in the formation of a layer of amorphous palladium on the surface. This explains why the valence band photoelectron spectra of this sample were different from the spectra of palladium deposited on a gold substrate. Not only were the fine details of the photoelectron spectrum changed, but also the intensity of the Fermi level was decreased. It has been found that the valence band photoelectron spectrum of palladium undergoes a greater modification than the spectra of some other metals such as copper, silver, and gold. For nickel, however, the trend of spectra behavior is similar to that observed in the case of palladium.

In the IEE-15 spectrometer, because of its construction, the sample is cylindrical with a diameter of 11 mm and length 20 mm. When the preparation of appropriately shaped bulk samples is not possible, and when nonconductive specimens are to be investigated, use is made of the powdered material deposited on an adhesive tape, which is then placed on a metallic cylinder.

In conclusion, it should be noted that X-ray photoelectron spectroscopy allows the study of the electron structure, not only of solid materials, but also of gases, metal vapors, and liquids. In the present work, however, we shall limit ourselves to discussion of problems related to the study of solids. X-ray photoelectron spectroscopy as applied to the study of gases has been treated in detail in a monograph by Siegbahn *et al.* [2], although at present the study of metal vapors and liquids is still at an early stage. The first metal vapors studied by X-ray photoelectron spectroscopy were bismuth and lead [59]. The metal samples were placed in a molybdenum crucible having an opening of 1 mm diameter, and were heated in a high-temperature furnace. The experimental arrangement was such that the exciting X-ray beam was perpendicular to the stream of molecular metal vapor. The photoelectrons were then admitted into the analyzer of the electron spectrometer. In the study of liquids [60], it has been found that it is important to produce a nonturbulent stream of liquid material, having a low diameter and passing very near to the entrance slit of the spectrometer.

2

Physical Principles of Electron Spectroscopy

In the field of X-ray photoelectron spectroscopy there exists at present a large amount of experimental data that requires a complete theoretical analysis. Before the first papers reporting experimental data on photoelectron spectra were published, it was considered that the energy eigenvalues determined for multielectron systems by applying the Hartree–Fock or Hartree–Fock–Slater methods were an adequate approximation for the electron binding energies. However, it has been shown that the magnitude of the binding energies of deeply lying core electrons is affected significantly not only by relativistic effects but also by relaxation effects. Creation of holes in the photoionization process results in a modification of the electron wave functions of the system, and this introduces a significant contribution to the values of binding energies. However, in the case of core electrons, both the relativistic and the relaxation effects are not very sensitive to the environment of the given atom incorporated in a molecule or in a solid material, and consequently the magnitude of core-level chemical shifts is determined mainly by modifications of the density-of-state distribution of the valence electrons.

Another interesting effect observed in the photoelectron spectra, that is also related to the existence of a hole in the core levels of atoms with incomplete valence shells, is the multiplet splitting of photoelectron lines. This effect is due to the interaction of the spin and orbital momenta of the hole with the spin and orbital momenta of the valence band electrons. In some cases, as a result of the photoionization process, one additional electron may be emitted, or an excitation of the system of electrons may occur.

For the interpretation of experimental photoelectron spectroscopy data in the study of valence states of molecules and solid state materials, it is in general necessary to take into account the dependence of the photoionization cross section on the nature of the particular state under investigation. For solid state materials, the different magnitude of the photoionization cross sections corresponding to electrons of different symmetry types is the main factor that determines the difference observed experimentally between the energy distributions of photoelectrons and the density-of-state distributions of valence band electrons in crystals. However, even in these cases, the valence band photoelectron spectra reflect the van Hove singularities in the density-of-states distributions.

At present, there exists a number of methods for the calculation of energy bands and electron wave functions in crystals, the most popular ones being the method of orthogonal plane waves (OPW), the method of augmented plane waves (APW), the method of Green's functions (KKR), and the pseudopotential method. In this chapter, we will discuss briefly the basic ideas of these methods. Subsequently, spectra calculated on the basis of the various theoretical models will be compared with the experimental results. First, it should be noted that the energy positions of the van Hove singularities, as determined experimentally from electron spectra, may be used as parameters in the determination of pseudopotential form factors in the pseudopotential method.

In recent years, special attention has been devoted to the study of such disordered systems as the alloys of transition and noble metals. We have therefore included a discussion of one of the most important methods used in the study of disordered systems, namely, the coherent potential approximation.

The method of X-ray photoelectron spectroscopy also allows the determination of the angular distribution of emitted photoelectrons. Study of angular distributions gives information about the partial density of states of electrons of various types of symmetry.

The results obtained by the use of X-ray photoelectron spectroscopy may in a number of cases be supplemented by results obtained from other techniques. In the present work, we will therefore discuss the present and future possibilities of a supplementary use of the method of X-ray photoelectron spectroscopy in combination with X-ray emission spectroscopy and, in some cases, with data obtained by the use of Mössbauer spectroscopy or nuclear magnetic resonance.

Binding Energies and Chemical Shifts of Core Levels

In order to evaluate the binding energy $E_B(k) = E^f(k) - E^i$, it is necessary to calculate the total energy of the multielectron system corresponding to the initial and the final states. Total energies may be calculated by the methods of Hartree–Fock or Dirac–Fock, in the nonrelativistic and relativistic approximations, respectively. Let us consider the simplest case, when the wave function Ψ of the system of $N = 2n$ electrons may be represented as a single Slater determinant

$$\Psi = \frac{1}{\sqrt{N!}} \begin{vmatrix} \varphi_1(1)\alpha(1)\,\varphi_1(1)\beta(1) & \ldots & \varphi_n(1)\beta(1) \\ \ldots & \ldots & \ldots \\ \varphi_1(N)\alpha(N)\,\varphi_1(N)\beta(N) & \ldots & \varphi_n(N)\beta(N) \end{vmatrix}, \tag{10}$$

in which each spin-orbital, relativistic effects being neglected, is represented as the product of functions that depend only on space coordinates and functions that depend only on the spin. In the given spin-orbital, the space parts of the wave functions are here assumed to be the same for both spin orientations, which gives the so-called restricted Hartree-Fock approximation.

The single-electron wave functions $\varphi_i(\mathbf{r})$ are determined from the Hartree-Fock equations. For the Hamiltonian

$$H = -\frac{1}{2}\sum_{i=1}^{N} \nabla_i^2 - \sum_{i=1}^{N}\sum_{l=1}^{P} \frac{Z_l}{r_{il}} + \sum_{i>j} \frac{1}{r_{ij}} + \sum_{l>s} \frac{Z_l Z_s}{R_{ls}},$$

which describes the system consisting of N electrons and P nuclei, the Hartree-Fock equations may be written as follows:

$$\left[-\frac{1}{2}\nabla_1^2 - \sum_{l=1}^{P} \frac{Z_l}{r_{1l}}\right] \varphi_i(\mathbf{r}_1) + \sum_{j=1}^{n} [2J_j - K_j]\, \varphi_i(\mathbf{r}_1) = \varepsilon_i \varphi_i(\mathbf{r}_1), \quad (11)$$

where

$$J_j \varphi_i(\mathbf{r}_1) = \int \varphi_i^*(\mathbf{r}_2) \frac{1}{r_{12}} \varphi_i(\mathbf{r}_1)\, \varphi_j(\mathbf{r}_2)\, dV_2,$$

$$K_j \varphi_i(r_1) = \int \varphi_j^*(\mathbf{r}_2) \frac{1}{r_{12}} \varphi_i(\mathbf{r}_2)\, \varphi_j(\mathbf{r}_1)\, dV_2.$$

Representation of the wave function Ψ by the expression (10) is generally valid only for systems with completely filled shells, since in this case the function (10) is also an eigenfunction of the operators S^2 and S_z^2, which commute with the Hamiltonian H. For systems with incompletely filled shells, the wave function Ψ is usually represented as a sum of Slater determinants.

After finding the self-consistent solution of the system of equations (11), the orbital energies ϵ_i may be determined by using the formula:

$$\varepsilon_i^{HF} = \varepsilon_i^0 + \sum_{j=1}^{n} (2J_{ij} - K_{ij}), \quad (12)$$

where

$$\varepsilon_i^0 = \left\langle \varphi_i \left| -\frac{1}{2}\nabla_1^2 - \sum_{l=1}^{P} \frac{Z_l}{r_{1l}} \right| \varphi_i \right\rangle,$$

and J_{ij} and K_{ij} represent the Coulomb and exchange integrals, respectively:

$$J_{ij} = \langle \varphi_i | J_j | \varphi_i \rangle = \iint \varphi_i^* (\mathbf{r}_1) \varphi_j^* (\mathbf{r}_2) \frac{1}{r_{12}} \varphi_i (\mathbf{r}_1) \varphi_j (\mathbf{r}_2) \, dV_1 dV_2,$$

$$K_{ij} = \langle \varphi_i | K_j | \varphi_j \rangle = \iint \varphi_i^* (\mathbf{r}_1) \varphi_j^* (\mathbf{r}_2) \frac{1}{r_{12}} \varphi_i (\mathbf{r}_2) \varphi_j (\mathbf{r}_1) \, dV_1 dV_2.$$

The total initial energy E^i is not equal to the sum of the single-electron energy values:

$$E^i = \langle \Psi_i | H | \Psi_i \rangle = 2 \sum_{i=1}^{n} \varepsilon_i^0 + \sum_{i,j=1}^{n} (2J_{ij} - K_{ij}). \tag{13}$$

It is to be noted that this expression does not include the term that describes the nuclear repulsion. With nuclei in fixed positions, inclusion of the additive term does not raise difficulties. Calculation of the energy of the system after the X-ray photoionization process has taken place represents a more difficult problem, since creation of holes results in relaxation of the wave functions $\varphi_i(\mathbf{r})$ which enter into the expression (10) for Ψ_i. In order to avoid the difficulties related to such hole effects, the calculation of binding energies in the past has often made use of the Koopmans' theorem [61]. The Koopmans' theorem is valid under the condition that the one-electron wave functions $\varphi_i(\mathbf{r})$ that enter into the expression for the electron wave function Ψ_i of the whole system in the initial state coincide with the corresponding wave functions $\varphi_i(\mathbf{r})$ that enter into the expression for the electron wave function Ψ_f of the system in the final state. In this case, the expression for the energy $E^f(k)$ may be obtained from formula (13) by eliminating the terms related to one of the two electrons in the kth orbital:

$$E^f(k) = \varepsilon_k^0 + 2 \sum_{i \neq k} \varepsilon_i^0 + \sum_{i,j=1}^{n} (2J_{ij} - K_{ij}) - \sum_{i=1}^{n} (2J_{ik} - K_{ik}).$$

It then follows that

$$E_B^V(k) = -\varepsilon_k^0 - \sum_{i=1}^{n} (2J_{ik} - K_{ik}),$$

and, by taking into account the expression (12), one obtains

$$E_B^V(k) = -\varepsilon_k^{HF},$$

which represents the Koopmans' theorem. In deriving this expression, it has been assumed that in the photoionization process the nuclei do not change their positions. This assumption is based on the fact that the oscillation periods of the nuclei (of the order of 10^{-13} sec) are much longer than the time interval in which the photoionization process takes place.

In some cases, particularly in the study of energy band structures of crystals, the wave functions φ_k and the energies E_k are deduced from the one-electron equation

$$H\varphi_k = E_k \varphi_k, \tag{14}$$

where

$$H = -\frac{1}{2}\nabla^2 + V(\mathbf{r}). \tag{15}$$

Here, it cannot be assumed that the energy eigenvalues E_k are close to the binding energies of electrons in the k state. Therefore, in order to obtain the binding-energy values, it is necessary to add a correction term δE_k to the energy E_k:

$$-E_B^V(k) = E_k + \delta E_k,$$

where

$$\delta E_k = \left\langle k \left| -\sum_{l=1}^{P} \frac{Z_l}{r_{1l}} - V(\mathbf{r}) \right| k \right\rangle + \sum_{j=1}^{n} (2J_{kj} - K_{kj}).$$

The term δE_k was introduced for the first time by Lindgren [62]. This correction improves appreciably the agreement with the one-electron energy values calculated by the Hartree-Fock method for atoms [1]. The expression for the total energy E^i, including the correction, may be written as follows:

$$\begin{aligned} E^i &= 2\sum_{i=1}^{n} \varepsilon_i^0 + \sum_{i,j=1}^{n} (2J_{ij} - K_{ij}) = 2\sum_{i=1}^{n} \varepsilon_i - \sum_{i,j=1}^{n} (2J_{ij} - K_{ij}) = \\ &= 2\sum_{i=1}^{n} E_i - \sum_{i,j=1}^{n} (2J_{ij} - K_{ij}) + 2\sum_{i=1}^{n} \delta E_i = \\ &= 2\sum_{i=1}^{n} E_i + \sum_{i=1}^{n} \left\langle i \left| -\sum_{l=1}^{P} \frac{Z_l}{r_{1l}} - V(\mathbf{r}) \right| i \right\rangle + \sum_{i=1}^{n} \delta E_i. \end{aligned}$$

Snow *et al.* [63] have calculated the energies E^i for atoms, by taking into account the Slater-type exchange potential $V_{ex}^S = -6\,[(3/8\pi)\,\rho]^{1/3}$, but without including the term $\Sigma_{i=1}^{n}\, \delta E_i$. The agreement with the energy values E^i obtained by the Hartree–Fock method is significantly improved if one takes into account the contribution of this term. Lindgren has also proposed the use of a modified Slater-type exchange potential:

$$V_{ex}^{L} = -C\left[\frac{81}{32\pi^2}\right]^{1/3} r^{n/3-1}\tilde{\rho}(r)^{m/3}, \tag{16}$$

where $\tilde{\rho} = 4\pi r^2 \rho$ is the radial density of electrons, and the parameters C, n, m, which are equal to unity in the Slater approximation, can be determined from the condition of minimum total energy of the electron system.

In the calculation of binding energies for heavy elements, account should be taken of relativistic effects. The relativistic equation for one particle, as proposed by Dirac, is

$$H\Psi_i = [c\boldsymbol{\alpha} \cdot \mathbf{p} + \beta m_0 c^2 + V(\mathbf{r})]\,\Psi_i = \varepsilon_i \Psi_i, \tag{17}$$

where c is the velocity of light,

$$\boldsymbol{\alpha} = \begin{pmatrix} \boldsymbol{\sigma} & 0 \\ 0 & \boldsymbol{\sigma} \end{pmatrix},$$

and

$$\beta = \begin{pmatrix} I & 0 \\ 0 & -I \end{pmatrix};$$

$\boldsymbol{\sigma}$ represents the Pauli matrices, I is a unitary matrix of dimension 2×2, and Ψ_i represents a bispinor:

$$\Psi_i = \begin{pmatrix} \psi_{i1} \\ \psi_{i2} \\ \psi_{i3} \\ \psi_{i4} \end{pmatrix}.$$

If the energy E is measured against the rest energy as reference, then equation (17) may be written as follows:

$$[c\boldsymbol{\alpha} \cdot \mathbf{p} + (\beta - 1)\,c^2 + V(\mathbf{r})]\,\Psi_i = E_i \Psi_i. \tag{18}$$

The approximate multielectron Hamiltonian that includes relativistic effects, but does not include magnetic interactions and retardation effects, may be written

$$H = \sum_i \left[c\boldsymbol{\alpha}_i \cdot \mathbf{p}_i + (\beta_i - 1)\,c^2 - \frac{Z}{r_i}\right] + \sum_{i>j} \frac{1}{r_{ij}}\,.$$

By using this Hamiltonian, Rosén and Lindgren [64] obtained the following expression for the total energy of the electron system:

$$E^i = \sum_i \langle \Psi_i | f | \Psi_i \rangle + \sum_{i<j} \langle ij | g | ij \rangle,$$

where

$$\langle \Psi_i | f | \Psi_i \rangle = \int \Psi_i(1) \left[c\alpha \cdot \mathbf{p} + (\beta - 1) c^2 - \frac{Z}{r} \right] \Psi_i(1)\, d\tau_1,$$

$$\langle ij | g | ij \rangle = \iint \Psi_i^*(1) \Psi_j^*(2) \frac{1 - P_{12}}{r_{12}} \Psi_i(1) \Psi_j(2)\, d\tau_1 d\tau_2.$$

Here, P_{12} represents the operator of permutation of the variables that characterize the first and the second particles. The Coulomb and exchange terms enter into both the nonrelativistic and the relativistic equations. In the relativistic equation, the nonlocal exchange potential may be replaced by the local Slater potential. The total energy is in this case given by

$$E^i = \sum_i E_i - \frac{1}{2} \sum_i \left\langle i \left| V(\mathbf{r}) + \frac{Z}{r} \right| i \right\rangle + \frac{1}{2} \sum_i \delta E_i, \tag{19}$$

where $V(\mathbf{r})$ represents the sum of the Coulomb and exchange potentials,

$$V(\mathbf{r}) = -\frac{Z}{r} + \frac{1}{r} \int_0^r \rho(r')\, dr' + \int_r^\infty \frac{\rho(r')\, dr'}{r'} + V_{ex}^S; \tag{20}$$

and

$$\delta E_i = \sum_j \langle ij | g | ij \rangle - \left\langle i \left| V(\mathbf{r}) + \frac{Z}{r} \right| i \right\rangle.$$

Ψ_i and E_i are the eigenfunctions and the eigenenergies, respectively, of the relativistic one-electron equation (18) with the potential (20). As exchange potential, use may also be made of the type of exchange potential given by formula (16). The parameters that enter into this potential are determined by minimization of the total energy of the system, as given by equation (19). For relatively light elements (i.e., $Z \leqslant 29$), $C \sim 0.8$–0.9, and $n \sim 1.1$–1.3, while for heavy elements ($Z > 36$), $C \sim 0.7$–0.8, and $n \sim 1.10$–1.20. In all of the cases considered, $m = 1$. It is worth noting that, by choosing the parameters $C = 1$ and $n = 1$, one obtains the Slater-type exchange potential.

In the discussion of the binding energy we should distinguish two different models. One, which we can denote by method A, is built upon identifying the binding energy with the corresponding Hartree–Fock energies. The more appropriate procedure is to recognize that the levels seen in the spectrum are the levels of the ion left behind, i.e., the ($N-1$)-particle system. In this method, referred to as B, one obtains the energies by solving the self-consistent problem for the ionic state. Figure 35 shows the comparison with experimental data for 1s electrons. For valence electrons, where the hole wave function is extended, the two methods agree rather well, whereas for strongly bound electrons only method B gives a satisfactory agreement.

The existing disagreement between the theoretical and experimental values of binding energies may be attributed to the fact that the correlation effects, the magnetic retardation effects, and the quantum-electrodynamical effects have not been taken into account. Since the large majority of experimental data have been obtained from solid specimens, it is also necessary to take into account the energy shift resulting from the transition from single atom states to solid or molecular states. The magnitude of this shift (5–10 eV) is of the same order as the chemical shift. Magnetic and quantum-electrodynamical effects are particularly important in the case of heavy atoms.

The total energy E_T may be written as follows:

$$E_T = E_T^{HF} + E^C + E^R. \tag{21}$$

The first term in this expression represents the nonrelativistic Hartree–Fock total energy, the second represents the electron–electron correlation energy, and the third represents the relativistic energy including also the quantum-electrodynamical effects. In order to calculate the correlation energy, it is necessary to

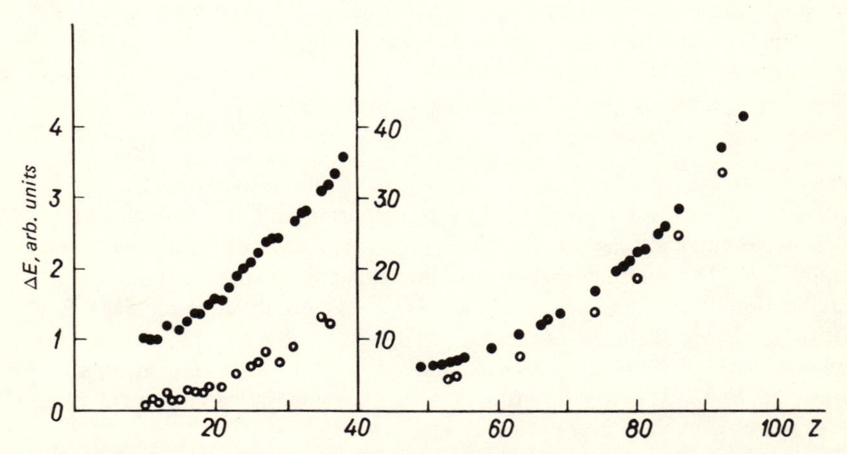

Figure 35. Discrepancy between theoretical values of binding energy of 1s electrons as calculated by methods A(●) and B(○). The corresponding experimental data are also shown.

express the wave function of the electron system as a linear combination of functions, each of them expressed by a Slater determinant. In this linear combination of determinants, besides the determinant corresponding to nonexcited orbitals, it is also necessary to include the Slater determinants that contain singly and multiply excited molecular orbitals. This method of calculating energies is called the method of configuration interaction (CI). By taking into account expression (21), the binding energy may be written as follows:

$$E_B^V = E^f - E^i = E^{f,HF} - E^{i,HF} + \Delta E^C + \Delta E^R,$$

where ΔE^C and ΔE^R represent the differences between the final-state and initial-state correlation and relativistic energies, respectively. By making use of the Koopmans' theorem, it is found that $E^{f,HF} - E^{i,HF}$ is equal to $-\epsilon^{HF}$. A more accurate expression for the binding energy should account for the relaxation effects, and therefore

$$E_B^V = -\varepsilon^{HF} + E^{R-n} + \Delta E^C + \Delta E^R. \tag{22}$$

Table 4 shows the values of the energies in equation (22), for the case of water molecules, as reported by Meyer [65]. The values of E_B^{CI} include the oscillation energy and the relativistic effects.

The theoretical values of binding energies obtained by the method of configuration interaction agree well with the experimental data. Good agreement between calculated and experimental values of binding energies is also obtained when account is taken of the relaxation, correlation, and relativistic energies. The value of E_B^{HF}, calculated by taking into account the relativistic effects and the oscillation energies, is lower than the experimental value. This indicates that the total correlation energy is somewhat larger for the initial state of the $2N$-electron system than for the final state of the $(2N - 1)$-electron system. This condition is, however, not fullfilled for $2a_1$ electrons, which may perhaps be explained by the fact that it is difficult to perform Hartree-Fock calculations for electron systems with an incompletely filled shell, if in these systems there

TABLE 4. Comparison between the Theoretical and Experimental Values of the Binding Energies of the Water Molecule

Orbital	$E_{B(\mathrm{exp})}$	E_B^{CI}	E_B^{HF}	$-\epsilon^{HF}$	$-E^{R-n}$	ΔE^C	ΔE^R
$1a_1$ (01s)	540.2 ± 0.2	540.0	539.5	559.5	−20.4	0.5	0.4
$2a_1$	32.2 ± 0.2	32.4	34.2	36.8	—2.5	−1.9	0.0
$1b_2$	18.6 ± 0.2	18.7	17.4	19.5	−1.9	1.3	0.0
$3a_1$	14.7 ± 0.1	14.6	13.2	15.9	−2.5	1.4	0.0
$1b_1$	12.6	12.4	11.0	13.9	−2.8	1.4	0.0

exist deeply lying states of the same type of symmetry, characterized by close energy values [66]. The relaxation energy for the 1*s* states of oxygen in the water molecule is large (20.4 eV). For valence electrons, it is an order of magnitude lower, but can vary for different states by approximately 1 eV. Consequently, the orbital energies calculated without taking into account ΔE^{R-n} are not always in the correct sequence for the closely spaced valence orbitals.†

By using equation (22), one may write for the magnitude of the chemical shift of the core level *i* the following expression:

$$\Delta E_i = -\Delta\varepsilon_i^{HF} + \Delta E_i^{R-n} + \delta\Delta E_i^C + \delta\Delta E_i^R.$$

The magnitudes of the energies ΔE_i^{R-n}, $\delta\,\Delta E_i^C$, and $\delta\,\Delta E_i^R$ are much lower than the corresponding individual contributions to the binding energies of electrons in the *i*th state for the atom in the first and in the second compound.

The energy $\Delta\epsilon_i^{HF}$ is determined by the ground state of molecules and solids, while ΔE_i^{R-n} is determined by a more complicated function that reflects the dynamics of the photoionization process itself. In formula (22), the relaxation energy E^{R-n} is determined as

$$-\varepsilon^{HF} = E_B^{HF} - E^{R-n},$$

where

$$E_B^{HF} = E^{f,HF} - E^{i,HF}.$$

E^{R-n} is negative, since $|E_B^{HF}| < |\epsilon^{HF}|$, and $\epsilon^{HF} < 0$.

The relaxation energy for molecules and crystals may be expressed as the sum of two terms‡:

$$E_A^{R-n} = E_A^{1,R-n} + E_A^{2,R-n},$$

where the energy $E_A^{1,R-n}$ is determined by the increase of the local electron density in the vicinity of atom *A*, and $E_A^{2,R-n}$ is determined by the additional redistribution of charge over the whole molecule or crystal. In the redistribution of the electron density, a tendency towards an increase of electron density in the direction of atom *A* is observed. Tables 5 and 6 show the values of energies

†*Editors' note*: One might remark that the configuration-interaction approach is useful only for molecules and is not applicable to valence electrons in solids.

‡*Editors' note*: The relation is only approximately valid. In practice it is limited to the ΔSCF approximation.

TABLE 5. Values of the Relaxation Energies for Atomic States Obtained by the Modified Hartree-Fock-Slater Method (Nonrelativistic Calculations)

Atom	1*s*	2*s*	2*p*
He	1.5		
Li	3.8		
Be	7.0	0.7	
B	10.6	1.6	0.7
C	13.7	2.4	1.6
N	16.6	3.0	2.4
O	19.3	3.6	3.2
F	22.0	4.1	3.8

E_A^{R-n} for the series of elements from helium to copper, taken from the data of Gelius *et al.* [67].

Snyder [68] has shown that the relaxation energy of an isolated atom may be represented as the sum of the contributions of different electronic shells. The

TABLE 6. Values of the Relaxation Energies for Atomic States Obtained by the Modified Hartree-Fock-Slater Method (Relativistic Calculations)

Atom	1*s*	2*s*	2*p*	3*s*	3*p*	3*d*	4*s*
F	22.1	4.1	3.9				
Ne	24.8	4.8	4.7				
Na	23.3	4.1	4.7	0.3			
Mg	24.6	5.2	6.0				
Al	26.1	6.1	7.1	1.0	0.2		
Si[a]	27.1	7.0	8.0				
P	28.3	7.8	8.8				
S	29.5	8.5	9.6	1.4	0.9		
Cl	30.7	9.3	10.4				
Ar	31.8	9.9	11.1	1.8	1.4		
K	31.2	9.1	10.5				
Ca	32.0	9.6	11.1				
Sc	33.8	11.5	12.9				
Ti	35.4	13.0	14.4	3.6	3.4	2.0	0.3
V	37.0	14.5	16.0				
Cr[a]	38.6	15.9	17.4				
Mn	40.1	17.2	18.8			3.6	0.4
Fe	41.6	18.5	20.0	5.7	5.3		
Co	43.2	19.8	21.6			4.1	
Ni	44.7	21.1	22.9	6.7	6.3		
Cu[b]	48.2	23.7	25.7	7.7	7.2	5.3	0.3

[a]Interpolated values.
[b]Values taken from [64].

contribution of each shell is proportional to the occupation number. These arguments are applied by Gelius [66] to molecules. For $E_A^{1,R-n}$, one may write

$$E_A^{1,R-n} = k'q_A + l'_A. \tag{23}$$

The constant k' is approximately equal to 2.5 eV for the 1s electrons of elements in the second period and to 1.5 eV for those of elements in the third period; l'_A is a nearly constant term, approximately equal to E_A^{R-n} for the free atom. According to Table 5, the electron relaxation energy for the 1s state of the oxygen atom is -19.5 eV. By introducing this value into equation (23), together with the approximate value of the charge $q_A = 0.3$, one obtains $E_A^{1,R-n} = -18.5$ eV. From Table 4, it can be seen that the total relaxation energy for this case is 20.4 eV. Therefore, $E_A^{2,R-n}$ represents only a small part of the relaxation energy. Since the relaxation energy is always negative, it follows that the magnitude of $\delta\Delta E_A^{2,R-n}$ is even smaller. The magnitude of $\delta\Delta E_A^{2,R-n}$ in molecules (or solid state materials) depends on the degree of localization of the molecular orbitals (or orbitals of the solid), on the type of symmetry, and on the distance to neighboring atoms. Knowledge of the magnitude of $E_A^{2,R-n}$ for molecules is at present limited.

Hedin and Johansson [69] have expressed ΔE^{R-n} in terms of the polarization potential arising from the presence of the hole on the kth orbital. The Hamiltonian was written as

$$H = h + V = h + \sum_i V_i,$$

where h is the one-electron operator, and the operator V_i describes the Coulomb and exchange interactions. The Hamiltonian H^* corresponds to the system in which one electron has been removed from the kth orbital:

$$H^* = h + V^* = h + V - V_K + V_P.$$

The polarization potential is given by the expression

$$V_P = \sum_{i \neq k} (V_i^* - V_i).$$

The binding energy may be written as follows:

$$\begin{aligned} -E_B^V(k) &= E(N) - E^*(N-1, k) = \\ &= \varepsilon_k^{HF} - \{E^*(N-1, k) - E(N-1, k)\} = \\ &= \varepsilon_k^{HF} + \frac{1}{2}\langle \varphi_k | V_P | \varphi_k \rangle + \frac{1}{2}\sum_i{}' \langle \delta\varphi_i | H^* - H - \delta\varepsilon_i^{HF} | \delta\varphi_i \rangle, \end{aligned}$$

where the function $\delta\varphi_i$ is determined from the expression $\varphi_i^* = \varphi_i + \delta\varphi_i$.

Theoretical calculations have shown that the magnitude of the last term in the above expression is small, compared to the first two terms, so that

$$- E_B^V(k) \simeq \varepsilon_k^{HF} + \frac{1}{2}\langle k | V_P | k \rangle. \tag{24}$$

The model polarization potential proposed by Hedin [69] has given good results in the calculation of binding energies of electrons in core orbitals.

For sodium and potassium in the atomic state and in the single-ionized state, calculation of the contributions of various orbitals to the magnitude of binding energy, as given by formula (24), has shown that the relaxation inside the given shell is small compared to that corresponding to outer shells, while for the more deeply lying shells it is negligibly small.

Shirley [70] has used the method of polarization potential for the calculation of binding energies of electrons in neon, argon, krypton, and xenon. It follows from equation (24) that in order to determine $E_B^V(k)$, it is necessary to know the polarization potential V_P. This can be calculated directly for atoms having a hole on the given level. Rather good results, however, are also obtained here by using the approximation of equivalent cores in which it is assumed that the electron lying on an inner shell almost completely screens the outer shell electrons from the nuclear charge equal to unity. Therefore, the outer orbital of an atom having a nuclear charge Z and a hole on an inner level can be approximated fairly well by the outer orbital that describes the ground state of the next element, i.e., the element with nuclear charge $Z + 1$. Although this approximation is no longer valid for deeply lying electronic states, this does not appreciably affect the magnitude of those Coulomb and exchange integrals that are determined by the wave functions of the deeply lying electrons, and of the valence electrons of the atom.

It can therefore be assumed that the magnitude of the integrals that describe the hole states of atoms with nuclear charge Z will be approximately equal to the integrals corresponding to the ground state of the atom with nuclear charge $Z + 1$. By using Slater's approach [71], this approximation, for $n' > n$, may be expressed as follows:

$$F_k(nl,\ n'l';\ Z(\overline{nl})^*) \approx F_k(nl,\ n'l';\ Z+1);$$
$$G_k(nl,\ n'l';\ Z(\overline{nl})^*) \approx G_k(nl,\ n'l';\ Z+1).$$

As we have already mentioned, the relaxations of the inner orbitals, $n' < n$, and of the orbital from which the electron has been removed, $n' = n$, are small and may therefore be neglected. Slater [71] has derived expressions for the interaction energy of electrons with angular momenta l and l'. Summation over the external shells enables one to write

$$\langle nl | V_R | nl \rangle = \sum_{l'} \frac{N(n'l')}{2(2l'+1)} \Big\{ f(ll') \Delta [F_0(nl, n'l')] - $$
$$- \Delta \sum_k [g_k(ll') G_k(n', n'l')] \Big\},$$

where

$$\Delta F_0(nl, n'l') = F_0(nl, n'l'; Z+1) - F_0(nl, n'l'; Z);$$
$$\Delta G_k(nl, n'l') = G_k(nl, n'l'; Z+1) - G_k(nl, n'l'; Z);$$

$N(n'l')$ represent the occupation numbers of the $n'l'$ subshells. Therefore, the quantity $N(n'l')/2(2l' + 1)$ determines the degree of filling of the given subshell. For completely filled subshells, it is equal to 1. The factors $f(ll')$ and $g(ll')$ are given by other expressions also deduced by Slater [71].

The approximation discussed above has been used for the calculation of relaxation energies of the noble gases neon, argon, krypton, and xenon. The method allows quite an accurate determination of binding energy values. The poor agreement obtained for the binding energies of 1*s* electrons in xenon is due to quantum-electrodynamic effects [64]. Figure 36 shows the difference between the theoretical values of binding energies, calculated for noble gases on the basis of various approximations, and the corresponding experimental values.

Shirley [70] has studied the problem of relaxation in condensed systems. He pointed out the fact that the binding energy of electrons in atoms belonging to condensed systems is lower than for the corresponding free atoms. For example, the binding-energy values in graphite and in free carbon atoms differ by 10 eV. Shirley explained this effect by the presence in solids of a superatomic relaxation related to the effect of a hole in a given atom on the neighboring atoms. This results in an additional redistribution of the electronic charge.

In order to get an idea of the magnitude of the correlation energy, let us consider the results for neon, an atom with completely filled shells [72]. The magnitude of correlation corrections ΔE^C for the 1*s* level of neon may be represented as the sum of paired-electron correlation energies corresponding to the 1*s* electron coupled to all of the other electrons of the atom. These energies depend on the overlapping and orientation of the spins, since the exchange correlation partly takes into account the correlation in the movement of electrons with parallel spins. For the 1*s* level of neon, we can write

$$\Delta E^C = \varepsilon(1s\alpha, 1s\beta) +$$
$$+ \varepsilon(1s\alpha, 2s\alpha) + \varepsilon(1s\alpha, 2s\beta) +$$
$$+ 3\varepsilon(1s\alpha, 2p\alpha) + 3\varepsilon(1s\alpha, 2p\beta), \tag{25}$$

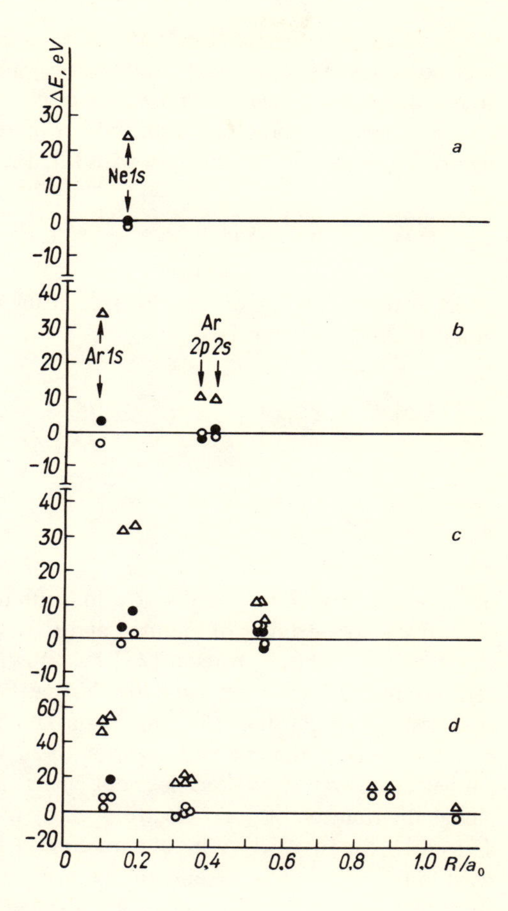

Figure 36. Discrepancy between theoretical and experimental values of binding energy for neon (a), argon (b), krypton (c), and xenon (d). The data are obtained by taking into account the orbital energies (△), by the Lindgren method (•) [62, 64], and by the Shirley method (○) [70].

where the quantities $\epsilon(i,j)$ have the values

$$\varepsilon(1s\alpha,\ 1s\beta) = 1.09\ \text{eV},\quad \varepsilon(1s\alpha,\ 2s\alpha) = 0.07\ \text{eV},$$
$$\varepsilon(1s\alpha,\ 2s\beta) = 0.06\ \text{eV},\quad \varepsilon(1s\alpha,\ 2p\alpha) = 0.11\ \text{eV},$$
$$\varepsilon(1s\alpha,\ 2p\beta) = 0.15\ \text{eV}.$$

We observe that the $\epsilon(i,j)$ take lower values for electrons with parallel spins. Since the wave functions of 1s electrons overlap strongly, it follows that they are characterized by large values of paired correlation energies. The experimental value of the binding energy is $E_B^V(1s) = 870.2$ eV [73]. The correlation energy, as calculated by using equation (25), is 1.4 eV, and the value of the binding energy $E_B^V(1s)$ obtained by Verhaegen *et al.* [72] is 870.8 eV. More accurate calculations performed later by Moser *et al.* [73] have given values of $E_B^V(1s) =$ 870.0 eV and $\Delta E^C = 0.6$ eV.

Regarding relativistic effects, it can be stated that the magnitude of ΔE^R is approximately 0.5–1.0% of the values of E_B^V [73, 74]. The relativistic effects do not depend appreciably on the type of chemical bond.

Let us consider now the magnitude of the shifts determined by the one-electron energies ϵ_i^{HF} for systems with filled shells. In this case

$$\varepsilon_i^{HF} = \left\langle \varphi_i \left| -\frac{1}{2}\nabla_1^2 \right| \varphi_i \right\rangle + \sum_P \left\langle \varphi_i \left| -\frac{Z}{r_{1P}} \right| \varphi_i \right\rangle + \sum_j (2J_{ij} - K_{ij}),$$

where φ_i represents the orbital centered at the nucleus of atom A. This expression can also be written [66]

$$\varepsilon_i^{HF} = \left\langle \varphi_i \left| -\frac{1}{2}\nabla_1^2 - \frac{Z_A}{r_{1A}} \right| \varphi_i \right\rangle + \sum_{j=C_A} (2J_{ij} - K_{ij}) -$$

$$- \sum_{j \neq C_A} K_{ij} + \sum_{j=C_B} 2J_{ij} + \sum_{j=v} 2J_{ij} + \sum_{B \neq A} \left\langle \varphi_i \left| -\frac{Z_B}{r_{1B}} \right| \varphi_i \right\rangle. \tag{26}$$

In this expression, $j = C_A$ and $j = C_B$ indicate that the summation is performed over all the core orbitals of atoms A and B, while $j = v$ indicates summation over all the valence orbitals. Watson [75] has shown, in a series of binding-energy calculations performed by using the Hartree–Fock method, that the first term in equation (26) is constant to within approximately 0.03 eV for $1s$, $2s$, and $2p$ core electrons, in the case of ionization of $3d$ valence electrons. The shape of the core-electron wave functions does not change significantly by changing the charge distribution of outer electrons, or by ionization of outer electrons. Therefore, the first two terms in equation (26) are almost constant. The third term in the equation represents the exchange integrals between two core electrons localized at different nuclei. The exchange integrals between core and valence electrons are rather small. The core electrons of other atoms efficiently screen the nuclear charge. The interaction of an inner electron of an atom with the nucleus and core electrons of another atom may be considered as being equivalent to an interaction between two point charges. Therefore, for the chemical shift, the following expression is valid:

$$\Delta E_i^{HF} \approx -\Delta\varepsilon_i^{HF} = \Delta\left(-\sum_{j=v} 2J_{ij} + \sum_{B \neq A} \frac{Z_B^*}{R_{AB}} \right), \tag{27}$$

where Z_B^* represents the reduced effective charge of atom B, i.e., the nuclear charge minus the charge corresponding to the total number of core electrons. The second term in this expression contains the two last terms in equation (26). Equation (27) shows that the chemical shift may be rather well described by an electrostatic potential that accounts for the distribution of valence electrons. By using the approximation suggested by Pople *et al.* [76]:

$$\left\langle \mu_A \nu_B \middle| \frac{1}{r_{\mu\nu}} \middle| \mu_A \nu_B \right\rangle \approx \frac{1}{R_{AB}}$$

(which is rather accurate if ν_B represents a core orbital), the expression (27) may be written as follows:

$$\Delta E_i \approx \Delta \left[- P_A \left\langle i\mu_A \middle| \frac{1}{r_{i\mu}} \middle| i\mu_A \right\rangle - \sum_{B \neq A} P_B \frac{1}{R_{AB}} + \sum_{B \neq A} \frac{Z_B^*}{R_{AB}} \right], \quad (28)$$

where μ_A represents the valence orbital of atom A. On the basis of the LCAO method, P_A is given by

$$P_A = \sum_j \sum_{\mu_A} C_{\mu j}^2.$$

By introducing the atomic charges q_A, we obtain

$$q_A = Z_A^* - P_A.$$

Equation (28) may then be written as follows:

$$\Delta E_i \approx \Delta \left(k_i q_A + \sum_{B \neq A} \frac{q_B}{R_{AB}} \right) - \Delta (k_i Z_A^*),$$

where k_i is a constant equal to the integral of the Coulomb interaction between core and valence electrons. The variation $\Delta(k_i Z_A^*)$ may be considered as being practically equal to zero. Since the chemical shift is measured relative to a given compound for which the energy of the corresponding level is taken as reference, the following expression is obtained for ΔE_i:

$$\Delta E_i = k_i q_A + \sum_{B \neq A} \frac{q_B}{R_{AB}} + l. \quad (29)$$

Here, l is a constant determined by the position of the reference level on the energy scale. Equation (29) expresses the basic idea of the potential model in the ESCA method. In practice, the quantities k_i and l are considered as adjustable parameters that may be determined by the method of least squares to give the best fit to a large series of chemical shifts for a given element. If the atomic charges in molecules are calculated, then by using these theoretical values and the experimental data on chemical shifts, the values of the constants k and l can be determined. If, instead, the magnitude of the chemical shifts and the constants k and l are determined, then by using equation (29) the magnitude of the charges can be obtained.

Another variant of the potential model for the determination of chemical shifts is a model that does not use adjustable parameters. If it is assumed that the core orbitals are localized in the immediate vicinity of nucleus A, then from equation (27) we obtain

$$\Delta E_{\iota} \approx \Delta U_A^v = \Delta\left(-2\sum_{j=v}\left\langle\varphi_j\left|\frac{1}{r_{jA}}\right|\varphi_j\right\rangle + \sum_{B\neq A}\frac{Z_B^*}{R_{AB}}\right). \quad (30)$$

In this expression, the one-electron integral $\langle\varphi_j|2/r_{jA}|\varphi_j\rangle$ is already included. Since the potential U_A^v does not contain any core orbitals it can be calculated by semiempirical methods. Such a model will evidently give good results only if it is used for the calculation of chemical shifts of the most deeply lying levels.

A further variant can be developed from the potential model if use is made of the condition of electroneutrality:

$$q_A = -\sum_{B\neq A} q_B.$$

By choosing the reference level so that $l = 0$, equation (29) leads to

$$\Delta E_A = \sum_{B\neq A}\left(\frac{1}{R_{AB}} - k_A\right) q_B.$$

In this expression, the summation is performed first over the nearest-neighbor atoms around the given atom A, and then over all of the groups of such atoms:

$$\Delta E_A = \sum_G\left[\sum_{B\in G}\left(\frac{1}{R_{AB}} - k_A\right) q_B\right] = \sum_G \Delta E_G. \quad (31)$$

ΔE_G is called the group shift. Introduction of group shifts will be useful if the shift due to the given group of atoms situated around the atom A is not affected by the presence of the other groups of atoms. This assumption is, of course, only approximately valid. In fact, both the charge of the given atom itself and the charges of the other atoms in the group will influence the atoms belonging to other groups, so that a single group cannot be considered as being isolated.

Multiplet Splitting in Core-Electron Spectra

In considering the problems related to binding energies and chemical shifts of core electrons, we have limited the discussion to the systems having completely filled valence shells. In this case the total angular and spin momenta of the system in the initial state are equal to zero ($L = S = 0$). If a system having incompletely filled shells is ionized, the created ion is characterized by an orbital momentum L' and a spin momentum S', which satisfy the condition [77]

$$|L-l| \leqslant L' \leqslant L+l, \quad S' = S \pm \frac{1}{2}, \quad S' \geqslant 0, \tag{32}$$

where L and S represent the orbital and the spin momenta, respectively, of the atom with incompletely filled shell, while l represents the orbital momentum of the emitted electron.

For systems with completely filled shells, only one ionic final state is possible, namely, $L' = l$ and $S' = \frac{1}{2}$. Therefore, in the electron spectrum only one line will appear, corresponding to the binding energy of the emitted electron. For systems with incompletely filled shells, at least one of the momenta L and S is different from zero. The rule of addition of momenta (32) predicts that, in this case, several final states are possible and, according to equation (1), it can be expected that a number of lines will appear in the electron spectrum. This effect is called multiplet splitting of electron lines. This process can be represented symbolically as follows:

$$(nl)^q (n'l')^p \xrightarrow{h\nu} (nl)^{q-1} (n'l')^p + \text{photoelectron.} \tag{33}$$

Here, the filled subshell $(nl)^q$ contains q electrons, while the unfilled valence subshell $(n'l')^p$ contains p electrons. Since for the $(nl)^q$ shell both the spin momentum and the orbital momentum are equal to zero, the momenta L and S correspond to the orbital and spin momenta of the valence subshell $(n'l')^p$. There exist several possible types of systems with incompletely filled shells. The majority of free atoms have incompletely filled outer valence shells. Such unfilled shells also exist in atoms of compounds of the transition and rare-earth metals, in free radicals and in a number of covalent molecules.

The unpaired d and f electrons are, to a great degree, localized around the given atom, and, therefore, to identify these orbitals, it is possible to use the atomic quantum numbers, as well as the notations from the atomic theory of multiplets. The simplest case occurs when one electron from the s shell is emitted. In this case, only two final states are possible, corresponding to $S' = S \pm \frac{1}{2}$, and the intensity ratio of the peaks corresponding to them will be equal to the ratio of their multiplicity, i.e.,

$$\frac{I\left(L, S+\frac{1}{2}\right)}{I\left(L, S-\frac{1}{2}\right)} = \frac{S+1}{S}. \tag{34}$$

The energy interval between these components will be given by the following expressions:

$$\left.\begin{aligned}\Delta\,[E_B\,(ns)] &= E\left(L,\; S-\tfrac{1}{2}\right)_f - E\left(L,\; S+\tfrac{1}{2}\right)_f,\\ \Delta\,[E_B\,(ns)] &= (2s+1)\,K_{ns,n'l'} \quad \text{for} \quad S\neq 0,\\ \Delta\,[E_B\,(ns)] &= 0 \quad \text{for} \quad S=0,\end{aligned}\right\} \tag{35}$$

where $K_{ns,\,n'l'}$ represents the $ns - n'l'$ exchange integral,

$$K_{ns,n'l'} = \frac{1}{2l'+1}\int\limits_0^{\infty}\int\limits_0^{\infty}\frac{r_<^{l'}}{r_>^{l'+1}}\,P_{ns}(r_1)\,P_{n'l'}(r_2)\,P_{ns}(r_2)\times \times P_{n'l'}(r_1)\,dr_1dr_2, \tag{36}$$

and $r_<$ and $r_>$ represent the smaller and the greater of the radii r_1 and r_2, respectively; P_{ns}/r and $P_{n'l'}/r$ represent the radial wave functions of the ns, and $n'l'$ electrons, respectively. A characteristic feature of the doublet obtained is that the more intensive peak corresponds to the lower binding energy. As can be seen from formula (36), the exchange integral $K_{ns,\,n'l'}$ is determined by the space distribution of the core and valence electrons.

In the one-electron approximation, as will be shown later on, the transition probability is determined by the square modulus of the transition matrix element $|\langle\varphi_{nl}|r|\varphi_{ph}\rangle|$, where φ_{nl} and φ_{ph} represent the one-electron orbitals corresponding to the initial and final states, respectively. Therefore, the orbital momentum of the electron is equal to $l_{ph} = l \pm 1$. Conservation of spin and orbital angular momenta imposes the following selection rules:

$$\left.\begin{aligned}\Delta S &= S' - S = \pm\tfrac{1}{2};\\ \Delta L &= L' - L = 0, \quad \pm 1, \; \pm 2, \; \ldots, \; \pm l.\end{aligned}\right\} \tag{37}$$

As has been shown by Cox and Orchard [78], in the spin-orbit approximation, the intensity $I(L', S')$ is given by

$$I\,(L',\; S') \sim (2S'+1)\,(2L'+1). \tag{38}$$

Let us consider as an example the photoemission spectra corresponding to electronic transitions from the $3s$ and $3p$ levels of the free Mn^{2+} ion. The ground state of the ion in the L-S coupling is $3d^5\ {}^6S$ ($S = \frac{5}{2}, L = 0$), with all five $3d$ electron spins oriented parallel to each other. After the emission of one $3s$ electron, two final states are possible, namely, $(3s)\,(3d)^5\ {}^5S$ ($S = 2, L = 0$), or $(3s)\,(3d)^5\ {}^7S$ ($S = 3, L = 0$). The difference between them is that for the ${}^5S$

state, the spin of the remaining $3s$ electron is oriented antiparallel to the spin of the five $3d$ electrons, while for the ^7S state, the spins of the $3s$ and $3d$ electrons are parallel to each other. Since the exchange interaction exists only for electrons with parallel spins, it follows that the energy of the ^7S state is lower than the energy of the ^5S state. The magnitude of energy splitting, according to formula (35), is proportional to the $3s$–$3d$ exchange integral:

$$\Delta [E_B(3s)] = \Delta E(3s3d^5) = 6K_{3s,3d} =$$

$$= \frac{6}{5} \int_0^\infty \int_0^\infty \frac{r_<^2}{r_>^3} P_{3s}(r_1) P_{3d}(r_2) P_{3s}(r_2) P_{3d}(r_1)\, dr_1 dr_2.$$

More complicated is the process of X-ray photoemission of $3p$ electrons of Mn^{2+}. In this case, $(nl)^{q-1} = 3p^5$, $(n'l')^p = 3d^5$, and the initial state is, as above, ^6S $(S = \frac{5}{2}, L = 0)$. The selection rules (37) allow only those final states corresponding to ^7P $(S = 3, L = 1)$ and to ^5P $(S = 2, L = 1)$. However, the ^5P state can be obtained in three different ways from the interaction of the $3p^5$ and $3d^5$ configurations:

$$3p^5\,(^2P)\,[3d^5\,(^6S)]\,^5P,$$
$$3p^5\,(^2P)\,[3d^5\,(^4D)]\,^5P,$$
$$3p^5\,(^2P)\,[3d^5\,(^4P)]\,^5P.$$

Consequently, these three states will interact with each other, and in the method of configuration interaction they will enter into the operation of diagonalization of the interaction matrix. The eigenfunctions that describe the ^5P states will be expressed as linear combinations of the functions:

$$\Psi(^5P_1) = c_{11}\Psi(^6S) + c_{12}\Psi(^4D) + c_{13}\Psi(^4P),$$
$$\Psi(^5P_2) = c_{21}\Psi(^6S) + c_{22}\Psi(^4D) + c_{23}\Psi(^4P),$$
$$\Psi(^5P_3) = c_{31}\Psi(^6S) + c_{32}\Psi(^4D) + c_{33}\Psi(^4P).$$

With the energy values corresponding to these eigenfunctions, it is possible to calculate the energy difference between the $5p$ states. The eigenfunctions and the eigenenergies are determined by diagonalization of (3×3) matrices, in which every matrix element is expressed as a linear combination of I_{3d-3d}, K_{3d-3d}, I_{3p-3d} and K_{3p-3d}. The ratio of the total intensities of ^5P and ^7P states is given by

$$I_T(^5P)/I_T(^7P) = [I(^5P_1) + I(^5P_2) + I(^5P_3)]/I_T(^7P) = {}^5/_7.$$

The intensity ratio corresponding to two electronic $3s$ emission lines is also equal to 5/7: $I_T(^5S)/I_T(^7S) = 5/7$. The intensity of each component is determined by the square modulus $|c_{il}|^2$. The position of the calculated multiplet lines and their intensities are schematically shown in Figure 37.

A detailed discussion of the most important experimental data on the multiplet splitting of photoelectron lines will be given in Chapter 8. Here we only mention that particular care should be exercised in the interpretation of experimental data.

Other phenomena can produce additional structures in X-ray photoelectron spectra that can be erroneously interpreted as being the result of multiplet splitting. Common among those are: multielectron excitations, Auger electrons, structures due to the presence of satellites of the excitation line or to the presence of small quantities of impurities in the sample. It should be pointed out that the Auger electrons are characterized by constant kinetic energy, while the intensity and the position of the satellites of the X-ray excitation line depend on the anode material of the X-ray tube.

Inelastic diffusion processes also generate satellites. These are situated on the high-energy side of the main lines on the binding energy scale. Lines generated by inelastic electron scattering are separated from the corresponding main photoelectron lines by an approximately similar interval regardless of the binding energy.

Lines caused by chemical reactions on the sample surface can hardly be distinguished from multiplet splitting effects. In this case, it is necessary to analyze

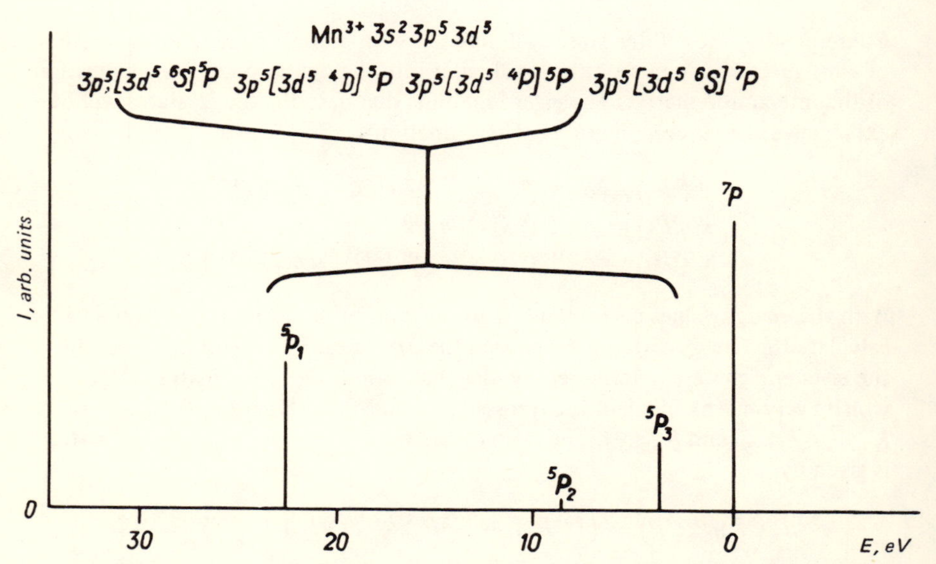

Figure 37. Structure of the photoelectron spectrum generated by photoionization of the 3p shell of the Mn^{2+} ion.

the positions, shapes, and relative intensities of the core and valence photoelectron lines of all the chemical compounds that can possibly be formed on the sample surface.

It is particularly difficult to distinguish between multiplet splitting and the effects due to multielectron excitation. This is due to the fact that multielectron excitations often possess their own multiplet splittings.

Multielectron Excitation Effects in Photoelectron Spectra

Multielectron processes in X-ray photoelectron spectroscopy were first observed and studied by Krause and Carlson [79–85]. Gaseous neon and argon were irradiated with X rays in the energy range 270 eV–1.5 keV. The charge distribution of the resulting ions and the kinetic energy of the emitted electrons were measured. The authors demonstrated that, as a result of X-ray absorption, not only one-electron, but also two-electron, and three-electron transitions are observed, with a total probability of approximately 20% for each absorbed photon. The two-electron process has about a 10-times greater probability than the three-electron process. Therefore, we will limit our discussion to the two-electron processes. There exist two possible types of two-electron transitions depending on whether the second electron is excited up to a higher bound state ("shake-up") or to an unbound state situated in the continuum spectrum ("shake-off"). These two types of transition may be symbolically represented as follows:

$$\text{shake-up} \quad (nl)^q\,(n'l')^p \xrightarrow{h\nu} (nl)^{q-1}\,(n'l')^{p-1}\,(n''l'')^1 + \text{photoelectron},$$

$$\text{shake-off} \quad (nl)^q\,(n'l')^p \xrightarrow{h\nu} (nl)^{q-1}\,(n'l')^{p-1}\,(E_{\text{kin}}l'')^1 + \text{photoelectron}.$$

The second electron is excited from the subshell $(n'l')^p$, which can be completely or partially occupied. In the shake-up and shake-off processes, excitation or emission of the additional electron requires an additional amount of energy. As a result, a satellite structure appears on the high-energy side of the main line on the binding energy scale. Increasing the excitation energy from the value of the electron binding energy up to a value several times greater increases the probability of shake-off-type processes from zero up to some constant value. The existing theoretical calculations of the position and intensity of the lines generated by multielectron excitations are based on the quantum-mechanical model of "sudden perturbations" [86]. This model is based on the assumption that, the emission of the photoelectron taking place very rapidly, the valence electrons do not have enough time to feel the potential change in the region of core levels. In this sense, the initial excitation may be considered as instantaneous.

The theoretical criterion for the validity of the "sudden perturbation" approximation has been expressed by Åberg [87] with the following inequality:

$$[E^f(nl,\ n'l') - E^f(nl)]\,\tau \ll \hbar, \tag{39}$$

where $E^f(nl, n'l')$ is the total energy of the final state after the shake-up process has taken place, $E^f(nl)$ is the total energy of the final state for the one-electron transition, and τ is the time interval during which the nl electron, leaving the atom, passes the $n'l'$ subshell.

An estimation of how well the inequality (39) is fulfilled for a typical case has been performed by Fadley [88]. At a photoelectron energy $E_{\text{kin}} = 1000$ eV, the electron velocity v is approximately 2×10^9 cm/sec, and, in the case of an atom of approximate dimension 2 Å, the value of τ is $\tau \sim 10^{-7}$ sec. In this case, $\tau/h \approx 1/65$ eV^{-1}, and therefore the above approximation is valid if the distance between the satellites and the main line in the electron spectrum is greater than 65 eV. Thus, the criterion (39) imposes significant limits on the use of the "sudden perturbation" method for calculation of the probability of shake-up processes. However, it has been shown that such calculations are in reasonable agreement with experiment, even in cases when the criterion is evidently not fulfilled [87].

Let us consider an arbitrary system of N electrons in the ground state. It is described by a wave function $\Psi_0(N)$, a solution of the Schrödinger equation:

$$H(N)\,\Psi_0(N) = E_0^N \Psi_0(N).$$

Let us express $\Psi_0(N)$ as the antisymmetrized product of the one-electron orbital nl and the function $\Psi_R(N-1)$:

$$\Psi_0(N) = A\varphi_{nl}\Psi_R(N-1), \tag{40}$$

where A represents the antisymmetrization operator. In the photoionization process, the N-particle Hamiltonian $H(N)$ undergoes a sudden change: $H(N) \to H(N-1)$. In the "sudden perturbation" approximation, one can consider that the orbitals of the remaining electrons do not have time to relax. Consequently, the electron system now containing $(N-1)$ electrons is left in a state described by the wave function $\Psi_R(N-1)$, which, however, is not an eigenfunction of the $(N-1)$-particle Hamiltonian $H(N-1)$.

The new Hamiltonian has the eigenfunctions $\Psi^f(N-1)$:

$$H(N-1)\,\Psi^f(N-1) = E^f(N-1)\,\Psi^f(N-1).$$

The superscript f is used in order to distinguish between the various final states generated by the excitation of the second electron, or between the various multiplet states appearing after the one-electron or two-electron transitions have taken place. The wave function $\Psi_R(N-1)$ represents a mixture of the wave functions $\Psi^f(N-1)$:

$$\Psi_R(N-1) = \sum_f C_f \Psi^f(N-1), \tag{41}$$

where the coefficients C_f are determined by the integrals

$$C_f = \langle \Psi^f(N-1) \mid \Psi_R(N-1) \rangle.$$

The probability that the system of $(N-1)$ electrons will be left in the state f after the photoionization process has taken place is given by the expression

$$P_f = |\langle \Psi^f(N-1) \mid \Psi_R(N-1) \rangle|^2. \tag{42}$$

From this expression it can be seen that the operator that determines the transition probability is a unitary operator. In this case, the transitions are called monopole transitions. From equations (41) and (42) it follows that only transitions between states with the same symmetry are allowed:

$$\Delta J = \Delta L = \Delta S = \Delta M_J = \Delta M_L = \Delta M_S = 0. \tag{43}$$

As an example of shake-up-type processes, let us consider the photoelectron spectrum generated by the X-ray ionization of the 1*s* electrons of neon. Figure 38 shows the photoelectron spectrum of neon over a wide range of kinetic en-

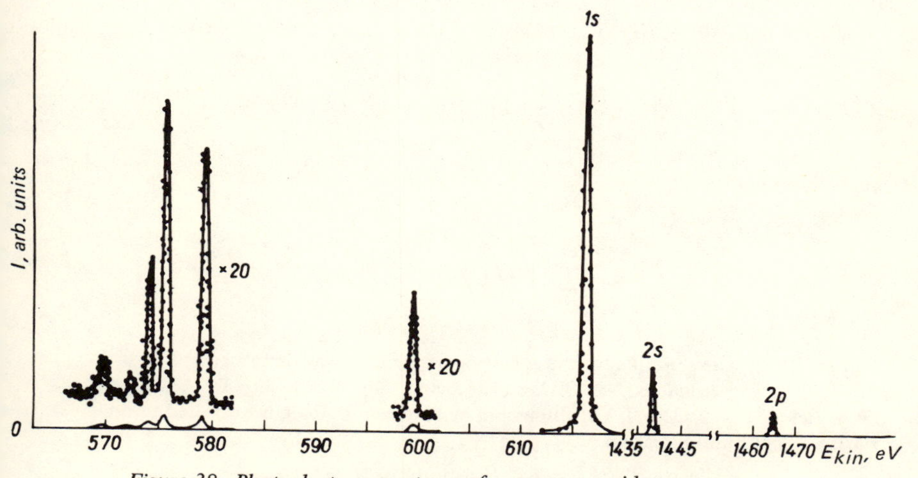

Figure 38. Photoelectron spectrum of neon over a wide energy range.

ergies [66]. The group of lines with kinetic energy values situated in the interval 570–580 eV corresponds to the shake-up transitions. The intensity of the line corresponding to a kinetic energy of 600 eV is pressure dependent, indicating that this line is generated by inelastic scattering processes. The two most intense peaks in the group of shake-up lines correspond to the final states originating from the excitation of a $2p$ electron into a $3p$ state. Therefore, the kinetic energy of these lines with respect to the $1s$ line of neon is determined by the relation

$$E_{\mathrm{kin}}^{ik} = h\nu - E_B^i - \Delta E_B^{ik},$$

where $h\nu - E_B^i$ is the kinetic energy of the main peak corresponding to the ground state of the ion and ΔE_B^{ik} is the additional energy necessary to move one valence electron up to an excited orbital of the ion. The selection rule (43) for monopole transitions ensures that the final states of the neon ion (Ne^+) will have a symmetry of the ^2S type. For each $(1s)^2(2s)^2(2p)^5np$ configuration, there exist two possible schemes of momenta summation that lead to the state ^2S, namely, $1s[2p^5np(^1S)]$ and $1s[2p^5np(^3S)]$. Therefore, two states having the same configuration will interact with one another. This process can be treated by using the multiconfiguration approximation, in which both the coefficients of the wave functions and the orbitals in the Slater determinant are simultaneously optimized. The calculated difference ΔE_B^{ik} for the two most intense peaks is equal to 35.6 eV and 39.5 eV, respectively. To this energy should be added the correlation energy, which amounts to 1.4 eV. The resulting values (37.0 eV and 40.9 eV) are in good agreement with the experimental data (37.3 eV and 40.7 eV). The less intensive lines in Figure 38 correspond to transitions from the $2p$ level to the $4p$, and $5p$ levels.

Manne and Åberg [89]† have studied the expression for $E_R(N-1)$:

$$E_R(N-1) = \langle \Psi_R | H(N-1) | \Psi_R \rangle = \sum_{f=0}^{\infty} E^f(N-1) |\langle \Psi^f | \Psi_R \rangle|^2.$$

Since the equality

$$\sum_{f=0}^{\infty} |\langle \Psi^f | \Psi_R \rangle|^2 = 1$$

†*Editors' note:* The sum rule for the spectrum of the hole was first discussed by B. I. Lundqvist, *Phys. Kondens. Mat.* **9**, 236 (1967) and fully discussed by D. C. Langreth, *Phys. Rev. B* **1**, 471 (1970). The discussion by Manne and Åberg is restricted to the Hartree–Fock approximation.

is valid, it follows that

$$E_{BR} = E_R(N-1) - E_0(N) = \sum_{f=0}^{\infty} |\langle \Psi^f | \Psi_R \rangle|^2 I_f.$$

If the wave function $\Psi_0(N)$ is calculated by using the Hartree–Fock approximation, it is found that the magnitude of E_{BR} coincides with the binding energy value in the Koopmans approximation E_B^{KT} given by the relation

$$E_B^{KT} = I_0 + \sum_{f=1}^{\infty} |\langle \Psi_i | \Psi_R \rangle|^2 (I_f - I_0), \tag{44}$$

where $I_0 = E_0(N-1) - E_0(N)$ is the binding energy value obtained by taking into account the relaxation of electron states.

Photoionization Cross Section in the One-Electron Approximation

Problems involved in the calculation of photoionization cross sections for free atoms and molecules in the one-electron approximation have been treated by a number of authors [90–96]. Usually, the calculations are performed with several simplifying assumptions, namely:

(1) The action of X-ray photons is treated as a perturbation.

(2) The photon wavelength λ is assumed to be greater than the dimension of the subshells from which the electrons are emitted. This assumption is completely justified if the $K\alpha_{1,2}$ lines of magnesium or aluminum are used as the X-ray excitation source, since in these cases $\lambda \sim 10$ Å.

(3) The wave function of the initial and final states may be represented as antisymmetric products of the one-electron wave functions φ_{nl}, φ_{ph} with the functions $\Psi_R(N-1)$, $\Psi^f(N-1)$, respectively, where $\Psi_R(N-1)$ and $\Psi^f(N-1)$ describe the $(N-1)$ passive electrons.

The wave functions $\Psi^i(N)$ and $\Psi^f(N)$ are often written as Slater determinants. Use of these approximations implies that the main effect in the X-ray photoelectron process consists in the excitation of the electron from the state φ_{nl} to the state φ_{ph}.

(4) In the process of photoelectron emission only one electron changes its orbital, while the passive electron orbitals remain unchanged. This assumption is equivalent to using the Koopmans' theorem.

(5) In order to describe the initial and final states, the L–S scheme of momenta summation is utilized.

The photoionization cross section is, by definition, given by the following formula:

$$\sigma_{nl} = \sigma_{nl}(h\nu) = \sigma_{nl}(E_{\text{kin}}) = \frac{4\pi\alpha a_0^2}{3} h\nu \left\langle \Psi^f(N) \left| \sum_{i=1}^{N} \mathbf{r}_i \right| \Psi^i(N) \right\rangle, \quad (45)$$

where α is the fine-structure constant and a_0 is the Bohr radius.

With the wave function (40), expression (45)–as will be demonstrated in the next section–may be written as follows:

$$\sigma_{nl} = \frac{4\pi\alpha a_0^2}{3} (h\nu) \,|\langle \varphi_{ph} | \mathbf{r} | \varphi_{nl} \rangle|^2 \,|\langle \Psi_f(N-1) | \Psi_R(N-1) \rangle|^2. \quad (46)$$

Use of the symmetry properties of the functions entering the matrix elements of this expression leads to the selection rule (37). The functions $\Psi_f(N-1)$ and $\Psi_R(N-1)$ should have the same type of symmetry. If the wave functions $\Psi^i(N)$ and $\Psi^f(N)$ can be expressed as Slater determinants, then equation (46) may be written as follows:

$$\sigma_{nl} = \frac{4\pi\alpha a_0^2}{3} (h\nu) \,|\langle \varphi_{ph} | \mathbf{r} | \varphi_{nl} \rangle|^2 \prod_{i \neq nl} |\langle \varphi_i' | \varphi_i \rangle|^2. \quad (47)$$

If the final state relaxation is neglected then σ_{nl} is given by a simpler formula

$$\sigma_{nl} = \frac{4\pi\alpha a_0^2}{3} (h\nu) \left| \int \varphi_{ph}^* \mathbf{r} \varphi_{nl} dV \right|^2 \quad (48)$$

obtained from equation (47) with $\varphi_i' \equiv \varphi_i$.

In order to calculate the ionization cross section for the whole nl subshell from (48), it is necessary to perform a summation over all of the final states and to take the average over all of the orbitals φ_{nl}. These operations are performed for the two possible orbital momenta $(l+1)$ and $(l-1)$ of the photoelectron, as well as for all of the values (initial and final) of the magnetic quantum numbers. The photoionization cross section for the nl subshell is finally determined by the expression obtained by performing the integration in formula (48):

$$\sigma_{nl} = \frac{4\pi\alpha a_0^2}{3} (h\nu) \,[l R_{\varepsilon,l-1}^2 + (l+1) R_{\varepsilon,l+1}^2], \quad (49)$$

where

$$R_{\varepsilon,l\pm 1} = \int_0^\infty P_{nl}(r)\, r P_{\varepsilon,l\pm 1}(r)\, dr;$$

$P_{nl}/r = R_{nl}$ and $P_{\epsilon,l\pm1}/r$ are the radial functions of the orbitals φ_{nl} and φ_{ph}, and $\epsilon = E_{\text{kin}}$.

A number of authors [92, 94, 98–100] have calculated photoionization cross sections and, with the exception of the regime at or near absorption edges, the values obtained agree quite well with experiment. If the photon energy $h\nu$ is greater than the edge, then the term $(l+1)R^2_{\epsilon,l+1}$ in formula (49) has a greater value than $(l-1)R^2_{\epsilon,l-1}$. The ionization cross section σ_{nl} is a function that decreases with increasing $h\nu$ or with increasing photoelectron kinetic energy E_{kin}. However, when $h\nu$ takes values near the edge, large oscillations and even zeros may appear in σ_{nl}. These oscillations are due to variations in the degree of overlapping of the functions $P_{nl}(r)$ and $P_{\epsilon,l\pm1}(r)$. The magnitude of σ_{nl} may be different for different subshells, for the same value of excitation energy.

The total photoionization cross section σ_{nl} is expressed by the integral over all of the electron emission angles. The angular dependence of photoemission may be determined by using the differential photoionization cross section $d\sigma_{nl}/d\Omega$. Cooper and Manson [94] have derived the following expression for the differential photoionization cross section:

$$\frac{d\sigma_{nl}}{d\Omega} = \left[\frac{\sigma_{nl}}{4\pi}\right]\left[1 - \frac{1}{2}\beta P_2(\cos\theta)\right] = \\ = \left[\frac{\sigma_{nl}}{4\pi}\right]\left[1 + \frac{\beta}{2}\left(\frac{3}{2}\sin^2\theta - 1\right)\right], \tag{50}$$

where β is the asymmetry parameter, θ is the angle between the photon and electron directions, and $P_2(\cos\theta)$ is the second-order Legendre polynomial: $P_2(\cos\theta) = \frac{1}{2}(3\cos^2\theta - 1)$. The asymmetry parameter β, like the photoionization cross section σ_{nl}, depends on the electron kinetic energy E_{kin}, and is given by the formula

$$\beta = \Big\{ l(l-1)R^2_{\varepsilon,l-1} + \\ + \frac{(l+1)(l+2)R^2_{\varepsilon,l+1} - 6l(l+1)R_{\varepsilon,l+1}R_{\varepsilon,l-1}\cos(\delta_{l+1} - \delta_{l-1})}{(2l+1)[lR^2_{\varepsilon,l-1} + (l+1)R^2_{\varepsilon,l+1}]}\Big\}. \tag{51}$$

Here, δ_{l+1} and δ_{l-1} are the phase shifts, dependent on E_{kin}. The range of variation of β is determined by the inequality $-1 \leqslant \beta \leqslant 2$. Positive values of β indicate that the electrons are preferentially emitted in directions perpendicular to the directions of the photons ($\theta = 90°$), while negative β values indicate that photoelectron emission takes place preferentially in parallel or antiparallel directions with respect to the direction of the exciting photons ($\theta = 0°$, or $\theta = 180°$). For s electrons $l = 0$, and therefore only waves with $l = 1$ are possible. In this case $\beta = 2$, and

$$\frac{d\sigma_{ns}}{d\Omega} = \frac{\sigma_{ns}}{4\pi} \sin^2 \theta.$$

Maximum intensity is obtained at $\theta = 90°$, and zero intensity at $\theta = 0°$ and $\theta = 180°$. For $\beta = -1$,

$$\frac{d\sigma_{ns}}{d\Omega} = \frac{\sigma_{ns}}{4\pi} \cos^2 \theta,$$

and the intensity of photoelectron emission is zero at $\theta = 90°$ and maximum at $\theta = 0$ and $\theta = 180°$.

Equation (50) is equivalent to

$$\frac{d\sigma_{ns}}{d\Omega} = A + B \sin^2 \theta, \tag{52}$$

where A and B are constants,

$$A = \frac{\sigma_{nl}}{4\pi}\left(1 - \frac{\beta}{2}\right) \quad \text{and} \quad B = \left(\frac{\sigma_{nl}}{4\pi}\right).$$

If A and B are determined empirically, β can be found by using the formula $\beta = 4B/(3A + 2B)$, obtained on the assumption that the radiation is unpolarized. It is worth mentioning that here we have only discussed the case of unpolarized radiation since, at present, efficient sources of polarized radiation do not exist in the X-ray region of the spectrum. For molecules and for unpolarized radiation, formulas similar to (49)–(51), have been derived by Cooper and Zare [96], Tully *et al.* [101], and Grimm [102].

Gelius [103] has shown that, in the case of molecules, one may correlate the relative photoionization cross sections of molecular orbitals with the photoionization cross sections of atomic valence subshells. In his calculations he used the Born–Oppenheimer approximation, "frozen orbitals," for the description of the ion and plane waves as the wave functions of the excited electrons. In this case, for a given molecular orbital, labeled φ_j, the photoionization cross section is given by the expression

$$\sigma_j^{MO} \sim |\langle \varphi_j | P \rangle|^2, \tag{53}$$

where P is the electron momentum. Since the variation of electron kinetic energy in the energy range corresponding to the localization of the j orbital is not large, the energy dependence of the photoionization cross section may be neglected. Since a typical value of the de Broglie wavelength is of the order of

0.35 Å, it follows that the most significant contributions to the integral (53) come from regions in which the wave function φ changes rapidly. To a first approximation, the photoionization cross section does not depend on the shape of the molecular orbital in the interatomic region, and therefore expression (53) may be written as

$$\sigma_i^{MO} = \sum_A \sigma_{Ai}. \tag{54}$$

Since the electron in the molecular orbital φ_j is localized mainly in the vicinity of the atom A, the molecular orbital φ_j can be expressed as follows in the LCAO approximation:

$$\varphi_i = \sum_{A,\lambda} C_{A\lambda_i} \Phi_{A\lambda},$$

where the summation is performed over all atoms A and over all types of symmetry λ for each atom. The LCAO approximation describes the shape of the orbital near the nucleus particularly well, since in this region the orbital shape is determined mainly by the condition of orthogonality of the given molecular orbital and core orbital, both of which are rather atomlike. Therefore, each term in equation (54) may be expressed in terms of the photoionization cross section of atomic subshells. The following expression is then valid:

$$\sigma_{Ai} = \sum_\lambda P_{A\lambda i} \sigma_{A\lambda}^{AO}, \tag{55}$$

where $P_{A\lambda j}$ is the contribution of the atomic $A\lambda$ orbital to the jth molecular orbital, and $\sigma_{A\lambda}^{AO}$ is the photoionization cross section for the atomic $A\lambda$ subshell. Using equations (54) and (55), the final expression for the photoionization cross section is obtained:

$$\sigma_i^{MO} = \sum_{A,\lambda} P_{A\lambda i} \sigma_{A\lambda}^{AO}. \tag{56}$$

In the interpretation of experimental results, it is convenient to use not the absolute but the relative values of the photoionization cross section. Therefore, if all of the photoionization cross sections are referred to the given subshell $A_0\lambda_0$, the intensity I_j^{MO} is given by

$$I_i^{MO} \sim \sum_{A,\lambda} P_{A\lambda i} \cdot \sigma_{A\lambda}^{AO} / \sigma_{A_0\lambda_0}^{AO}. \tag{57}$$

Gelius has performed calculations using formula (57) for a large number of molecules, namely, C_6H_6, C_4H_4O, C_4H_4S, CF_4, SF_6, and C_3O_2. Figure 39

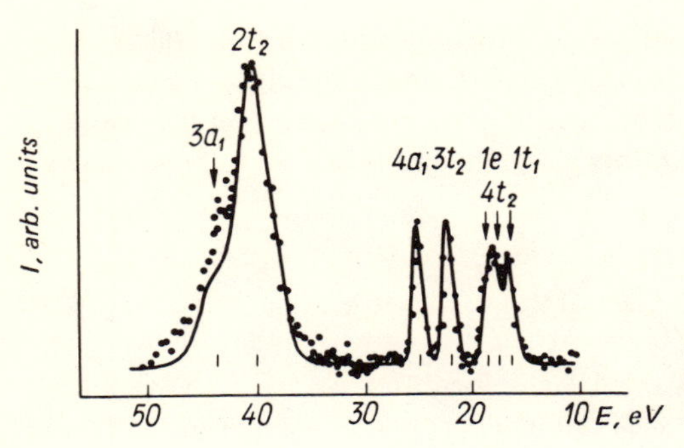

Figure 39. Photoelectron spectrum of valence electrons in CF_4 excited by the $K\alpha_{1,2}$ radiation of magnesium.

shows the experimental and theoretical results for the CF_4 molecule. The calculated values agree very well with the experimental ones. In the calculations, the contribution of each molecular orbital to the electron spectrum was approximated by a Gaussian curve with a half-width determined experimentally, and an area proportional to the calculated transition intensity.

Nefedov *et al.* [104] have used formulas (56) and (57) to calculate the valence electron spectra for a large group of molecules, viz., CO, CO_2, C_4H_5N, $LiNO_3$, and a few others. For $P_{A\lambda j}$, the authors used previously published values, while $\sigma_{A\lambda}^{AO}$ values were calculated by using the atomic relativistic electron wave functions. Their theoretical calculations agree well with the experimental results. The authors have also measured experimentally the photoionization cross sections for the atomic subshells of the series of elements from carbon to calcium. The relative photoionization cross sections (theoretical and experimental) for the inner atomic subshells, normalized to one electron, are shown in Figure 40.

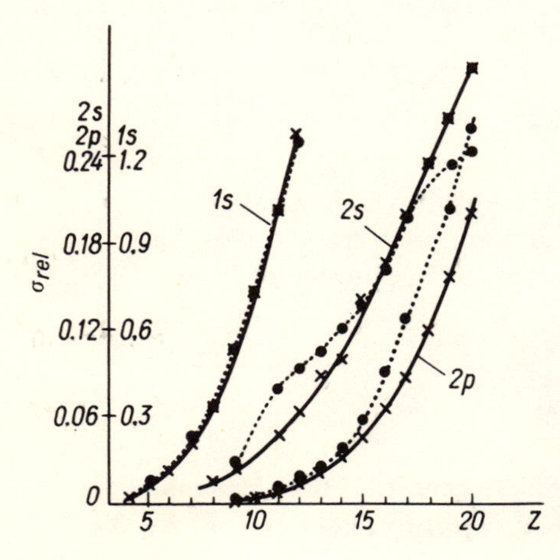

Figure 40. Comparison between the relative photoionization cross sections of internal shells. The cross sections are normalized to a single electron, and the value of the photoionization cross section of Na 1s electrons is taken as unity: ●—experimental data; ×—theoretical results.

The theoretical and experimental values of the relative photoionization cross sections are in good agreement in the case of $1s$ electrons. Large disagreement is observed for the $2s$ electrons and especially for the $2s$ electrons of sodium, for which the experimental value is about twice as great as the theoretical one.

Influence of Multielectron Transitions and Multielectron Processes on the Magnitude of the Photoionization Cross Section

The problem of the influence of multielectron transitions on the magnitude of the photoionization cross section, as determined by the approximation of the "sudden perturbation" theory, has been treated by Fadley [105]. As the wave functions of the initial and final states, he used functions of the type given by formula (40):

$$\Psi^i(N) = A\varphi_{nlm_lm_s}(1)\Phi(N-1),$$

$$\Psi^f(N) = A\varphi_{\varepsilon,l''',m_l''',m_s'''}(1)\Psi^f(N-1),$$

where $\varphi_{nlm_lm_s}$ is the one-electron orbital describing the state of the electron before the transition, $\Phi(N-1)$ is the wave function describing the initial state of those electrons that do not participate in the transition, $\varphi_{\epsilon l''' m_l''', m_s'''}$ is the wave function of the electron with kinetic energy ϵ and angular momentum $l''' = l \pm 1$, and $\Psi^f(N-1)$ is the wave function of the electrons of the ionic core. The author treated both the case of one-electron and of two-electron transitions, which may be represented schematically as follows:

one-electron transition:

$$\begin{array}{ccc} \ldots (nl)^q \ldots (n'l')^p\, L,\, S & \xrightarrow{h\nu} & \ldots (nl)^{q-1} \ldots (n'l')^p\, L',\, S' + \varphi_{\varepsilon_\beta, l\pm 1} \\ \Psi^i(N)_\alpha,\ E^i(N) & & \Psi^f(N-1)_\beta,\ E^f(N-1)_\beta; \end{array}$$

two-electron transition of the shake-up, or shake-off type:

$$\begin{array}{ccc} \ldots (nl)^q \ldots (n'l')^p\, L,\, S & \xrightarrow{h\nu} & \ldots (nl)^{q-1} \ldots (n'l')^{p-1} (n''l'')^1\, L',\, S' + \\ & & + \varphi_{\varepsilon_\gamma, l\pm 1} \\ \Psi^i(N)_\alpha,\ E^i(N) & & \Psi^f(N-1)_\gamma,\ E^f(N-1)_\gamma. \end{array}$$

The subshell $(n'l')^p$ is completely or partially occupied. In the final state, the second electron which participates in the transition is labeled $(n''l'')^1$. In the case of processes of the shake-off type, n'' is replaced by ϵ_γ; it is assumed that the initial state is the same for all of the possible transitions. The final state may be characterized by different energies, including the possible structure due to multiplet splitting. The indexes α, β, and γ indicate the magnetic quantum numbers and the scheme of addition of momenta for the subshells.

The probability of transition from the initial state α to one of the final states β or γ is determined by the expression

$$\sigma_{nl,\alpha-\beta(\gamma)} \approx \left| \left\langle A\varphi_{\varepsilon,l''',m_l''',m_s'''} \Psi^f (N-1)_{\beta(\gamma)} \left| \sum_{i=1}^{N} \mathbf{r}_i \right| \times \right.\right.$$
$$\left.\left. \times A\varphi_{nlm_l m_s} \Phi (N-1)_\alpha \right\rangle \right|^2. \quad (58)$$

The representation of electron wave functions by formula (40):

$$\Psi_0 (N) = \frac{1}{\sqrt{N}} \sum_{i=1}^{N} (-1)^i \varphi_{nl} (x_i) \Psi_R (x_1, \ldots, x_{i-1}, x_{i+1}, \ldots, x_n)$$

implies that correlations between the electron and the rest of the system are negligible. The functions φ_{nl} and $\Phi_R(N-1)$ are usually restricted by an additional condition [106]:

$$\int \varphi_{nl} (x_1) \Psi_R (x_1, x_2, \ldots, x_{N-1}) \, dx_1 = 0. \quad (59)$$

When $\Psi_0(N)$ is represented by a Slater determinant, this condition is satisfied automatically. Let us write expression (58) as

$$\sigma_{nl,\alpha-\beta(\gamma)} \sim |(I_1 + I_2)|^2,$$

where

$$I_1 = \langle \varphi_{\varepsilon\beta(\gamma),l''',m_l''',m_s'''} (x_N) \Psi^f (x_1, x_2, \ldots, x_{N-1}) | \mathbf{r}_N | \times$$
$$\times A\varphi_{nlm_l m_s} \Phi (N-1)_\alpha \rangle,$$
$$I_2 = \left\langle \varphi_{\varepsilon\beta(\gamma),l''',m_l''',m_s'''} (x_N) \Psi^f (x_1, x_2, \ldots, x_{N-1}) \left| \sum_{i=1}^{N-1} \mathbf{r}_i \right| \times \right.$$
$$\left. \times A\varphi_{nlm_l m_s} \Phi (N-1)_\alpha \right\rangle.$$

Due to the orthogonality condition (59), I_2 is equal to zero and the photoionization cross section becomes

$$\sigma_{nl,\alpha-\beta(\gamma)} \sim |\langle \varphi_{\varepsilon\beta(\gamma),l''',m_l''',m_s'''} | \mathbf{r} | \varphi_{nlm_l m_s} \rangle|^2 \times$$
$$\times |\langle \Psi^f (N-1)_{\beta(\gamma)} | \Phi (N-1)_\alpha \rangle|^2. \quad (60)$$

Overlapping of the wave functions $\Psi^f(N-1)$ and $\Phi(N-1)_\alpha$ is nonzero only when both of these states have the same values of the quantum numbers L, S, M_L, M_S, while the one-electron orbitals in both of the functions are characterized by the same values of the orbital momenta l. These conditions lead to the known monopole-selection rules (43), which comprise the condition $\Delta l = l' - l'' = 0$.

In order to determine the photoionization cross section of the subshell, it is necessary to take the average of the expression (60) over all the degenerate initial states and to perform the summation over all of the final states related to the given configuration. Finally, the following expressions are obtained for the ionization cross sections for the one-electron and two-electron transitions $[\sigma_{nl}$ and $\sigma_{nl}(n'l' - n''l'')]$, where the initial and final states of the second electron are indicated in parentheses:

$$\sigma_{nl} \sim [C_{l+1}R^2_{\varepsilon,l+1} + C_{l-1}R^2_{\varepsilon,l-1}] \, | \langle \Psi^f (N-1)_0 \, | \, \Phi (N-1) \rangle_r |^2, \quad (61)$$

$$\sigma_{nl(n'l'-n''l'')} \sim [C_{l+1}R^2_{\varepsilon,l+1} + C_{l-1}R^2_{\varepsilon,l-1}] \times \\ \times | \langle \Psi^f (N-1)_{n'l'-n''l''} \, | \, \Phi (N-1) \rangle_r |^2, \quad (62)$$

where $\epsilon = E_{\text{kin}}$ represents the kinetic energy of the electron; the subscript 0 refers to the final-state configuration of the one-electron transition, and r indicates that only the overlap due to the radial integrals remains to be calculated. The wave function $\Phi(N-1)_\alpha$ may be represented as a series (41):

$$\Phi(N-1)_\alpha = \sum_\beta \langle \Psi^f (N-1)_\beta \, | \, \Phi (N-1)_\alpha \rangle \, \Psi^f (N-1)_\beta + \\ + \sum_\gamma \langle \Psi^f (N-1)_\gamma \, | \, \Phi (N-1)_\alpha \rangle \, \Psi^f (N-1)_\gamma + \ldots$$

The terms included in the sum over γ correspond to transitions of the shake-up and shake-off type. The state $\Phi(N-1)_\alpha$ is normalized, and therefore,

$$| \langle \Psi \, (N-1)_0 \, | \, \Phi (N-1) \rangle_r |^2 + \\ + \sum_{n'l', n''l''} | \langle \Psi^f (N-1)_{n'l'-n''l''} \, | \, \Phi (N-1) \rangle_r |^2 + \cdots = 1. \quad (63)$$

As already discussed, the binding energy $E_B^{KT}(nl)$ satisfies the relation

$$E_B^{KT}(nl) = \sum_{\delta=0}^{\infty} | \langle \Psi^f (N-1)_\delta \, | \, \Phi (N-1) \rangle |^2 \, E_B (nl)_\delta.$$

Therefore, the contribution of each binding energy to $E_B^{KT}(nl)$ is proportional to the magnitude of the corresponding photoionization cross section.

The expression for the photoionization cross section, without taking into account relaxation processes, is

$$\sigma^u_{nl,\alpha-\alpha} \sim \left| \left\langle \Psi^f(N)^u_\alpha \left| \sum_{j=1}^{N} \mathbf{r}_j \right| \Psi^i(N)_\alpha \right\rangle \right|^2 =$$

$$= \left| \left\langle A\varphi^u_{\varepsilon_\alpha, l''', m'''_l, m'''_s} \Phi(N-1)_\alpha \left| \sum_{j=1}^{N} \mathbf{r}_j \right| A\varphi_{nlm_lm_s} \Phi(N-1)_\alpha \right\rangle \right|^2 =$$

$$= |\langle \varphi^u_{\varepsilon_\alpha, l''', m'''_l, m'''_s} | \mathbf{r} | \varphi_{nlm_lm_s} \rangle|^2.$$

Performing the necessary summation and averaging gives us

$$\sigma^u_{nl} \sim [C_{l+1} R^2_{\varepsilon, l+1} + C_{l-1} R^2_{\varepsilon, l-1}]. \tag{64}$$

and combining expressions (61)-(64):

$$\sigma^u_{nl} = \sigma_{nl} + \sum_{n'l', n''l''} \sigma_{nl(n'l'-n''l'')} + \ldots \tag{65}$$

Therefore, without taking into account relaxation processes, the photoionization cross section is represented by the total photoionization cross section for one-electron and multielectron processes. This result may be generalized to take into account multiplet splitting:

$$\sigma^u_{nl} = \sum_{\delta=0}^{\infty} \sigma_{nl,\delta}, \tag{66}$$

where the ionization cross section $\sigma_{nl,\delta}$ corresponds to all final states of the given configuration having the same values of L', S' and the same energy. By using the ionization cross sections σ^u_{nl}, $\sigma_{nl,\delta}$ determined by equations (64)-(66), $E^{KT}_B(nl)$ $[= E^u_B(nl)]$ can be written

$$E^{KT}_B(nl) = \frac{\sum_{\delta=0}^{\infty} \sigma_{nl,\delta} E_B(nl)_\delta}{\sigma^u_{nl}}.$$

The "sudden perturbation" method is less appropriate for the description of the process of valence electron ionization at energies near to the absorption edge. There exist more efficient theoretical descriptions of the process of electron photoemission. These treatments take into account multielectron effects in the photoionization process. However, because of their sophisticated character, they have not yet been used extensively. The photoionization cross sections calculated by Amusia [10], Wendin [108, 109], and Lundqvist and Wendin [110] using the random-phase approximation and taking into account exchange processes are in quite a good agreement with the experimental data. We mention here that inclusion of multielectron effects is particularly important for the outer subshells, since in this case all the electrons of the subshell take part in the photoionization process [107, 111].

Energy Distribution of Valence Electrons in Crystals

The problem of the structure of the photoelectron spectra of the valence electrons of crystals has been treated by Hedin *et al.* [112]. After excitation, the final-state electron wave function may be written

$$|\Psi_f\rangle = a^+_{\tilde{\mathbf{k}}} | N-1, \; s\rangle,$$

where $\tilde{\mathbf{k}}$ is the momentum of the emitted electron. The transition probability is given by the formula

$$I(\varepsilon_{\tilde{\mathbf{k}}}) \sim \sum_f |\langle \Psi_f | P | \Psi_i\rangle|^2 \, \delta(\omega - E_f + E_i) =$$

$$= \sum_{\tilde{\mathbf{k}},s} |\langle N-1, \; s | a_{\tilde{\mathbf{k}}} \sum_{\mathbf{k},\mathbf{k}'} P_{\mathbf{k}\mathbf{k}'} a^+_{\mathbf{k}'} a_{\mathbf{k}} | N\rangle|^2 \, \delta(\omega - \varepsilon_{\tilde{\mathbf{k}}} + \varepsilon_s).$$

Since $E_f = \varepsilon_{\tilde{\mathbf{k}}} + E(N-1)$, $E_i = E(N)$, follows that

$$\varepsilon_s = E(N) - E(N-1, \; s).$$

In this formula, the matrix elements $P_{\mathbf{k}\mathbf{k}'}$ of the momentum operator $P = -i\hbar\nabla$ have been used. If $a_{\tilde{\mathbf{k}}}|N\rangle = 0$, then

$$I(\varepsilon_{\tilde{\mathbf{k}}}) = \sum_{\tilde{\mathbf{k}},s} |\langle N-1, s | \sum_{\mathbf{k}} P_{\tilde{\mathbf{k}}\mathbf{k}} a_{\mathbf{k}} | N\rangle|^2 \, \delta(\omega - \varepsilon_{\tilde{\mathbf{k}}} + \varepsilon_s) =$$

$$= \sum_{\tilde{\mathbf{k}}} \sum_{\mathbf{k}\mathbf{k}'} P^*_{\tilde{\mathbf{k}}\mathbf{k}} P_{\tilde{\mathbf{k}}\mathbf{k}'} A_{\mathbf{k}\mathbf{k}'}(\varepsilon - \omega),$$

where

$$A_{\mathbf{k}\mathbf{k}'}(\varepsilon - \omega) = \sum_s |\langle N-1, \; s | a^+_{\mathbf{k}} | N-1, \; s\rangle \times$$

$$\times \langle N-1, \; s | a_{\mathbf{k}'} | N\rangle | \, \delta(\omega - \varepsilon_s).$$

The number of vectors $\tilde{\mathbf{k}}$ for which the energy is $\epsilon_{\tilde{\mathbf{k}}} = \epsilon$ is proportional to $\epsilon^{1/2}$, and therefore

$$I(\varepsilon) \sim \sqrt{\varepsilon} \sum_{\mathbf{kk'}} A_{\mathbf{kk'}}(\varepsilon - \omega) P^{*}_{\tilde{\mathbf{k}}\mathbf{k}} P_{\tilde{\mathbf{k}}\mathbf{k'}}. \tag{67}$$

When a core electron is emitted, then by neglecting the excitation of the remaining electrons and by taking into account the fact that

$$a_c \mid N\rangle = a_c a_c^{+} \mid N_v,\ N_c - 1\rangle,$$

the following is obtained:

$$I(\varepsilon) \sim \sum_{\tilde{\mathbf{k}},s} |\langle N_v,\ s \mid N_v\rangle P_{\tilde{\mathbf{k}}c}|^2 \times \times \delta(\omega - \varepsilon_{\tilde{\mathbf{k}}} + \varepsilon_c),$$

where $|N_v, s>$ describes the state of the valence electrons including the effect of the core hole. This formula gives the radiation intensity in the "sudden perturbation" approximation. If the nondiagonal terms in the spectral function $A_{\mathbf{kk'}}$ and the dependence of $P_{\tilde{\mathbf{k}}\mathbf{k}}$ on the momenta $\tilde{\mathbf{k}}$ and $\mathbf{k}$ can be neglected, then instead of formula (67) the following is obtained:

$$I(\varepsilon) \sim \sqrt{\varepsilon}\, P_{eff}^2 N(\varepsilon - \omega), \tag{68}$$

since

$$N(E) = \sum_{\mathbf{kk'}} A_{\mathbf{kk'}}(E) = SpA(E).$$

If the dependence of $P_{\tilde{\mathbf{k}}\mathbf{k}}$ on the momenta $\tilde{\mathbf{k}}$ and $\mathbf{k}$ cannot be neglected, then calculations should be based on formula (67). In the one-electron approximation, $A_{\mathbf{kk'}}(\epsilon - \omega)$ is given by the following expression:

$$A_{\mathbf{kk'}}(\varepsilon - \omega) \sim \delta_{\mathbf{kk'}} \delta(\varepsilon - \omega - \varepsilon_{\mathbf{k}}).$$

In discussing the problem of the X-ray photoelectron emission of crystals, we have not considered the processes that determine the emission of the excited electron with momentum $\tilde{\mathbf{k}}$ outside the sample. For simplicity, it has been considered that the process of photoemission takes place through the following stages:

(1) The valence electron is excited into a state in the conduction band.

(2) The electron travels toward the surface (and possibly undergoes inelastic scattering processes).

(3) The electron passes through the surface (or is possibly reflected backwards).

Before considering the experimental and theoretical calculations of photoelectron spectra, we will discuss briefly the basic methods for the calculation of energy bands in crystals.

Methods of Calculation of Energy Bands in Crystals

In order to be able to calculate the valence band photoelectron spectra of crystals it is necessary, as indicated by equations (67) and (68), to know the wave functions and the energy values corresponding to one-electron states.

At present, the methods that are most extensively used for calculation of the structure of energy bands and of electron wave functions are the methods of augmented plane waves (APW), the method of Green's functions, the method of orthogonal plane waves (OPW), and the method of the pseudopotential. These methods are described in detail in a number of monographs devoted to the band theory of solids [113-119, 5], as well as in a number of review articles [119-125]. Some of these works are entirely devoted to just one of these methods. Thus, the monographs of Slater [113] and Loucks [117] deal only with the APW method, the book of Harrison [115] considers the method of the pseudopotential, the review article of Woodruff [124] discusses the OPW method, and that of Dimmock [125] the APW method.

Of course, the question arises why have so many different methods been developed to describe the electron energy spectra of crystals? Can it be explained by the fact that each method should be used only for a limited number of solids? To answer these questions, attention should be paid to the period in which the methods were developed. The first method for calculation of energy bands of crystals was proposed by Bloch at the end of the 1920s in those years when the quantum theory of matter was founded [126]. It was immediately used to obtain results of a general character on the dynamics of the motion of electrons in crystals. The development of the APW and OPW methods is associated with the names of Slater [127] and Herring [128], respectively. However, specific band-structure calculations for various materials did not immediately follow the development of these methods. The reason for the delay in application and progress of these methods arises from their great mathematical difficulty and the necessity to perform tedious calculations. The situation only changed in the 1950s when the first computers appeared. Even though they were far from advanced, their arrival initiated a period of intensive calculations of the band structure of crystals. Since the 1960s progress in our knowledge of the band structure of the most important solid materials has been closely related to advances in computer techniques.

A number of distinct directions in the development of band-structure theories have now emerged, and groups of scientists involved in the use and development of each particular method have formed. Thus, Herman has achieved great success in the development of the OPW method, Slater has made the APW method one of the most extensively used in band calculations, while Harrison, Heine, and Cohen have made valuable contributions to the development of the pseudopotential method, proposed at the end of the 1950s. These groups of scientists are still involved in development of improved methods.

The first calculations for a given crystalline material using different methods yielded quite different results. Usually, the ordering of the energy bands and the main characteristics of their structure are well reproduced. The quantitative differences that are observed may be partly attributed to differences in the approximations used, the character of these approximations, and the different convergence properties (dependence of calculated energy values on the number of functions in the basis).

It is well known that the crystal potential varies rapidly in the vicinity of atomic nuclei and only slowly in the region between the atoms. Consequently, the wave function of electrons in crystals is also characterized by rapid and abrupt changes (with characteristic critical points) in the vicinity of nuclei and a relative smooth behavior in the region between the atoms. This property of the potential and of wave functions is expressed differently in the different methods, and this leads to different final results. Thus, in the APW and Green's function methods, some simplifying assumptions are made concerning the shape of the crystal potential. The potential is chosen to have a muffin-tin shape.

Outside a sphere of given dimensions, the potential is assumed to be constant. This assumption is justified since in the region between atoms the potential varies slowly (see Figure 41).

In the OPW method, it is not necessary to use a muffin-tin (MT) potential. For this reason alone the results obtained by the APW, Green's function, and OPW methods may differ. This discrepancy might be reduced by using a poten-

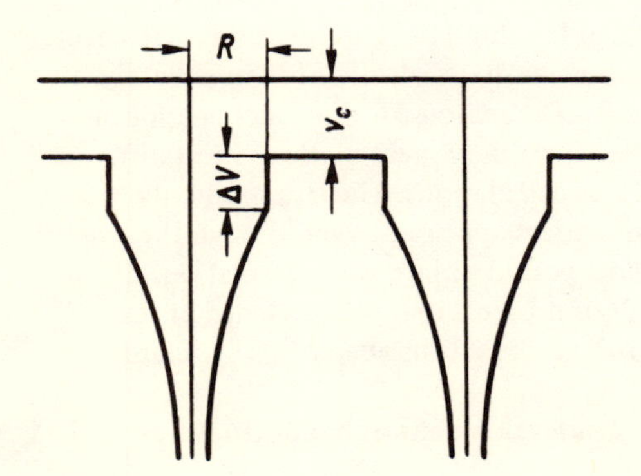

Figure 41. MT potential for a monatomic crystal.

tial deviating from the muffin-tin shape. This, however, could not be easily done at the beginning of the 1960s, since at that time the memory and calculation speed of computers were not high enough for the large volume of calculations necessary. By the end of the 1960s, when computers with calculation speeds of the order of 10^6 operations/sec became available, it was possible to perform calculations based on a smaller number of simplifying assumptions, and with a greater accuracy. As a consequence, use has begun to be made of equivalent crystal potentials and certain other approximations, so that the final results obtained by using different methods have become quantitatively similar.

Nevertheless, the different methods provide different opportunities to perform empirical corrections, to vary parameters, or to include relativistic effects. It has been shown that the calculation of the band structures of insulators and semiconductors is easier to perform by using the OPW method or the pseudopotential method, while for the transition and rare-earth metals, it is easier to use the APW method or the method of Green's functions. For transition metal compounds and intermetallic compounds, it is preferable to use the APW method. At present no universal method exists for the calculation of the band structures of all types of crystals. However, the group of existing methods allows an approach to most problems of interest in solid state physics.

The basic equation of the band theory is the one-electron Schrödinger equation:

$$\{-\nabla^2 + V(\mathbf{r})\}\,\Psi(\mathbf{r}) = E\Psi(\mathbf{r}), \tag{69}$$

where $V(\mathbf{r})$ is the crystal potential. For crystals, $V(\mathbf{r})$ is a periodic function in the space of lattice vectors $V(\mathbf{r} + \mathbf{R}_n) = V(\mathbf{r})$. The solutions of equation (69) that are of interest in the energy-band theory should satisfy the boundary condition

$$\Psi(\mathbf{r} + \mathbf{R}_n) = \exp i\mathbf{k}\cdot\mathbf{R}_n\Psi(\mathbf{r}), \tag{70}$$

where k is the wave vector or the electron quasimomentum; this takes N different values in the vector space of the reciprocal lattice in the Brillouin zone. (N is equal to the number of elementary cells in the crystal.) A consequence of condition (70) is that the solutions of equation (69) are quasimomentum dependent and that it is possible to represent them as Bloch functions: $\Psi(\mathbf{k}, \mathbf{r}) = \exp i\mathbf{k}\cdot\mathbf{r}U_k(\mathbf{r})$, where $U_k(\mathbf{r})$ is a periodic function: $U_k(\mathbf{r} + \mathbf{R}_n) = U_k(\mathbf{r})$.

The quasimomentum dependence of the energies $E_\mathbf{k}$ and the wave functions $\Psi(\mathbf{k}, \mathbf{r})$ has enabled an interpretation to be performed in terms of energy-band structure of crystals. For each $\mathbf{k}$, equation (69) has an infinite number of solutions, which may be ordered so that $E_n(\mathbf{k}) \leqslant E_{n+1}(\mathbf{k})$ for all of the n values. The totality of the energy values $E_n(\mathbf{k})$, when k takes all values in the Brillouin zone, is called the nth energy band, and the system of energy bands constitutes the energy-band structure of the crystal. In each band, the energy $E_n(\mathbf{k})$ is a continuous function of the vector $\mathbf{k}$. According to the Pauli principle, there are two electrons on each energy level $E_n(\mathbf{k})$, and $2N$ electrons in each energy band. Therefore, in insulators, several of the first energy bands are completely occupied, while the remaining bands are empty. In metals, there exist energy bands that are only partially occupied. In this case, the energy value of the highest level that separates the occupied states from the empty ones is called the Fermi energy E_F, and the surface in the $\mathbf{k}$ space determined by the equation $E(\mathbf{k}) = E_F$ is called the Fermi surface. Before proceeding to a treatment of the basic symmetry properties of energy bands and of wave functions, we will briefly discuss the methods for their determination.

At present, in the most widely used methods for the determination of the energy eigenvalues and eigenfunctions, in equation (69) the wave function $\Psi_n(\mathbf{k}, \mathbf{r})$ is represented as a series:

$$\Psi_n(\mathbf{k},\ \mathbf{r}) = \sum_i C_i^n(\mathbf{k})\, \varphi_i(\mathbf{k},\ \mathbf{r}), \tag{71}$$

where each of the basis functions $\varphi_i(\mathbf{k}, \mathbf{r})$ satisfies the Bloch condition (70). In this series the coefficients $C_i^n(\mathbf{k})$ of the basis functions are unknown and may be determined by solving the following system of linear homogeneous equations

$$\sum_j (H_{ij} - ES_{ij})\, C_j^n = 0, \tag{72}$$

where

$$H_{ij} = \int_V \varphi_i^*(r)\, H\varphi_j(r)\, dV, \quad S_{ij} = \int_V \varphi_i^*(r)\, \varphi_j(r)\, dV.$$

Since the series (71) is limited to a finite number of terms, it follows that the system of linear homogeneous equations (72) is also finite. The energy values $E_n(\mathbf{k})$ are determined by the condition that the determinant of the system is equal to zero:

$$\det |\, H_{ij} - ES_{ij} \,| = 0.$$

In this way, the difficult problem of solving the differential equation (69) is reduced to the much simpler problem of solving the system of algebraic equations (72). These can be obtained from the condition of minimization of the functional

$$E = \frac{\int\limits_V \Psi^* H \Psi dV}{\int\limits_V \Psi^* \Psi dV}.$$

In some cases, equation (69) is transformed into an integral equation, obtained by varying in a determined way the chosen functional. By introducing expression (71) for the functions $\Psi(\mathbf{k}, \mathbf{r})$ into this functional, one obtains instead of the integral equation, a system of algebraic equations for the energies $E_n(\mathbf{k})$ and the coefficients $C_i^n(\mathbf{k})$. When applying the variation principle to the integral equations, the treatment can be limited to a single elementary cell, since knowledge of the Bloch function for a single elementary cell is sufficient to allow determination of its values over the whole crystal by simply applying relation (70). In this case, the wave function in each elementary cell should satisfy some boundary conditions. An elementary cell represents a geometrical figure determined by the intersection of pairs of parallel planes such that to each point $\mathbf{r}_1$ situated on one boundary, there always corresponds a point $\mathbf{r}_2$ situated on the second boundary, at a distance from $\mathbf{r}_1$ given by a translation vector $\mathbf{a}$, i.e., $\mathbf{r}_2 = \mathbf{r}_1 + \mathbf{a}$. Here $\mathbf{a}$ is a constant vector, characteristic for each pair of parallel planes and having different magnitude for different pairs. Therefore, the following relation is valid: $\Psi(\mathbf{k}, \mathbf{r}_2) = \exp(i\mathbf{k} \cdot \mathbf{a})\Psi(\mathbf{k}, \mathbf{r}_1)$. Since on opposite boundaries of the elementary cell the external normals are oriented in opposite directions, $\mathbf{n}_1 = -\mathbf{n}_2$,

$$\frac{\partial}{\partial \mathbf{n}_2} \Psi(\mathbf{k}, \mathbf{r}_2) = -\exp i\mathbf{k} \cdot \mathbf{a} \frac{\partial}{\partial \mathbf{n}_1} \Psi(\mathbf{n}, \mathbf{r}_1),$$

where $\partial/\partial n$ denotes differentiation with respect to the direction of the external normal to the boundary of the elementary cell.

The same simple choice of the basis functions $\varphi_i(\mathbf{k}, \mathbf{r})$ is made in the methods of tight binding and of plane waves. In the tight-binding method, which is sometimes called the method of linear combination of atomic orbitals (LCAO), the functions $\varphi_i(\mathbf{k}, \mathbf{r})$ are represented as linear combinations of the atomic orbitals $a_i \equiv a_{nlm}(\mathbf{r}) = R_{nl}(\mathbf{r}) Y_{lm}(\Theta, \varphi)$:

$$\varphi_i(\mathbf{k}, \mathbf{r}) = \frac{1}{\sqrt{N}} \sum_{\mathbf{R}_n} \exp i\mathbf{k} \cdot \mathbf{R}_n \cdot a_i(\mathbf{r} - \mathbf{R}_n). \tag{73}$$

The LCAO method has not been used extensively because the calculation of the overlapping integrals for atomic functions centered at different points of the lattice proves to be very difficult. Recently, some modifications of the LCAO-method have been proposed [129, 130], but they have as yet only been used for band-structure calculations of crystals of the light elements lithium and carbon. We mention here that the functions (73) may be used for describing the core electrons of atoms in crystals.

The method of plane waves is based on the use of the plane waves

$$\varphi_i\ (\mathbf{k},\ \mathbf{r}) = \frac{\exp i\,(\mathbf{k} + \mathbf{K}_i)\cdot\mathbf{r}}{\sqrt{N\Omega_0}}$$

as basis functions, where Ω_0 is the volume of the elementary cell, and $\mathbf{K}_i$ is the vector of the reciprocal lattice. Plane waves are conveniently used as basis functions in the determination of the matrix elements of the Hamiltonian in equations (72). In the method of plane waves, the solution depends on the Fourier components of the potential:

$$V(\mathbf{K}_i) = \frac{1}{V}\int_V \exp i\mathbf{K}_i\cdot\mathbf{r}V(\mathbf{r})\,dV,$$

which may be determined with great accuracy. However, $V(\mathbf{K}_i)$ decreases rapidly only at large values of the vectors $\mathbf{K}_i$, and therefore the energy values obtained depend strongly on the order of the equations. In order to determine the valence-electron wave functions and energies with a sufficient accuracy, it is necessary to choose several thousand basis functions $\varphi_i(\mathbf{k}, \mathbf{r})$. Consequently, this method, which is simple in principle, becomes very tedious when used for performing practical calculations.

The method of plane waves has been considerably improved by Herring [128]. In order to reproduce the rapid variation of the wave functions in the neighborhood of atomic centers in the crystal, he has suggested the construction of functions orthogonal to the wave functions of the core electrons. Such functions, constructed from LCAO functions and plane waves, are called orthogonal plane waves:

$$\varphi_i\ (\mathbf{k},\ \mathbf{r}) = \frac{1}{\sqrt{N\Omega_0}}\exp i\,(\mathbf{k} + \mathbf{K}_i)\cdot\mathbf{r} - \sum_{nlm} B_{nlm}\,(\mathbf{k} + \mathbf{K}_i)\,\Psi_{nlm}\,(\mathbf{k},\ \mathbf{r}), \quad (74)$$

where B_{nlm} is given by $B_{nlm}(\mathbf{k} + \mathbf{K}_i) = \delta_{m0}A_{nl}(|\mathbf{k} + \mathbf{K}_i|)$ in the system of coordinates having the z axis parallel to the vector $\mathbf{k} + \mathbf{K}_i$; $A_{nl}|\mathbf{k} + \mathbf{K}_i|$ is the orthogonality coefficient determined by the atomic wave functions $P_{nl}(\mathbf{r})$. The functions $\varphi_i(\mathbf{k}, \mathbf{r})$ are constructed to be orthogonal to $\Psi_{nlm}(\mathbf{k}, \mathbf{r})$:

$(\varphi_i, \Psi_{nlm})=0$, for all of the quantum numbers n, l, m over which the summation in expression (74) is performed, provided that the atomic functions entering into $\Psi_{nlm}(\mathbf{k}, \mathbf{r})$, localized at nearby points in the lattice, do not overlap appreciably. This condition is quite accurately satisfied, and therefore the orthogonal plane waves constitute a convenient system of basis functions to be used for the determination of the wave functions of valence electrons, although it is not an orthogonal system in itself. The energies $E_n(\mathbf{k})$ and the wave functions $\Psi_n(\mathbf{k}, \mathbf{r})$ are determined by the coefficients of the Fourier components of the potential, by the values of the energies of the core electrons E_{nl}, and by the coefficients of orthogonality. In this method, the coefficients $C_i^n(\mathbf{k})$ decrease more rapidly than in the method of plane waves, and 150–200 OPW functions are usually sufficient. In accurate calculations, the core functions should be determined by taking into account the crystal potential field.

The OPW method also has drawbacks. The convergence of series (71) is considerably worsened if the orthogonal plane waves (74) do not contain core functions of a given symmetry type. Thus, at the point $\mathbf{k} = 0$ of diamond, in the determination of valence band p-state energies the series (71) contains only plane waves because the carbon atoms have only 1s-core states. The wave function for $\mathbf{k} = 0$, constructed from symmetrized plane waves, is characterized by a p-type symmetry, and therefore, being orthogonal to the ground 1s state, it expresses to some extent the characteristics of the structure of valence-band p states. The convergence of p states in diamond is very slow. Euwema *et al.* [131] consider that a good convergence is obtained if the number of basis functions is of the order of 5000.

In contrast, silicon has core states of both s and p type, which improves the convergence considerably. For silicon the convergence is usually satisfactory with only about 150–200 OPW basis functions. The OPW method is not recommended for the band-structure calculations of transition metals, since in the transition metals of the first transition period there are no core states of d-type symmetry.

It is, of course, possible to add to the expansion in OPW functions some localized wave functions of d-type symmetry, determined by the crystal potential field. In doing so, a boundary condition may be used, namely that the d functions should be equal to zero outside a sphere of radius approximately equal to the radius of the muffin-tin sphere for the given atom of the crystal. Due to the complexity of this procedure, the OPW method is usually applied to band-structure calculations of the compounds of type A_3B_5 and A_2B_6. The convergence in the OPW method for these compounds is affected by the following factors: the presence or the absence of the corresponding core function, the relative dimensions of the cores of the various atoms in the elementary cell, and the degree of localization of the valence wave function. Studies performed by Euwema *et al.* [131] have shown that the convergence for the p-type states is determined to a greater extent by the dimensions of the anion than by those of the cation. More localized valence states are characterized by a worse convergence. Thus, solid argon is situated in the same line of the periodic system as silicon, but for argon it is necessary to use twice as many orthogonal plane waves as for silicon. Poor convergence is also characteristic for compounds with strong ionic character, such as NaCl.

Better results are obtained if self-consistent calculations are performed. In order to improve the non-self-consistent calculations, Herman *et al.* [132] have suggested that empirical corrections may be used. Such corrections are easy to introduce into the crystal potential and the energy values of core electrons. By making these corrections, the potential V is transformed into a new potential $V + \Delta V$, where the magnitude of ΔV is chosen so that the calculated energy values agree as well as possible with the experimental ones. By using empirical corrections, it is possible to compensate to some extent for the neglected relativistic and correlation effects, and for the insufficient number of basis functions. Thus, for example, in calculating the band structure of diamond-type crystals, Herman *et al.* [132] used three parameters: $\Delta V(111)$, $\Delta V(220)$, $\Delta V(311)$. For compounds with a crystal lattice of the sphalerite type, Herman *et al.* [133] as well as Shay *et al.* [134] used a somewhat different three-parameter scheme. This included the following parameters: $\Delta V_S(111)$ determining the change of the symmetrical Fourier component of the potential $V_S(111)$, ΔE_c^a and ΔE_c^c representing the core shifts of the anion and cation, respectively. The magnitude of the shifts was chosen identically for all of the core states of both the cation and the anion.

The APW method, which has been applied extensively for the calculation of the energy bands in transition metals and their compounds, is also based on the use of the variation principle. It also makes use of plane waves and localized functions. It differs from the OPW method, however, in a number of ways. First, in the APW method the potential usually has a muffin-tin shape:

$$V(\mathbf{r}) = \begin{cases} V(|\mathbf{r}|), & \text{for } r \leqslant R_{\mathrm{MT}}; \\ V_c, & \text{for } r > R_{\mathrm{MT}}. \end{cases}$$

As a consequence of the fact that the elementary cell is divided into two regions, each of the basis functions consists of two parts. Outside the muffin-tin sphere, the basis functions $\varphi_i(\mathbf{k}, \mathbf{r})$ behave like plane waves, while inside the muffin-tin sphere they are represented as linear combinations of the functions $R_l(r)$, which are solutions of the radial Schrödinger equation for the potential $V(\mathbf{r})$:

$$-\frac{1}{\rho^2}\frac{d}{d\rho}\left(\rho^2\frac{dR_l}{d\rho}\right) + \left[\frac{l(l+1)}{\rho^2} + V(r)\right]R_l(\rho) = E'R_l(\rho). \quad (75)$$

$R_l(\rho)$ should satisfy a boundary condition; that is, it should be regular at $\rho = 0$. Since there is only one boundary condition, it follows that E' in equation (75) may take arbitrary values, and in particular it may be equal to the initial value of the energy $E' = E$. The magnitude of the coefficients A_{lm} in the series expansion of the functions $\varphi_i(\mathbf{k}, \mathbf{r})$ inside the muffin-tin sphere,

$$\varphi_i(\mathbf{k}, \mathbf{r}) = \sum_{l=0}^{\infty} \sum_{k=-l}^{l} A_{lm}(\mathbf{k} + \mathbf{K}_i) R_l(r) Y_{lm}(\Theta, \varphi),$$

are determined by equating this expression with the expression of the plane wave, $\exp i(\mathbf{k} + \mathbf{K}_i) \cdot \mathbf{r}$, at the surface of the sphere:

$$A_{lm} = \frac{4\pi i^l Y_{lm}(\widehat{\mathbf{k} + \mathbf{K}_i}) Y_{lm}(\widehat{\mathbf{r}})}{R_l(R_{\mathrm{MT}})}.$$

Therefore, the basis functions $\varphi_i(\mathbf{k}, \mathbf{r})$ in the APW method are given by the following formulas:

$$\varphi_i(\mathbf{k}, \mathbf{r}) = \begin{cases} \displaystyle\sum_{l=0}^{\infty} \sum_{m=-l}^{l} (2l+1) i^l \frac{j_l(|\mathbf{k} + \mathbf{K}_i| R_{\mathrm{MT}})}{R_l(R_{\mathrm{MT}})} R_l(\rho, E) Y^*_{lm}(\widehat{\mathbf{k} + \mathbf{K}_i}) Y_{lm}(\widehat{\rho}), \\ \exp i(\mathbf{k} + \mathbf{K}_i) \cdot \mathbf{r}. \end{cases} \tag{76}$$

These functions are constructed to be continuous at the surface of the sphere, but their derivatives exhibit discontinuities. This is the reason why a functional is used that takes into account the existence of these derivative discontinuities. If the region of the elementary cell inside the sphere is labeled I, and that outside the sphere II, then the functional in the APW method is given by the expression:

$$E \int_{\mathrm{I+II}} \Psi^* H \Psi dV = \int_{\mathrm{I+II}} \Psi^* H \Psi dV + \frac{1}{2} \int_S \Big[(\Psi_{\mathrm{II}} - \Psi_{\mathrm{I}}) \Big(\frac{\partial}{\partial \rho} \Psi^*_{\mathrm{II}} + \frac{\partial}{\partial \rho} \Psi^*_{\mathrm{I}} \Big) - (\Psi^*_{\mathrm{II}} + \Psi^*_{\mathrm{I}}) \Big(\frac{\partial}{\partial \rho} \Psi_{\mathrm{II}} - \frac{\partial}{\partial \rho} \Psi_{\mathrm{I}} \Big) \Big] dS,$$

where the derivative $\partial/\partial\rho$ is taken in the direction of the external normal of the first region. The values of the energies $E_n(\mathbf{k})$ and wave functions $\Psi_n(\mathbf{k}, \mathbf{r})$ in the APW method are determined mainly by the logarithmic derivatives $R_l'(\rho)/R_l(\rho)$

at the surface of the muffin-tin sphere. The convergence depends not only on the number of functions in the series (71), but also on the number of harmonics in the expansion (76). Typically, 50 basis functions are used for each atom situated inside the elementary cell, l_{max} being taken as 12. The convergence in the APW method is better than in the OPW method. It has been found that the muffin-tin potential represents a good first approximation to the crystal potential. If necessary, account can be taken of the nonspherical symmetry of the potential inside the muffin-tin sphere and its nonconstancy outside it. The latter feature has the greater importance. In this case, the secular equation comprises the Fourier components of the crystal potential outside the sphere.

A modification of the APW method including deviation of the crystal potential from the muffin-tin shape has been proposed by Slater and de Ciceo [135]. In the APW method, the series expansion of the wave function contains localized functions of $s, p, d, f, \ldots$ type and therefore no difficulties arise in calculating the energy values for any type of symmetry. This is the reason why the APW method is so efficient for the calculation of energy bands for a large class of materials, particularly for the transition metals and their compounds.

The results obtained by the method of Green's functions are similar to those obtained by the APW method since both methods use similar approximations for the potential. In the method of Green's functions, it is convenient to choose the potential with a muffin-tin shape. The equations used in the method of Green's functions also contain the logarithmic derivatives of the radial functions $R_l(r)$ at the surface of the muffin-tin sphere. For $r < R_{MT}$, the solution obtained by applying the method of Green's functions is expressed as a series

$$\sum_{l=0}^{\infty} \sum_{m=-l}^{l} C_{lm} R_l(r) Y_{lm}(\Theta, \varphi),$$

where the functions $R_l(r)$ are determined from equation (75). In the method of Green's functions, the following functional is varied:

$$\Lambda = \int_{\Omega_0} \Psi^*(\mathbf{r}) \mathbf{V}(\mathbf{r}) \Psi(\mathbf{r}) \, dV - \int_{\Omega_0} \int_{\Omega_0} \Psi^*(\mathbf{r}) V(\mathbf{r}) G(\mathbf{r}, \mathbf{r}') V(\mathbf{r}') \times$$

$$\times \Psi(\mathbf{r}') \, dV dV',$$

where $G(\mathbf{r}, \mathbf{r}')$ is the Green's function determined by the equation

$$(\nabla^2 + E) G(\mathbf{r}, \mathbf{r}') = \delta(\mathbf{r} - \mathbf{r}').$$

The convergence in this method is high. Usually it is sufficient to consider only 4-6 harmonics in the expansion of the wave function $\Psi(\mathbf{r})$. However, in using the method of Green's functions, as compared to the APW method, it is more difficult to solve the secular equation; the deviation of the potential from the muffin-tin shape is less readily accounted for, and it is also more difficult to determine the electron wave function outside the muffin-tin sphere.

Recently, particularly wide use in band-structure calculations has been made of the pseudopotential method, in which the wave functions are expanded in series of plane waves. In a number of cases, this method allows a considerable simplification of calculations of the energy-band structure for crystals. Phillips and Kleinman [136] have pointed out that, in the equations used in the OPW method, the terms containing the orthogonality coefficients compensate for the values of the Fourier components of the potential. This allows a determination of the pseudopotential, which contains the repulsion terms present in the equation of the OPW method. This potential compensation depends on the number of functions in the core. Since the system of core functions is far from being complete, it follows that the potential compensation is also incomplete. It should be noted that the pseudopotential represents in general a nonlocal operator. The action of V_{ps} on an arbitrary function $\varphi(\mathbf{r})$ may be expressed by the relation

$$V_{ps}\varphi(\mathbf{r}) = V(\mathbf{r})\varphi(\mathbf{r}) + \sum_c (E - E_c)\Psi_c(\mathbf{r}) \int \Psi_c(\mathbf{r}')\varphi(\mathbf{r}')\,dV'.$$

It has been shown that, on the basis of data obtained from the study of scattering properties, it is possible to construct pseudopotentials of a more general shape [137]. The pseudopotential wave function (Figure 42) is characterized by the fact that it does not contain the radial nodes that are characteristic for the real function. Since the scattering amplitude is determined by the logarithmic derivative of the radial wave function, the potential can be replaced

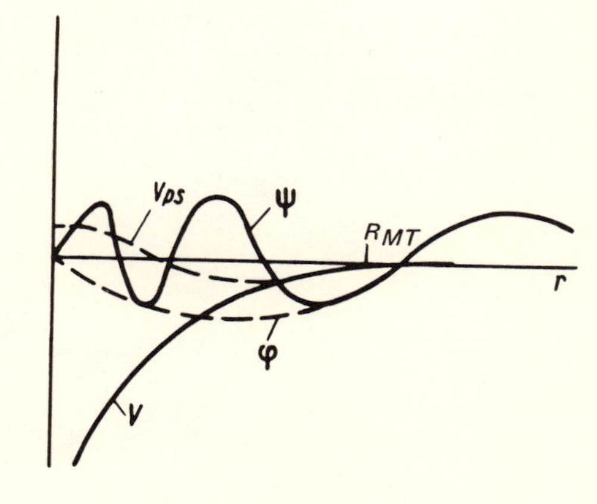

Figure 42. Wave function $\Psi(\mathbf{r})$ and pseudo-wave-function $\varphi(\mathbf{r})$ corresponding to the potential $V(\mathbf{r})$ and the pseudopotential $V_{ps}(\mathbf{r})$, respectively.

by a pseudopotential that has exactly the same scattering amplitude. For this, it is necessary that the logarithmic derivative of the wave function $R_l(r)$ be equal to the logarithmic derivative of the pseudowave function at the surface of a sphere of radius equal to the effective ion radius in the crystal. On the basis of these considerations, Abarenkov and Heine [137] have constructed a model potential, which, for an ion of charge Z, may be written as follows:

$$\tilde{V}_{ps} = \begin{cases} -\sum_l A_l(E)\, P_l(E), & r < R_{\mathrm{MT}}, \\ -\dfrac{Z}{r}, & r > R_{\mathrm{MT}}, \end{cases} \tag{77}$$

where E is the energy of the state, $A_l(E)$ is a factor that depends on E and l, and P_l is the projection operator that, when applied to wave functions, separates the components having the azimuthal quantum number l:

$$P_l f(r, \Theta, \varphi) = \sum_{m=-l}^{l} Y_{lm} \int_0^{2\pi} \int_0^{\pi} Y_{lm}^{*}(\Theta', \varphi')\, f(r, \Theta', \varphi') \sin\Theta' d\Theta' d\varphi'.$$

The factor $A_l(E)$ is determined experimentally, i.e., is chosen so that the pseudopotential (77) generates the spectroscopically observed one-electron energy levels. For the total collection of ions distributed at the nodes of the crystal lattice, the pseudopotential may be written as

$$\tilde{V}_{ps}(\mathbf{r}, \mathbf{r}', E) = \sum_j V_{ps}(\mathbf{r} - \mathbf{R}_j, \mathbf{r}' - \mathbf{R}_j, E). \tag{78}$$

Due to its nonlocal character, the potential depends on $\mathbf{r}$ and $\mathbf{r}'$. In calculations performed using the pseudopotential method, the secular equation contains the matrix element

$$\langle \mathbf{k} | \tilde{V}_{ps}(\mathbf{r}, \mathbf{r}', E) | \mathbf{k}' \rangle,$$

which may be written as follows:

$$\langle \mathbf{k} | \tilde{V}_{ps} | \mathbf{k}' \rangle = S(\mathbf{q})\, \tilde{V}_{ps}(\mathbf{q}),$$

where

$$\mathbf{q} = \mathbf{k} - \mathbf{k}', \quad S(\mathbf{q}) = \frac{1}{N} \sum_j \exp i\mathbf{q} \cdot \mathbf{R}_j$$

is a structure factor, and $\tilde{V}_{ps}(\mathbf{q})$ is the ionic pseudopotential form factor:

$$\tilde{V}_{ps}(\mathbf{q}) \equiv \tilde{V}_{ps}(\mathbf{q}, \mathbf{k}, \mathbf{k}', E) = \\ = \Omega_0^{-1} \int \exp(-i\mathbf{k} \cdot \mathbf{r}) \tilde{V}_{ps}(\mathbf{r}, \mathbf{r}') \exp(i\mathbf{k}' \cdot \mathbf{r}) \, dV dV'. \tag{79}$$

Often, in the treatment of electronic properties, only the local approximation is considered—the energy E in expression (79) is assumed to be equal to the Fermi energy E_F, and the vectors $\mathbf{k}$ and $\mathbf{k}'$, for $q = |\mathbf{k} - \mathbf{k}'| < 2k_F$, are taken as being equal to k_F and $q - k_F$, respectively, being oriented in opposite directions if $q > 2k_F$. In order to obtain the pseudopotential of the crystal from the pseudopotentials of the free ions situated at the nodes of the crystal lattice, account should be taken of the fact that the electrons in the conduction band are screening the pseudopotential (78). For the Fourier components, the following expression is then obtained:

$$V_{ps} = \frac{\tilde{V}_{ps}(q)}{\varepsilon(q)},$$

where $\epsilon(q)$ is the dielectric permeability.

The local approximation of the pseudopotential has been applied successfully in studies of the energy-band structure of a great number of crystals. Moreover, it can be stated that existing knowledge of the band structure of A_3B_5 and A_2B_6 compounds has been gained from the use of the pseudopotential method in its local-approximation form. It has been found that in order to describe the band structure of these crystals, six parameters are sufficient, namely, the Fourier components of the pseudopotential. In this case, the pseudopotential is represented by a fragment of a Fourier series:

$$V_{ps}(\mathbf{r}) = \sum_{\mathbf{K}} V_{ps}(\mathbf{K}) \exp i\mathbf{K} \cdot \mathbf{r}.$$

For crystals of the sphalerite type, it is convenient to set the origin of the coordinate system between the atoms, and then the following relation is valid:

$$V_{ps}(\mathbf{K}) = S^S(\mathbf{K}) V^S(\mathbf{K}) + iS^A(\mathbf{K}) V^A(\mathbf{K}),$$

where

$$S^S(\mathbf{K}) = 2\cos \mathbf{K} \cdot \boldsymbol{\tau}, \quad S^A(\mathbf{K}) = 2\sin \mathbf{K} \cdot \boldsymbol{\tau}, \\ V^S(\mathrm{K}) = \frac{1}{2}(V_1(\mathrm{K}) + V_2(\mathrm{K})), \quad V^A(\mathrm{K}) = \frac{1}{2}(V_1(\mathrm{K}) - V_2(\mathrm{K})), \\ \boldsymbol{\tau} = \frac{a}{8}(111).$$

Here, $V_1(\mathrm{K})$ and $V_2(\mathrm{K})$ are the Fourier components of the pseudopotentials corresponding to the atoms of type 1 and 2 [138].

To calculate the band structure of molecules and cluster groupings in solids, the X_α-cluster method proposed by Johnson and Smith [139] can be used. In this method, multicenter integrals that complicate the calculations for selection of a basis related to atomic orbitals do not appear. This fact also significantly reduces the computer operation time necessary for calculation of the electron band structure. When calculations are performed for clusters in solids, it is convenient to consider a single isolated cluster. The volume occupied by the cluster (see Figure 43) may be divided into three types of regions:

(I) the atomic region, consisting of spherical volumes centered at the atoms in the cluster;

(II) the interatomic region, between the inner atomic spheres and an outer sphere enclosing the whole cluster and centered at its central atom (the Watson sphere);

(III) the extra-atomic region, outside the Watson sphere.

If the extra-atomic region is neglected, the division of the space is analogous to that in the muffin-tin approximation, though here the muffin-tin spheres may intersect each other. The radii of the spheres depend on the particular Hartree–Fock model potential chosen as a first approximation in the self-consistent calculations.

For an arbitrary point in the cluster, the potential may be written as follows:

$$V(\mathbf{r}) = V^C(|\mathbf{r} - \mathbf{R}_0|) + \sum_{j=1}^{n} \tilde{V}(|\mathbf{r} - \mathbf{R}_j|), \tag{80}$$

where $\mathbf{R}_0$ defines the position of the central atom in the cluster, and $\mathbf{R}_j$ determines the position of the other atoms. This expression for the potential is aver-

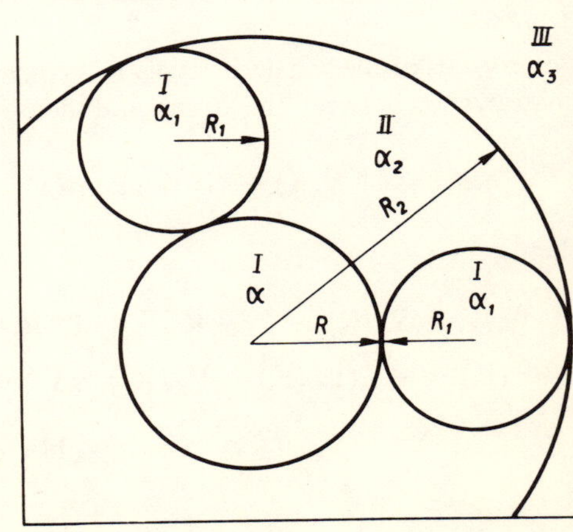

Figure 43. Atomic (I), interatomic (II), and extra-atomic (III) regions in the cluster.

aged inside each atomic sphere j and also over the extra-atomic region. For region II, the potential value averaged over the volume is used.

The Watson sphere is used in order to obtain stabilization of the cluster orbitals. Watson [140] used it first for the stabilization of orbitals in a Hartree–Fock calculation of the electron structure of the negative double-charged oxygen ion. The Watson sphere is assumed to have an electric charge opposite to the charge of the cluster. For each atomic center situated inside the corresponding atomic sphere, one can write the following series expansion of the wave function:

$$\Psi_{\mathrm{I}}^{j}(\mathbf{r}) = \sum_{L} C_{L}^{j} R_{l}^{j}(E, r) Y_{L}(\mathbf{r}) \quad (0 < r \leqslant R_{j}), \tag{81}$$

where $Y_L(\mathbf{r})$ is the real spherical harmonics of index $L \equiv (l, m)$. The functions $R_l^j(r, E)$ represent the solution of the radial Schrödinger equation for the potential (80), spherically averaged with respect to the center of the jth sphere. As in the APW method and in the method of Green's functions, it is assumed that the wave function is finite at the center of the sphere, for $r = 0$. A similar series expansion is also used for the partial wave representation of the wave function in region III:

$$\Psi_{\mathrm{III}}(r) = \sum_{L} D_{L}^{0} R_{l}^{0}(E, r) Y_{L}(\widehat{\mathbf{r}}) \quad (R_{W} < r < \infty), \tag{82}$$

where R_W is the radius of the Watson sphere. The functions $R_l^0(E, r)$ represent the solutions of the radial Schrödinger equation for the spherically averaged potential in region III. The radial wave functions of the molecular orbitals should decrease exponentially at large distances from the center of the cluster. This represents a boundary condition for the orbitals. In the interatomic region, one makes use of the multicenter partial wave representation of the wave function:

$$\begin{gathered}\Psi_{\mathrm{II}}(r) = \sum_{L} B_{L}^{0} j_{l}(\varkappa r_{0}) Y_{L}(\widehat{\mathbf{r}}_{0}) + \sum_{j=1}^{n} \sum_{L} A_{L}^{j} f_{l}(\varkappa r_{j}) Y_{L}(\widehat{r}_{j}) \\ (R_{0} \leqslant r_{0} \leqslant R_{W}), \quad (R_{j} \leqslant r_{j}), \quad (r_{0} \leqslant R_{W}).\end{gathered} \tag{83}$$

In this expression, $\varkappa \equiv (E - V_0)^{\frac{1}{2}}$;

$$f_{l}(\varkappa r) = \begin{cases} h_{l}^{(1)}(\varkappa r), \ \varkappa \text{ imaginary}, & E < V_{0} < 0, \\ n_{l}(\varkappa r), \ \varkappa \text{ real}, & V_{0} < E < 0; \end{cases}$$

V_0 is the mean value of the interatomic potential; j_l is the spherical Bessel function; $h_l^{(1)}$ is the spherical Hankel function of the first type; and n_l is the spherical Neuman function. The molecular orbital functions (81)-(83) and their corresponding first derivatives should be continuous when passing through the spherical surfaces. This condition can be satisfied by using the theory of multiple scattering [141]. The coefficients A_L^i and B_L^0 are found to be correlated to C_L^j and D_L^0. The equations obtained for the energy and for the coefficients A_L^i and B_L^0 include the functions $R_l^i(E, R_j), f_l(\kappa R_j)$, their first derivatives taken at the surface of the corresponding spheres, and the structure factors, i.e., factors of the type $f_l(\kappa R_{jj'})$, $Y_L(\mathbf{R}_{jj'})$, which depend on the interatomic distances and on the vectors $\mathbf{R}_{jj'}$.

One advantage of the method proposed by Johnson is the fact that it can also be used for the calculation of the band structure of crystals. In this case, such a cluster grouping that repeats itself periodically should be considered to give the whole crystal structure. Moreover, the cluster orbitals should satisfy the Bloch boundary conditions.

Let us consider the secular equations of the energy eigenvalues. In the methods of OPW, pseudopotential, APW, and Green's functions, the secular equation may be written as follows:

$$\det | \{(\mathbf{k} + \mathbf{K}_i)^2 - E\} \delta_{\mathbf{K}_i,\mathbf{K}_j} + F_{\mathbf{K}_i,\mathbf{K}_j}(\mathbf{k}, E) | = 0. \tag{84}$$

In the OPW method, the function $F_{\mathbf{K}_i,\mathbf{K}_j}(\mathbf{k}, E)$ is given by the expression

$$F^{OPW}_{\mathbf{K}_i,\mathbf{K}_j}(k, E) = V(\mathbf{K}_i - \mathbf{K}_j) - \sum_c E_c \langle \mathbf{k} + \mathbf{K}_i | c \rangle \langle c | \mathbf{k} + \mathbf{K}_j \rangle,$$

where $V(\mathbf{K}_i - \mathbf{K}_j)$ is the Fourier component of the crystal potential. For a local pseudopotential, $F^{PS}_{\mathbf{K}_i,\mathbf{K}_j}, (\mathbf{k}, E) = V^{PS}(\mathbf{K}_i - \mathbf{K}_j)$. Even for a nonlocal potential, however, the energy eigenvalues are determined from equations (84), with

$$F^{PS}_{\mathbf{K}_i,\mathbf{K}_j}(\mathbf{k}, E) = V(\mathbf{K}_i, \mathbf{K}_j).$$

The pseudopotential also depends on the quasimomentum $\mathbf{k}$ and on the energy E.

In the APW method, the function $F_{\mathbf{K}_i\mathbf{K}_j}(\mathbf{k}, E)$ is given by the following expressions:

$$F^{APW}_{\mathbf{K}_i,\mathbf{K}_j}(\mathbf{k}, E) = \frac{4\pi R^2_{\mathrm{MT}}}{\Omega_0} \left\{ - [(\mathbf{k} + \mathbf{K}_i)^2 - E] \frac{j_1(|\mathbf{K}_i + \mathbf{K}_j| R_{\mathrm{MT}})}{|\mathbf{K}_i - \mathbf{K}_j| R_{\mathrm{MT}}} + \right.$$

$$+ \sum_{l=0}^{\infty} (2l+1) P_l(\widehat{\mathbf{K}_i; \mathbf{K}_j}) j_l(|\mathbf{k} + \mathbf{K}_i| R_{\mathrm{MT}}) j_l(|\mathbf{k} + \mathbf{K}_j| R_{\mathrm{MT}}) \times$$

$$\left. \times \left[\frac{R_l'(R_{\mathrm{MT}}, E)}{R_l(R_{\mathrm{MT}}, E)} - \frac{j_l'(|\mathbf{k} + \mathbf{K}_j| R_{\mathrm{MT}})}{j_l(|\mathbf{k} + \mathbf{K}_j| R_{\mathrm{MT}})} \right] \right\},$$

where P_l are the Legendre polynomials, and j_l are the spherical Bessel functions. The convergence in the APW method is better than that in the OPW method, since in the former the wave function reproduces better the oscillations at the atomic core in the crystal.

In the method of Green's functions, the rows and the columns of the determinant that gives the energy values are numbered by the values of the orbital momenta, included in the series expansion of the wave function. This implies that the partial wave representation is utilized. Ziman [142] has shown that the matrix equations in the method of Green's functions may also be written in a similar way to equation (84), where

$$F^{GF}_{\mathbf{K}_i,\mathbf{K}_j}(\mathbf{k},\ E) = \frac{4\pi R^2_{\mathrm{MT}}}{\Omega_0} \sum_{l=0}^{\infty} (2l+1) P_l(\widehat{\mathbf{K}_i \cdot \mathbf{K}_j})\, j_l(|\mathbf{k}+\mathbf{K}_i|\, R_{\mathrm{MT}}) \times$$
$$\times\, j_l(|\mathbf{k}+\mathbf{K}_j|\, R_{\mathrm{MT}}) \left[\frac{R'_l(R_{\mathrm{MT}},E)}{R_l(R_{\mathrm{MT}},E)} - \frac{j'_l(\sqrt{E}R_{\mathrm{MT}})}{j_l(\sqrt{E}R_{\mathrm{MT}})} \right].$$

As can be seen from this expression, the function $F^{GF}_{\mathbf{K}_i,\mathbf{K}_j}(\mathbf{k}, E)$ is similar to $F^{APW}_{\mathbf{K}_i,\mathbf{K}_j}(\mathbf{k}, E)$, but differs from it by the terms that contain the Bessel functions j_l. In the APW method, they depend on $\mathbf{k}$, while in the method of Green's functions, such a dependence does not exist. This improves the l convergence in the method of Green's functions, owing to the compensation of the terms

$$\frac{R'_l(R_{\mathrm{MT}},E)}{R_l(R_{\mathrm{MT}},E)} \approx \frac{j'_l(\sqrt{E}R_{\mathrm{MT}})}{j_l(\sqrt{E}R_{\mathrm{MT}})}.$$

The slow l convergence in the APW method arises from the fact that the sum that contains the spherical functions describing the behavior of the crystal electron-wave function inside the muffin-tin sphere requires increasingly more terms for waves with high kinetic energies. The plane wave convergence in the method of Green's functions is lower than in the APW method. Johnson [143] has shown that the APW method has the optimal convergence among all the methods in which plane waves are used.

In the method of Green's functions, equation (84) may be more conveniently written in the partial wave representation:

$$\det |\sqrt{E}\, \delta_{ll'} \delta_{mm'} + \tan \eta_l \sum_{LM} C^{LM}_{lm;l'm'} D_{LM}(\mathbf{k},\ E)| = 0,$$

where η_l is the phase shift determined by the function $R_l(r)$; $C^{LM}_{lm,l'm'}$ are the coefficients given by

$$C^{LM}_{lm;l'm'} \equiv \int Y_{LM} Y_{lm} Y_{l'm'} d\Omega,$$

and D_{LM} are structure factors, which coincide with the coefficients in the partial wave expression of the one-electron Green's function:

$$G_E(R) = -\frac{1}{\Omega_0}\sum_{\mathbf{K}_i}\frac{\exp i(\mathbf{k}+\mathbf{K}_i)\cdot\mathbf{R}}{(\mathbf{k}+\mathbf{K}_i)^2 - E} =$$

$$= -\frac{\cos\sqrt{E}R}{4\pi R} + \sum_{LM} i^L D_{LM}(\mathbf{k}, E)\, j_L(\sqrt{E}R)\, Y_{LM}(\hat{\mathbf{R}}).$$

The good l convergence results in the fact that the secular equations in the method of Green's functions have a lower order, as compared to other methods. However, the matrix elements are more complicated, and it is difficult in this method to calculate the structure constants because of the bad plane-wave convergence. The convergence may be improved by using the method proposed by Ewald [144]. Here, the constants contain exponential functions, spherical harmonics, and integrals. In performing the band calculations with different methods, approximately similar efforts are involved, since the drawbacks of a given method are balanced by the advantages of some others, and conversely. Direct calculations have shown that, in performing similar calculations with the APW method and with the method of Green's functions, approximately the same computer time is necessary.

We would like to point out once again that in each of the methods discussed, it is possible to introduce empirical parameters that yield a better agreement between theory and experiment and make possible a more reliable interpretation of experimental data. In the OPW method, this is accomplished by varying some of the Fourier components of the potential and by making use of the core-level energy shifts, whereas in the APW method and in the method of Green's functions a change in the value of the crystal potential outside the atomic spheres is considered.

Special interpolation methods exist, to be used in calculations based on experimental values of the parameters, or on parameters determined by other calculation methods. First, there is the scheme based on the LCAO method [145]. In this case, the complicated overlapping integrals of atomic functions are not calculated exactly, but are considered to be parameters, determined from the condition of optimal agreement between the calculated and experimental results. Very good results for the interpolation of energy bands are achieved for pure transition metals with the method proposed by Hodges *et al.* [146] in which, besides the tight-binding functions describing mainly the d bands, use is also made of four orthogonal plane waves. The presence of hybridization between the localized functions and the plane waves allows a very accurate description of the energy band structure in transition metals. In this case, 15 parameters are used, these being partly obtainable from experimental data.

A number of other methods have also been developed [147–149], based on the use of the symmetry properties of the energy bands and on the Fourier series expansion of the dispersion functions of electrons in the crystal. By taking into account the symmetry properties, the number of the necessary parameters may be diminished. As a convenient semiempirical scheme, the pseudopotential method may also be used.

One of the fundamental problems is the determination of the crystal potential. The problem is to construct for a multielectron system a potential that takes into account the average effects of interactions between the electrons in the system. It must be admitted that the potentials used in energy-band calculations are in a high degree phenomenological. In the one-electron approximation, the crystal potential is represented by the sum of the Coulomb and exchange terms. Evaluation of the contribution of the Coulomb term in the total potential is not difficult, since it is determined by the density of electrons in the crystal. To a first approximation, the electron density in the crystal may be taken as equal to the sum of the electron densities of the free atoms. For the exchange term, Slater [150] has proposed an expression that we have already mentioned in the discussion of the magnitude of electron binding energies in atoms:

$$V_{ex}^{S} = -6\left[\frac{3}{8\pi}\rho\right]^{\frac{1}{3}} \tag{85}$$

Prior to 1965, when Kohn and Sham [151] proposed a new formula for the exchange potential, calculations were performed by using the Slater potential. The Kohn–Sham exchange potential may be obtained in an analogous way to the Slater potential; it is necessary to take the value of the exchange potential for the gas of free electrons at the Fermi surface, but not average it over the states inside the Fermi sphere. Subsequently, Slater and Johnson [152] have proposed the exchange potential X_α.

The charge density of electrons with spins oriented up and down can be expressed as follows:

$$\rho_+ = \sum_{j+} n_j \Psi_j^* \Psi_j, \quad \rho_- = \sum_{j-} n_j \Psi_j^* \Psi_j, \quad \rho = \rho_+ + \rho_-, \tag{86}$$

where the summation in the formula for ρ_+ is performed over the spin-orbitals of the electrons with spin oriented upward, while in the formula for ρ_- it is performed for electron spins oriented downward; n_j is the occupation number of the spin-orbital, equal to 0 or 1. Let us consider the total energy of the system:

$$\langle E_{X_\alpha}\rangle = \sum_i n_i \int \Psi_i^*(1) f_1 \Psi_i(1)\, dV_1 + \frac{1}{2}\int \rho(1)\rho(2)\, g_{12}\, dV_1 dV_2 + \\ + \frac{1}{2}\int [\rho_+(1)\, U_{X_\alpha +}(1) + \rho_-(1)\, U_{X_\alpha -}(1)]\, dV_1. \tag{87}$$

Here, f_1 is a one-electron operator, related to the first electron. The first term in this expression represents the sum of the kinetic and potential energies in the field of all the nuclei, while the second term represents the Coulomb interaction between two electron charge distributions, including also the interaction of the electron on the ith spin-orbital with itself. Of course, any terms that describe the interaction of electrons with themselves should not enter into the expression for the total energy of the system. The last term in equation (87) represents the exchange term, which not only compensates such terms in the expression for the total energy, but also takes into account the characteristic exchange effects. Thus, it would be more appropriate to call this last term an exchange-correlation term. A term for the Coulomb interaction between all of the pairs of nuclei should be added to equation (87) but, since this term is a constant, it can be put aside and added to the energy only at the end of the calculations. By doing so, the exchange potential $U_{X_\alpha^+}(1)$ representing a generalization of the potential (85) is given by the expression

$$U_{X_\alpha+}(1) = -9\alpha\left[\left(\frac{3}{4\pi}\right)\rho_+\right]^{\frac{1}{3}}. \qquad (88)$$

An analogous formula may also be written for $U_{X_\alpha^-}(1)$. By varying expression (87) with respect to the spin-orbital Ψ_i, the following equation is obtained:

$$[-\nabla^2 + V_{\text{Coul}}(1) + V_{X_\alpha+}]\,\Psi_i+(1) = \varepsilon_{iX_\alpha+}\Psi_i+(1), \qquad (89)$$

where $V_{\text{Coul}}(1)$ represents the Coulomb potential acting upon the electron. It is determined by the total density of electronic and nuclear charge, while the exchange potential in the X_α method is given by

$$V_{X_\alpha+}(1) = \frac{2}{3}U_{X_\alpha+}(1) = -\alpha\left[6\left(\frac{3}{4\pi}\rho_+\right)^{\frac{1}{3}}\right].$$

The eigenvalue ϵ_{iX_α} may be obtained by multiplying equation (89) on the left side by $\Psi_{i+}(1)$ and by subsequent integration over the electron coordinates, taking into account the condition of orthonormality of the functions Ψ_i. On the basis of equations (86)-(88), it can be demonstrated that

$$\varepsilon_{iX_\alpha} = \frac{\partial\langle EX_\alpha\rangle}{\partial n_i}, \qquad (90)$$

by differentiating expression (87) with respect to the occupation numbers for the spin-orbitals characterized by upward spin orientation. Equation (89) is solved self-consistently. It should be mentioned that the eigenvalues given by equation (90) do not coincide with the eigenvalues obtained by the Hartree–Fock method, which represent the difference between the total energies of the free atom and those of the ion in which the ith electron has been removed; that is:

$$\varepsilon_{iHF} = \langle EHF \rangle_{n_i=1} - \langle EHF \rangle_{n_i=0}. \tag{91}$$

The difference between the energy calculated by formula (90) and the Hartree–Fock value (91) becomes evident if the total energy $\langle E \rangle$ is considered, by making use of the fact that it depends on n_i in both the X_α method and the Hartree–Fock method. With a precision up to the terms of second order in n_i, one obtains

$$\langle E \rangle = \langle E \rangle_0 + \left.\frac{\partial \langle E \rangle}{\partial n_i}\right|_0 (n_i - n_{i0}) + \frac{1}{2} \left.\frac{\partial^2 \langle E \rangle}{\partial n_i^2}\right|_0 (n_i - n_{i0})^2 + \ldots \tag{92}$$

When $n_i = 0$ and $n_{i0} = 1$, the following equality is valid:

$$\langle E \rangle_{n_i=1} - \langle E \rangle_{n_i=0} = \left.\frac{\partial \langle E \rangle}{\partial n_i}\right|_0 - \frac{1}{2} \left.\frac{\partial^2 \langle E \rangle}{\partial n_i^2}\right|_0 .$$

Therefore, the eigenvalue obtained by the Hartree–Fock method is equal to the eigenvalue obtained by the X_α method minus the second derivative:

$$\frac{1}{2} \left.\frac{\partial^2 \langle E \rangle}{\partial n_i^2}\right|_0 .$$

It can be demonstrated that the second derivative of the energy with respect to the occupation number is positive, and therefore the energy eigenvalue in the Hartree–Fock method is greater in absolute value than the energy eigenvalue in the X_α method.

The value of α may be chosen so as to yield as good a description of the individual atoms as possible [153]. For this, it is necessary that the energy $\langle E_{X_\alpha} \rangle$ coincide exactly with the Hartree–Fock energy of the same atom (if the atom has completely occupied shells), or with the energy averaged over all the multiplet levels determined for the core-state configuration. Lindgren [154] has proposed a method for the determination of the magnitude of α in which the spin-orbitals, calculated with the X_α method using equation (89), are introduced into the Hartree–Fock expression of the total energy, after which its minimum is found as a function of α. In general, a single value of α is not completely satisfactory for all the spin-orbitals, and it is preferable to use various values of α for the various spin-orbitals. For molecules or crystals, α is determined inside each atomic sphere in a similar way as for the corresponding atoms, while in the interatomic region, a value averaged over different atoms is used.

An exchange potential, $V_{ex}^{\alpha,\beta}$, has been proposed by Herman *et al.* [155]:

$$V_{ex}^{\alpha,\beta} = \left[\frac{2}{3} + \beta G(\rho)\right] V_{ex}^{S},$$

where $\beta \ll \frac{2}{3}$,

$$G(\rho) = \frac{1}{\rho^{2/3}}\left[\frac{4}{3}\left(\frac{\nabla\rho}{\rho}\right)^2 - \frac{\nabla^2\rho}{\rho}\right].$$

Another expression for the exchange potential has been proposed by Liberman [156]:

$$V_{ex}^{L} = -\left(\frac{3}{8\pi}\rho(\mathbf{r})\right)^{1/3} F\left(\frac{\sqrt{2m(E_n - V(\mathbf{r}))}}{(3\pi^2\rho(\mathbf{r}))^{1/3}}\right).$$

This potential is characterized by the fact that different electron states have different exchange potentials, since in this expression E_n is included within the square root.

Orthenburger and Herman [157] have proposed the following exchange potential:

$$V_{X\alpha\beta\gamma}(\rho) = -6\left(\frac{3}{8\pi}\rho\right)^{\frac{1}{3}} [\alpha + \beta G(\rho) + \gamma_1 G_1(\rho) + \gamma_2 G_2(\rho) + \\ + \gamma_3 G_3(\rho)],$$

where $G(\rho)$, $G_1(\rho)$, $G_2(\rho)$, $G_3(\rho)$ are corrections for the nonhomogeneity of the exchange term, namely, $G(\rho)$ is the correction of the second order, while the others are corrections of the fourth order. The terms of fourth order exhibit a convergence behavior as a function of r. The use of this exchange potential, however, is perhaps not necessary since the X_α approximation may offer a sufficiently great accuracy in the description of the electron structure of atoms in crystals.

The problem of the inclusion of correlation effects in the one-electron approximation will not be discussed here, since it has not yet reached a sufficiently advanced stage.

In conclusion, the Hamiltonian used at present for determination of electron energy spectra has the shape

$$H = -\nabla^2 + V_c + V_{ex}. \tag{93}$$

As V_{ex}, one may use the exchange potential proposed by Slater, Kohn and Sham, Liberman, or the exchange potential X_α.

Symmetry Properties of the Energy Bands in Crystals

The fundamental problem of the theory of energy bands in crystals is to solve the Schrödinger equation with the Hamiltonian (93). Nevertheless, important conclusions about the properties of the solutions, i.e., the electron wave functions $\Psi(\mathbf{k}, \mathbf{r})$ and energies $E_n(\mathbf{k})$, may be drawn from a study of the symmetry properties. Thus the possible multiplicity of the energy eigenvalues can be determined, and it may be established that certain matrix elements of various types, in particular those related to electromagnetic radiation, are equal to zero. Study of the symmetry properties of the Hamiltonian (93) allows classification of the eigenvalues $E_n(\mathbf{k})$ and prediction of the possible behavior of $E_n(\mathbf{k})$ when $\mathbf{k}$ varies in the Brillouin zone. One can therefore realize how important it is to study the symmetry properties of the Schrödinger equation.

The Hamiltonian H remains invariant under the action of all the symmetry operations of the crystal space group. The space group consists of translations, rotations, and reflections, followed by translations either by the lattice vector, or by another vector different from the lattice vector. The elements of the space group may be written as $\{R_i | \mathbf{R}_n + \boldsymbol{\tau}_i\}$, where R_i represents a rotation or a reflection, $\mathbf{R}_n$ represents the lattice translation vector, and $\boldsymbol{\tau}_i$ represents a vector that is not a translation vector (in particular, the vector $\boldsymbol{\tau}_i$ may be equal to zero). The rotational parts of the elements of the space symmetry group $\{R_i|0\}$ form a group that is called the point-symmetry group. In crystals, there exist 32 point groups. The elements of the space group $\{R_i|\mathbf{R}_n + \boldsymbol{\tau}_i\}$ transform the vector $\mathbf{r}$ into the vector $\mathbf{r}' = \tilde{R}_i\mathbf{r} + \boldsymbol{\tau}_i + \mathbf{R}_n$, where $\tilde{R}_i$ is the matrix corresponding to the transformation R_i. Let us now see how the function $\Psi(\mathbf{r})$ is transformed by the operation $\{R_i|\mathbf{R}_n + \boldsymbol{\tau}_i\}$:

$$\{R_i \,|\, \mathbf{R}_n + \boldsymbol{\tau}_i\}\, \Psi(\mathbf{r}) = \Psi(\tilde{R}_i\mathbf{r} + \boldsymbol{\tau}_i + \mathbf{R}_n).$$

The groups for which the vector $\boldsymbol{\tau}_i$ is equal to zero, for all of the operations R_i of the point group, are called symmorphic groups. The requirement that the elements $\{R_i|\mathbf{R}_n + \boldsymbol{\tau}_i\}$ form a group imposes stringent limitations on the vectors $\mathbf{R}_n$ and $\boldsymbol{\tau}_i$. There exist only 230 space groups out of which 73 are symmorphic.

The most usual crystals have face-centered, body-centered, or hexagonal lattices. For example, such semiconductors as silicon and germanium have a diamond-type crystal structure. Their crystal lattice may be represented as two face-centered cubic lattices, displaced with respect to each other by the vector $(a/4)(1, 1, 1)$, where a is the lattice constant. The crystals GaAs, GaP, and GaSb are characterized by a zinc-blende-type structure. This also consists of two face-centered lattices, displaced with respect to each other by the vector $(a/4)(1, 1, 1)$. The crystals NaF, NaBr, and NaI have an NaCl-type structure, which consists of two face-centered lattices, displaced with respect to each other by the vector $(a/2)(1, 1, 1)$. The compounds CsBr and CsI are characterized by a structure of CsCl type, formed by two simple cubic lattices displaced with respect to each other by the vector $(a/2)(1, 1, 1)$. In these cases, one of the sublattices contains the atoms of one type, while the other contains the atoms of the other type. In the hexagonal crystals of wurtzite type, such as AlN, GaN, and ZnO, the elementary cell contains 4 atoms. Metals such as copper, palladium, silver, gold, platinum, and lead have a face-centered crystal lattice, while lithium, sodium, potassium, titanium, and vanadium have a body-centered crystal lattice. Several compounds and pure elements can exist in more than one type of crystal lattice. Thus, BN may crystallize in the cubic or in the hexagonal system, while iron can have a body-centered or face-centered lattice.

Let us label the unitary element of the point group R_1. By using the Bloch condition, it can be demonstrated that, for the elements of the group $\{R_1|\mathbf{R}_n\}$, the function $\{R_1|\mathbf{R}_n\}\ \psi\ (\mathbf{k}, \mathbf{r})$ is transformed under the irreducible representation of the translation group. For an arbitrary element of the group, the following relation is valid:

$$\{R_i\,|\,\boldsymbol{\tau}_i + \mathbf{R}_n\}\,\Psi_n(\mathbf{k}, \mathbf{r}) = \exp i\mathbf{k}\cdot(\boldsymbol{\tau}_i + \mathbf{R}_n)\exp i\,(R_i^{-1}\mathbf{k})\cdot\mathbf{r}\,\times$$
$$\times\,U_{n\mathbf{k}}(R_i\mathbf{r} + \boldsymbol{\tau}_i) = \exp i\,[(\tilde{R}_i^{-1}\mathbf{k})\cdot\mathbf{r}]\,U'_{R_i^{-1}\mathbf{k}}(\mathbf{r}),$$

which confirms that $\{R_i|\boldsymbol{\tau}_i + \mathbf{R}_n\}\ \Psi\ (\mathbf{k}, \mathbf{r})$ is a Bloch function with a wave vector $R_i^{-1}\mathbf{k}$ and an energy $E_n(R_i^{-1}\mathbf{k})$. Since the Hamiltonian H is invariant under the operation $\{R_i|\boldsymbol{\tau}_i + \mathbf{R}_n\}$ of the space group, it follows that $E_n(R_i^{-1}\mathbf{k}) = E_n(\mathbf{k})$. Each energy band has a complete point-symmetry group.

Among the elements $\{R_i|\boldsymbol{\tau}_i + \mathbf{R}_n\}$ of the space group, there exist some that lead again to Bloch functions with the same wave vector $\mathbf{k}$: $\tilde{R}_i^{-1}\mathbf{k} = \mathbf{k}$, or with the equivalent wave vector $\tilde{R}_i^{-1}\mathbf{k} = \mathbf{k} + \mathbf{K}_i$. These elements form a subgroup of the space group, and this is called the group of the wave vector $\mathbf{k}$. Such a subgroup determines the degeneration multiplicity and the symmetry of the energy band states belonging to the given point $\mathbf{k}$ of the Brillouin zone. Let us consider the wave vector groups for GaS and germanium crystals. The Brillouin zone for these crystals is shown in Figure 44. The wave vector group at the point $\mathbf{k} = (0, 0, 0)$ for the GaAs crystal is the whole symmorphic space group T_d^2. Let $E_n(0)$ be a p-times degenerated energy level with wave vector $\mathbf{k} = (0, 0, 0)$, and $\Psi_{n0j}(\mathbf{r})$ the basis functions of the m-dimensional irreducible representation of the wave vector group at the point Γ. Then

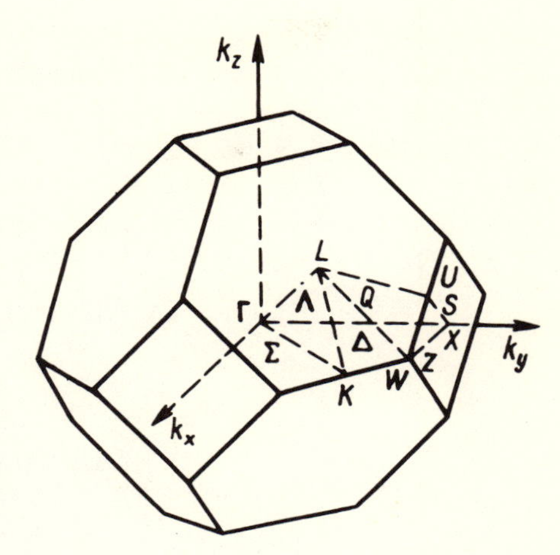

Figure 44. Brillouin zone for the bcc lattice.

$$\{R_l \mid \mathbf{R}_m\} \Psi_{n0j}(\mathbf{r}) = \sum_{s=1}^{p} M_{sj}^{lm} \Psi_{n0s}(\mathbf{r}).$$

On the other hand,

$$\{R_l \mid \mathbf{R}_m\} \Psi_{n0j}(\mathbf{r}) = \Psi_{n0j}(R_l \mathbf{r}) = R_l \Psi_{n0j}(\mathbf{r}).$$

For the elements $\{R_l|\mathbf{R}_m\}$, with l fixed and m variable, the following is obtained:

$$M^{lm} = M^{ls} = \cdots = M^l.$$

Therefore, all the elements of the space group having the same element $\{R_l|0\}$ of the point group are represented by the matrix M^l:

$$R_l \Psi_{n0j}(\mathbf{r}) = \sum_{s=1}^{p} M_{sj}^{l} \Psi_{n0s}(\mathbf{r}), \qquad l = 1, 2, \ldots, 24.$$

The Bloch functions form the basis of the p-dimensional irreducible representation of the point group T_d. The elements of the point group are represented by the matrix M^l. Therefore, in order to find the irreducible representations of the group of the wave vector $\mathbf{k}$, it is only necessary to find the irreducible representations of the group T_d, since each irreducible representation of T_d corresponds to an irreducible representation of the group of the wave vector $\mathbf{k} = 0$. The group

T_d has five irreducible representations: $\Gamma_1(1)$, $\Gamma_2(1)$, $\Gamma_{12}(2)$, $\Gamma_1(3)$, and $\Gamma_{25}(3)$, whose dimensions are indicated in parentheses.

Let us consider the direction $\mathbf{\Delta}$. Let $E_n(\mathbf{\Delta})$ be the p-times degenerated energy level with the wave vector oriented in the direction $\mathbf{\Delta}$ and the wave functions $\Psi_{nkj}(\mathbf{r})$ ($j = 1, 2, \ldots, p$). In this case, for all of the elements $\{R_l|\mathbf{R}_m\}$ in the group of the wave vector $\mathbf{k}$, the following equality is valid:

$$\{R_l\,|\,\mathbf{R}_m\}\,\Psi_{n\mathbf{k}j}(\mathbf{r}) = \exp(i\mathbf{k}\cdot\mathbf{R}_m)\,R_l\Psi_{n\mathbf{k}j}(r) = \sum_{s=1}^{p} M_{sj}^{lm}\Psi_{n\mathbf{k}s}(\mathbf{r}).$$

Therefore,

$$R_l\Psi_{nj}(\mathbf{k},\mathbf{r}) = \sum_{s=1}^{p} \exp(-i\mathbf{k}\cdot\mathbf{R}_m)\,M_{sj}^{lm}\Psi_{ns}(\mathbf{k},\mathbf{r}).$$

For the elements with different translation vectors, the following is obtained:

$$\exp(-i\mathbf{k}\cdot\mathbf{R}_m)\,M^{lm} = \exp(-i\mathbf{k}\cdot\mathbf{R}_s)\,M^{ls} = \cdots = M^l,$$

and consequently,

$$R_l\Psi_{n\mathbf{k}j}(\mathbf{r}) = \sum_{s=1}^{p} M_{sj}^{l}\Psi_{n\mathbf{k}s}(\mathbf{r}).$$

Therefore, the Bloch functions $\Psi_{nks}(\mathbf{r})$ ($s = 1, 2, \ldots, p$) form a selection of basis functions for the p-dimensional irreducible representation of the point group of the direction $\mathbf{\Delta}$.Consequently, in order to construct the irreducible representations of the group of the wave vector $\mathbf{\Delta}$, the irreducible representations of the point group should be used. The matrices of irreducible representations of the wave vector group may be obtained from the irreducible representations of the point group:

$$M^{ls} = M^l \exp i\mathbf{k}\cdot\mathbf{R}_s.$$

In the case of the GaAs crystal, the point group of the vectors $\mathbf{k}$ distributed along the direction $\mathbf{\Delta}$ has four irreducible representations, $\mathbf{\Delta}_1$, $\mathbf{\Delta}_2$, $\mathbf{\Delta}_3$, $\mathbf{\Delta}_4$ of dimension 1. Let us consider the wave vector group for the point $X = (2\pi/a)(1, 0, 0)$. The matrices representing the element $\{R_l|\mathbf{R}_m\}$ correspond to the irreducible representation of the wave vector group in the given point, and may be written as $\exp(ik_X R_m)M^l$, where M^l are the matrices of the irreducible representations of the point group in the point X. This point group has five irreducible representations X_1, X_2, X_3, X_4, X_5. Since the dimension of representation X_5 is 2, the states with symmetry X_5 will be twice degenerated.

The space group of the diamond crystal is nonsymmorphic, which makes the construction of its irreducible representations more difficult. In general, however, even in this case, the method of construction of the irreducible representations of the space group is analogous to that discussed above. At the point $\mathbf{k} = 0$, the classification of electronic states is performed by using 10 irreducible representations of the group 0_h: $\Gamma_1(1)$, $\Gamma'(1)$, $\Gamma_2(1)$, $\Gamma_2'(1)$, $\Gamma_{12}(2)$, $\Gamma_{12}'(2)$, $\Gamma_{15}(3)$, $\Gamma_{15}'(3)$, $\Gamma_{25}(3)$, and $\Gamma_{25}'(3)$. For diamond, the group of the wave vector $\mathbf{k}$ along the direction $\mathbf{\Delta}$ is characterized by five irreducible representations: $\mathbf{\Delta}_1$, $\mathbf{\Delta}_1'$, $\mathbf{\Delta}_2$, $\mathbf{\Delta}_2'$ and $\mathbf{\Delta}_5$, the last one two dimensional. The group of the wave vector at the point X has four irreducible representations, all two dimensional.

As we have already mentioned, from the orders of the irreducible representations the degree of degeneracy of the energy levels can be determined. However, in some cases, additional degeneration may occur. Besides the conditions of space symmetry, the time-reversal symmetry of the Schrödinger equation represents an additional condition, which may cause the degeneracy of energy bands along certain directions in the Brillouin zone. The equation which is the complex conjugate of (14) is

$$H\Psi^*_{n\mathbf{k}}(\mathbf{r}) = E_n(\mathbf{k})\,\Psi^*_{n\mathbf{k}}(\mathbf{r}).$$

Therefore, the wave function $\Psi^*_{n\mathbf{k}}(\mathbf{r})$ also represents an eigenfunction of the Hamiltonian H and corresponds to the same eigenvalue $E_n(\mathbf{k})$. However, since $\Psi^*_{n\mathbf{k}}(\mathbf{r})$ is a Bloch function corresponding to the wave vector $\mathbf{k}$, it follows that

$$\Psi^*_{n\mathbf{k}}(\mathbf{r}) = \exp(-i\mathbf{k}\cdot\mathbf{r})\,U_{n\mathbf{k}}(\mathbf{r}),$$

and therefore,

$$E_n(\mathbf{k}) = E_n(-\mathbf{k}).$$

In this case, the wave function

$$\Psi^*_{n\mathbf{k}}(\mathbf{r}, -t) = \exp(-iE_n(\mathbf{k})\,t)\,\Psi^*_{n\mathbf{k}}(\mathbf{r})$$

satisfies the same time-dependent Schrödinger equation as the function $\Psi_{n\mathbf{k}}(\mathbf{r}, t)$, namely

$$H\Psi_{n\mathbf{k}}(\mathbf{r}, t) = i\hbar\,\frac{\partial\Psi_{n\mathbf{k}}(\mathbf{r}, t)}{\partial t}.$$

For example, in the case of the compound GaAs, the states that transform under the irreducible representations $\mathbf{\Delta}_3$ and $\mathbf{\Delta}_4$ have the same energy $E(\mathbf{k}_\Delta)$.

The change of symmetry that occurs on passing from a point $\mathbf{k}$ to a neighboring point $\mathbf{k}_0$ may have a stepwise character. Therefore, some of the symmetry elements of the first point may be absent for the second point. Let us consider a transition from a point with a large number of symmetry elements to a point with a lower number of symmetry elements. The representations that were irreducible at the first point may be reducible at the second point. By studying the properties of the totality of irreducible representations, it is possible to determine the possible types of states between which a given irreducible representation (for a given energy) is divided, at points contiguous to $\mathbf{k}_0$. Let us consider, for example, that the representation at the point $\mathbf{k}_0$ is three dimensional. The compatibility relations made it possible to establish that at a point neighboring $\mathbf{k}_0$ this representation is divided into three one-dimensional representations. Unfortunately, however, these relations do not provide an answer to the question: which of the states has the highest or the lowest energy value? One representation may be compatible with representations of different types, and which of these will in fact occur can only be determined by calculation. Nevertheless, the compatibility relationships and the time-reversal symmetry properties provide a means of reducing considerably the number of possible band structures for the given crystal. Figure 45 shows several of the possible and impossible types of band structure for crystals of the diamond type and of the zinc-blende type.

In conclusion, the energies and electron wave functions of crystals may be classified by using the irreducible representations of the space group. In an

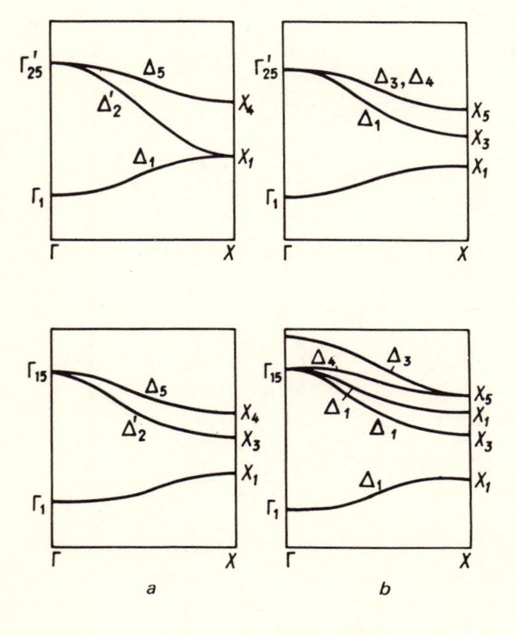

Figure 45. Possible and impossible (from the point of view of the group theory) band structures for crystals of (a) the diamond type and (b) the zinc-blende type.

arbitrary point **k** of the Brillouin zone, only translation symmetry exists, and therefore the wave functions should satisfy the Bloch condition. The series expansion of the valence electron wave functions in an arbitrary point $\mathbf{k}_0$ contains harmonics of $s, p, d, f, \ldots$ type. At points of high symmetry in the Brillouin zone, some of the harmonics do not appear in the series expansion of the wave function. Thus, at the point $\mathbf{k} = 0$, the principal types of symmetry for cubic crystals will be: s for the Γ_1 state, d for Γ_{12}, p for Γ_{15}, f for Γ_{25}, g for Γ'_{15} and d for Γ'_{25}. At the point X, the representation X_1 will have s- and d-type symmetry, and the representation X_5 will have p-type symmetry.

Electron States in Disordered Alloys

The essential difference between the theoretical treatment of disordered and of ordered systems is that, for disordered systems, the Bloch theorem is not valid. However, in disordered binary alloys, the atoms of the two components are situated at the lattice points of a periodic lattice. A more complicated situation is encountered for amorphous semiconductors, in which there exists short-range but not long-range atomic ordering. One of the most advanced methods for the calculation of the electron properties of disordered binary alloys is the method of coherent potential, proposed by Soven [158, 159]. The principle of this method is rather simple. Let us consider a point $\mathbf{R}_\nu$ of the crystal lattice, at which the probability of finding an atom of type A is c (c is identical to the concentration of the A component in the alloy), and the probability of finding an atom of type B is $(1 - c)$. Therefore, in the vicinity of the point $\mathbf{R}_\nu$, the alloy potential will be described either by the function U_A or by the function U_B.

Let us introduce the potential V_0 corresponding to some periodical structure. If $\overline{G}_0$ is the Green's function describing the electron behavior in the lattice with potential V_0, and G_0 the Green's function of a free electron, then

$$\overline{G}_0 = G_0 + G_0 \left(\sum_{\mathbf{R}_\nu} V_0 \right) \overline{G}_0.$$

In this model, the alloy is characterized either by the potential $V_A - V_0$, or by the potential $V_B - V_0$. The t matrix of the various centers in the alloy may be determined by the relation

$$t_i = (V_i - V_0) + (V_i - V_0) \overline{G}_0 t_i. \tag{94}$$

The Green's function of electrons in the alloy may be written as follows:

$$G = \overline{G}_0 + \sum_\alpha \overline{G}_0 t_\alpha \overline{G}_0 + \sum_\alpha \sum_{\beta \neq \alpha} \overline{G}_0 t_\alpha \overline{G}_0 t_\beta \overline{G}_0 + \cdots$$

By assuming that the average of $\langle t_i \rangle$ over all the configurations is zero,

$$\langle t_i \rangle = 0 = ct_A + (1 - c)\, t_B,$$

and using equation (94), the following expression for V_0 is obtained:

$$V - V_0 = (V_A - V_0)\, \bar{G}_0\, (V_B - V_0),$$

where $V = cV_A + (1 - c)V_B$.

The use of the above approximation in the determination of V_0 is called the method of coherent potential. This approximation implies that

$$\langle G \rangle = \bar{G}_0 + \sum_{\alpha} \sum_{\beta \neq \alpha} \sum_{\gamma \neq \beta} \sum_{\delta \neq \gamma} \langle \bar{G}_0 t_\alpha \bar{G}_0 t_\beta \bar{G}_0 t_\gamma \bar{G}_0 t_\delta \bar{G}_0 \rangle + \cdots$$

Frequently in practice only the first term of the expansion is retained, i.e., $\langle G \rangle \approx \bar{G}_0$. The reliability of this approximation has been confirmed in practice in a number of electron band-structure calculations for alloys.

Comparison between Electron Spectroscopy and Some Other Methods of Investigation of the Electron Structure of Crystals

Prior to the development of the method of X-ray photoelectron spectroscopy, the valence band structure in crystals was studied mainly by X-ray emission spectroscopy, by photoelectron spectroscopy in the far-ultraviolet range, and by optical-reflection and absorption spectroscopy.

The intensity of emitted X-ray radiation is given by the formula [5]:

$$\mathcal{T}(\nu) = \nu \sum_n \int_S \frac{|M_n(\mathbf{k})|^2}{|\nabla E_n(\mathbf{k})|}\, dS, \tag{95}$$

where the integration is performed over the isoenergetic surface $E = E_n(\mathbf{k}) - E_C$; $M_n(\mathbf{k})$ is the matrix element of the probability of transition from the valence band to the core levels, i.e.,

$$M_n(\mathbf{k}) = \langle \Psi_{n\mathbf{k}} | \nabla | \Psi_{c\mathbf{k}} \rangle. \tag{96}$$

This matrix element is determined by the wave functions $\Psi_{nk}(\mathbf{r})$ of the valence electrons, and $\Psi_{c\mathbf{k}}(\mathbf{r})$ of the core electrons. Aleshin and Smirnov [160] and Topol *et al.* [161] have demonstrated that the matrix elements of the transition probability, especially in semiconductors and in insulators, may change appre-

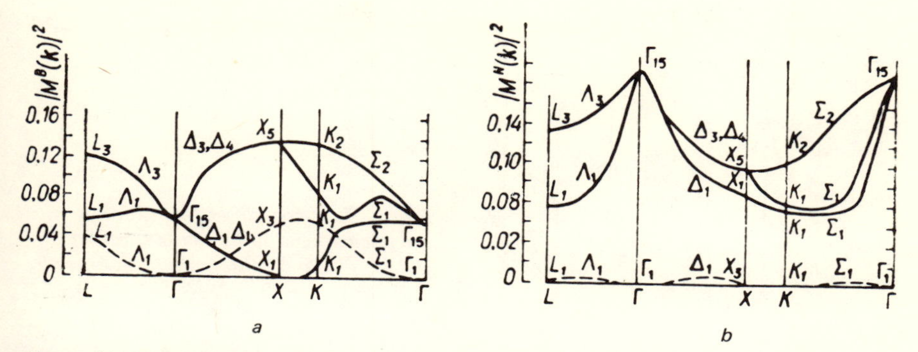

Figure 46. Probability of transition from the valence band to the 1s-core states of (a) boron and (b) nitrogen in BN.

ciably when **k** changes in the Brillouin zone, even on the same isoenergetic surface. Therefore, $|M_n(\mathbf{k})|^2$ cannot be extracted from the integral in equation (95). This means that the intensity $\mathcal{J}(\nu)$ cannot be considered to be proportional to the electron density of states, which is given by

$$\rho(E) = \int_S \frac{ds}{|\nabla E(\mathbf{k})|}, \tag{97}$$

since the proportionality factor varies significantly with energy. Such a matrix element will be different for the same **k** but different energy bands.

In the case of binary compounds, $|M_n(\mathbf{k})|^2$ may be significantly different for the transitions corresponding to the core levels of the different components. Figures 46 and 47 show the quasimomentum dependence of the square modulus of the matrix elements of transition probability for BN and GaP compounds, according to the data of Aleshin and Smirnov [160] and Topol *et al.* [161]. As a result of the difference in the magnitude of $|M_n(\mathbf{k})|^2$ for the various components in compounds, the emission bands will be significantly different, even for transitions on levels of the same type of symmetry. In some compounds, there exist valence subbands for which the electron wave functions are localized in the vicinity of the nuclei of one of the components in the compound. This usually results in great differences between the probabilities of transition from this subband to the core levels for different components.

The wave function of a valence electron in the vicinity of the atomic nucleus may be represented as a series expansion in spherical harmonics:

$$\Psi_n(\mathbf{k}, \mathbf{r}) = \sum_{lm} C_{lm}(\mathbf{k}) R_l Y_{lm}(\Theta, \varphi).$$

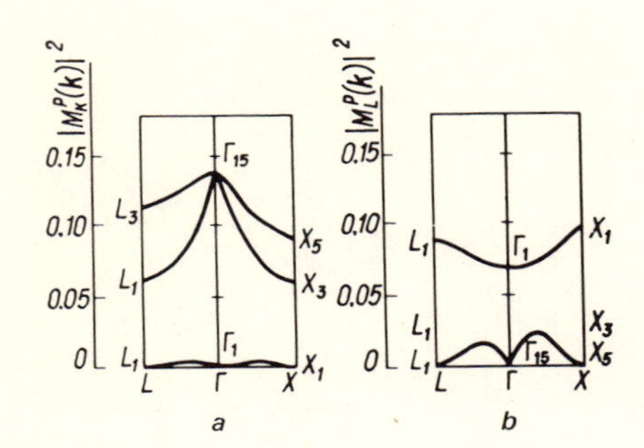

Figure 47. Probability of transition from the valence band to (a) the 1s core states and (b) the 2p core states of phosphorus in GaP.

Since the wave function of a core electron in the vicinity of the nucleus of any one of the components is approximately the same as the atomic function $\varphi_c(\mathbf{r}) = R_{cl'} Y_{l'm}(\Theta, \varphi)$, it follows that the matrix elements (96) will be different from zero if the dipole selection rules $\Delta l = \pm 1, \Delta m = \pm 1$ are satisfied. Therefore, X-ray emission spectroscopy allows the determination of the energy localization of states with different types of symmetry in the valence band of the crystal.

There exist points $\mathbf{k}$ in the Brillouin zone for which $|\Delta E(\mathbf{k})| = 0$. Consequently, singularities appear in the structure of the density-of-state curves of valence band electrons; these are the so-called "van Hove singularities" [162]. Comparison of formulas (95) and (97) indicates that these singularities should also appear in the curves of the X-ray intensity distribution, in spite of the fact that formula (95) contains the square modulus of the matrix element of the transition probability (96).

The X-ray emission spectra of the various components in compounds may differ significantly from each other because of the differences in the transition probability matrix elements. Therefore, a general criterion for identifying the X-ray emission bands of the different components of the crystal on the same energy scale does not exist. Such identification can be achieved efficiently by using X-ray photoelectron spectroscopy data. The states situated at the bottom of the conduction band may also be studied by using X-ray absorption spectroscopy. Figure 48 illustrates schematically the possibility to use the method of X-ray spectroscopy in the study of electronic states in crystals. It should be mentioned that, because of the influence of the transition probability on both the X-ray emission spectra and the X-ray absorption spectra, the determination of the width of the forbidden band in insulators by analysis of X-ray spectra of the different components may lead to different results. In this case, the lowest of these values should be accepted as the most reliable value of the forbidden band width.

In optical spectra, the shape of the lines corresponding to interband transitions is given by the following relation [163]:

$$\varepsilon_2(\omega) \sim \sum_{ii'} \bar{f}_{ii'} \frac{d\rho}{dE_{ii'}},$$

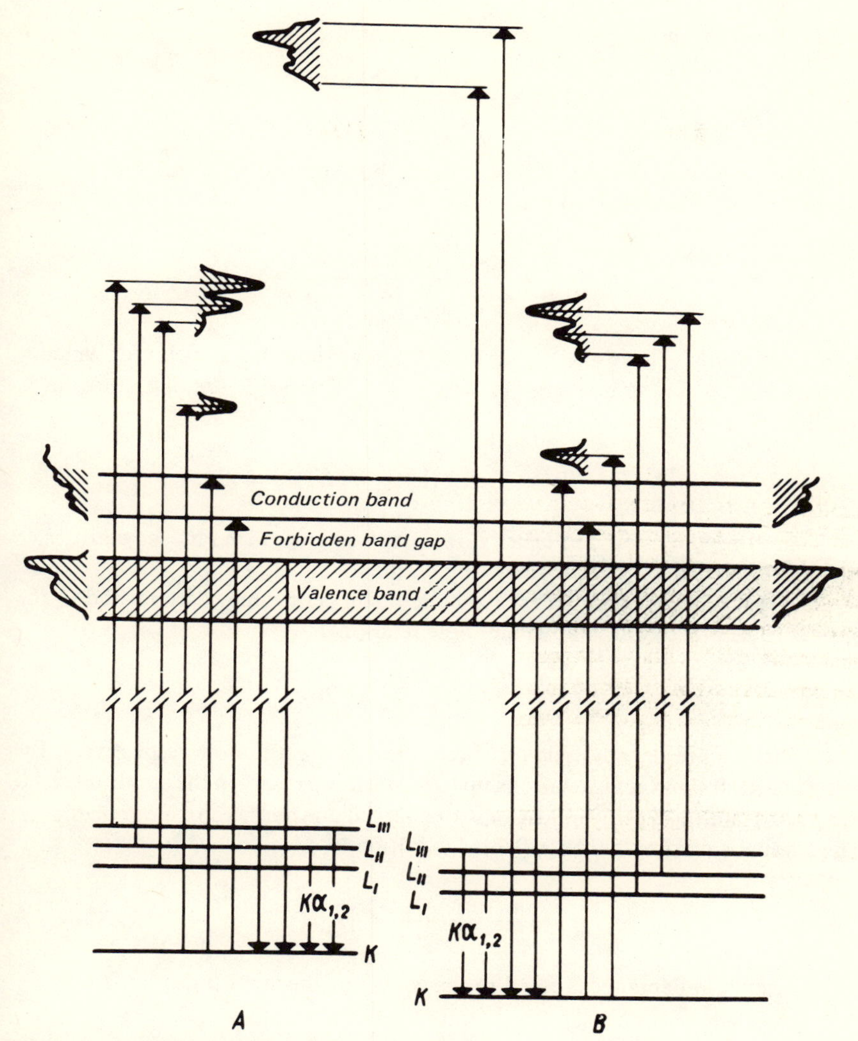

Figure 48. Use of X-ray photoelectron spectroscopy for the study of the electron structure in compounds of the AB type.

where $d\rho/dE_{jj'}$ represents the interband density of states, i.e.,

$$\frac{d\rho}{dE_{jj'}} = \int \frac{dS}{|\nabla_{\mathbf{k}}(E_j - E_{j'})|},$$

and $\bar{f}_{jj'}$ is the mean value of the oscillator strength:

$$f_{jj'}(\mathbf{k}) = \frac{2}{3} \frac{|\langle \mathbf{k}j | \mathbf{P} | \mathbf{k}j' \rangle|^2}{E_{j'}(\mathbf{k}) - E_j(\mathbf{k})}.$$

If the mean value of the oscillator strength $\bar{f}_{jj'}$ cannot be calculated, $\epsilon_2(\omega)$ can be obtained from the expression

$$\varepsilon_2(\omega) \sim \sum_{jj'} \int \frac{f_{jj'}(\mathbf{k})\, dS}{|\nabla_{\mathbf{k}}(E_j - E_{j'})|}.$$

$\epsilon_2(\omega)$ and $d\rho/dE_{jj'}$ have analytical singularities at those frequencies for which, at some point $\mathbf{k}$ on the isoenergetic surface $h\nu = E_{j'}(\mathbf{k}) - E_j(\mathbf{k})$, the following condition is satisfied:

$$\nabla_{\mathbf{k}} E_{jj'}(\mathbf{k}) = \nabla_{\mathbf{k}} E_{j'}(\mathbf{k}) - \nabla_{\mathbf{k}} E_j(\mathbf{k}) = 0.$$

Therefore, the interband density of states is characterized by singularities, which in this case are the van Hove singularities.

The real and imaginary parts of the dielectric permeability are related to the optical constants by the following simple relations:

$$\varepsilon_1 = n^2 - k^2, \quad \varepsilon_2 = 2nk,$$

where n and k are the coefficients of refraction and extinction, respectively. By using the Kramers–Kronig relationship (a relationship between the amplitude and the phase) and the Fresnel equations, the optical constants can be calculated. The Fresnel equation for the reflected radiation is

$$r = \frac{n - ik - 1}{n - ik + 1} = |r|\, e^{i\theta}.$$

The measured reflectance is equal to the square of the amplitude:

$$R = r^2 = \frac{(n-1)^2 + k^2}{(n+1)^2 + k^2}, \tag{98}$$

and the phase angle is determined by the expression

$$\theta(\omega) = \arctan\left(\frac{2k}{n^2 + k^2 - 1}\right). \tag{99}$$

The phase $\theta(\omega)$ for any frequency may be calculated from reflectance data by using the Kramers-Kronig relation:

$$\theta(\omega_0) = \frac{1}{2\pi} \int_0^\infty \frac{d \ln R}{d\omega} \ln \left| \frac{\omega + \omega_0}{\omega - \omega_0} \right| d\omega. \tag{100}$$

By solving the system of equations (98) and (99) together with relation (100), the dependence of the refraction and extinction coefficients on the energy of incident photons can be determined.

As far as the photoelectron emission in the far-ultraviolet range is concerned, Smith [164] and Brust [165] have shown that the energy distribution of electrons emitted without energy loss from the sample is given by the formula

$$I(E_i, \hbar\omega) \sim \sum_{j'j} \int d^3k \, | P_{j'j}(\mathbf{k}) |^2 \, \delta(E_{j'}(\mathbf{k}) - E_j(\mathbf{k}) - \hbar\omega) \times$$
$$\times \, \delta(E_i - E_j(\mathbf{k})) \, T(E_{j'}, \mathbf{k}),$$

where $T(E_{j'}, \mathbf{k})$ represents the exit function that determines the probability that an electron excited into the state $E_{j'}(\mathbf{k})$ will reach the sample surface and escape out of it. At sufficiently high energies, $\hbar\omega$, the magnitude of $I(E_i, h\omega)$ is approximately proportional to that part of $\epsilon_2(\omega)$ corresponding to transitions from the states E_i, i.e., $I(E_i, \hbar\omega) dE_i \sim \epsilon_2(\omega)$.

In conclusion, it can be stated that optical methods offer the possibility to study the structure of valence and conduction bands. In the majority of cases, however, it is just the structure of the conduction band at excitation photon energies of the order of 10–15 eV that hinders the determination of the characteristics of the crystal valence bands with the method of photoelectron spectroscopy.

Park *et al.* [166] have proposed a new method for the determination of the binding energies of electrons in crystals, namely, the so-called appearance potential spectroscopy (APS). The principle of this method is to apply a linearly and slowly increasing accelerating potential between a cathode and an anode, the latter being the sample itself. The increasing accelerating potential results in an increasing number of photons in the bremsstrahlung spectrum. When the edge potential corresponding to the excitation of a characteristic spectrum is reached the detector will record an additional number of photons. To improve the sensitivity of the method, the voltage derivative of the radiation intensity dI/dV is recorded rather than the intensity itself. Knowing the values of edge potentials, it is possible to determine the binding energies of core electrons belonging to atoms inside a surface layer of 5–10 Å thickness.

Information on the chemical shifts of core levels obtained with photoelectron spectroscopy can in some cases be complemented with data obtained from Mössbauer spectroscopy, in which the distribution of γ quanta emitted under nuclear transitions from excited to ground states is studied. It can be considered that, during the nuclear transition, the electron distribution remains unchanged. The total energy of the system includes the total electron and nuclear energy, and also the energy of Coulomb interaction between the electronic and nuclear charge distributions. It is determined by the magnitude of the overlap of nuclear and electronic charge distributions at the nucleus sites. The size of the nucleus in the excited state is different from its size in the ground state, and this affects the energy of emitted γ quanta. Of course, the variation of nuclear dimensions in the excited states does not depend on the chemical environment of the nucleus. However, the density of electronic charge at the nucleus site is sensitive to the chemical environment.

In spite of the fact that the main contribution to the density of electronic charge at the nucleus arises from core electrons, while variations related to the chemical environment only contribute approximately 10^{-3} of the total charge density at the nucleus, Mössbauer spectroscopy is characterized by a very high sensitivity, making possible the detection of extremely low changes in the chemical environment.

The isomer shift between two compounds A and B (B being the absorbant) is given by the formula

$$\delta_A = C\{|\Psi_s(0)_A|^2 - |\Psi_s(0)_B|^2\} \frac{\Delta\langle r^2\rangle}{\langle r^2\rangle}, \tag{101}$$

where $|\Psi_s(0)|^2$ is the charge density of s electrons at the nucleus, and $\Delta\langle r^2\rangle/\langle r^2\rangle$ is the relative change of the square of nuclear radius during the excitation. Equation (101) contains only the density of s electrons because electrons having nonzero orbital momenta will have singularities in their charge density distribution at the nucleus site. The magnitude of $\Delta\langle r^2\rangle$ depends on the isotope type and on the type of γ transition, and can be positive or negative. One way to determine the magnitude of $\Delta\langle r^2\rangle/\langle r^2\rangle$ is to calculate the charge density at the nucleus site for a particular group of molecules. It is then possible to determine C and the ratio $\Delta\langle r^2\rangle/\langle r^2\rangle$ using equation (101). Since the main contribution to the charge density at the nucleus arises from the core electrons, it follows that these calculations should take into account relativistic effects, as well as effects related to the correlation between the core electrons. In order to reduce energy losses arising from recoil effects, in Mössbauer spectroscopy the sample atoms or molecules are included in a solid state matrix. Consequently, the calculations should also include solid state effects. Because of their interdependence, each of these factors gives rise to tedious theoretical calculations. This explains the fact that for the majority of Mössbauer transitions the magnitude of $\Delta\langle r^2\rangle/\langle r^2\rangle$ is as yet not sufficiently well known. The accuracy in the determination of its value is at best ±20%, and in the common cases as poor as ±100%. Consequently, the information on electron structure obtained by the study of isomer shifts is as yet of qualitative rather than quantitative character.

In a qualitative description of the electron structure, the most important information is the sign of $\Delta\langle r^2\rangle$. If $\Delta\langle r^2\rangle$ is positive, the electron density at the nucleus site increases, while if it is negative, the electron density decreases. However, a change of sign of $\Delta\langle r^2\rangle$ is not always connected to a change in the density of *s* electrons at the nucleus of the given atoms. In fact, both an increase in the number of *s* electrons in atom *A* and a decrease in the number of its *p*, *d*, *f* valence electrons have the same effect, namely, to produce an increase in $|\Psi_s(0)|^2$. This is because the *p*, *d*, *f* valence electrons screen the *s*-core electrons from the nucleus, so that their removal causes a contraction of the *s* orbitals. Theoretical calculations performed for free atoms have shown that this effect is less important than the direct increase of the density of *s* electrons. Nevertheless, both effects should normally be taken into account. This is the reason the complementary use of photoelectron spectroscopy and Mössbauer spectroscopy would be useful in the study of the nature of chemical bonds in molecules.

The correlation between chemical and isomer shifts has been established by Barber *et al.* [167] and Adams *et al.* [168]. Figure 49 shows the relationship between chemical and isomer shifts for inorganic complexes of bivalent iron ^{57}Fe, for a transition energy of 14.4 keV [168]. In general, a correlation between the chemical and isomer shifts is not to be expected. However, it is possible to predict certain series of compounds, having a given type of hybridization, for which a linear relation between ΔE and δ_A would be valid. Possible deviations from such a linear dependence are of particular interest.

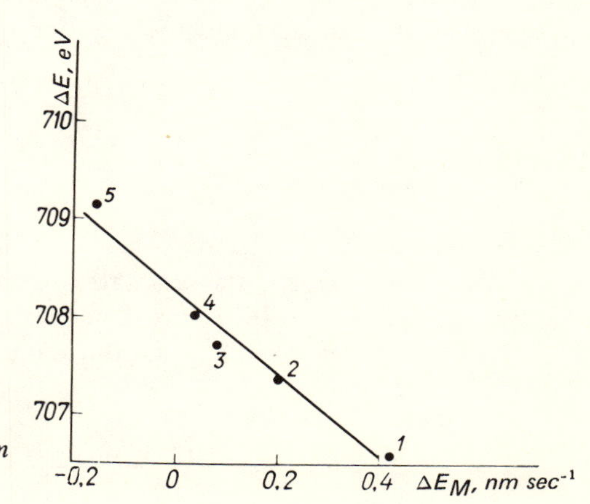

Figure 49. Chemical and isomer shifts for ^{57}Fe in the inorganic complexes of bivalent iron, according to the data from [168].

The advanced and rapid development of computational techniques will soon allow accurate determinations of the factor $\Delta\langle r^2\rangle/\langle r^2\rangle$, and interesting quantitative information will become possible.

In nuclear magnetic resonance experiments, the electron cloud in molecules in the initial state is already excited by the action of the strong magnetic field. As a reaction of the electron system to the field B, a screening field $B' = -\bar{\sigma}_A B$ is generated at the nucleus. The chemical shifts appearing in nuclear magnetic resonance arise from the different screening constants of the atomic nuclei situated in different chemical environments. The magnitude of $\bar{\sigma}_A$ is determined as the mean value:

$$\bar{\sigma}_A = \frac{1}{3}(\sigma_{xx} + \sigma_{yy} + \sigma_{zz}),$$

where σ_{xx}, σ_{yy}, σ_{zz} depend on molecular properties, namely, the molecular dimensions, the local electron density, and the stability of the electron distribution of the ground state with respect to the perturbations due to the magnetic fields. The change of $\bar{\sigma}_A$ for an element A in various environments is called the "NMR chemical shift." A more efficient magnetic screening corresponds to a positive value of the chemical shift. The screening constant $\bar{\sigma}_A$ may be represented as the sum

$$\bar{\sigma}_A = \sigma_A^d + \sigma_A^p,$$

where σ_A^d and σ_A^p are the diamagnetic and paramagnetic screening constants, respectively. The diagmagnetic screening describes the screening related to the spherically symmetric part of the charge distribution. The paramagnetic term appears as a result of the fact that in molecules the charge distribution is not spherically symmetric. Therefore, the paramagnetic contribution to the screening compensates the diamagnetic screening. For σ_A^d, one may write the following expression:

$$\sigma_A^d = \frac{e^2}{3mc^2}\left\langle \Psi_0 \left| \sum_i \frac{1}{r_{Ai}} \right| \Psi_0 \right\rangle.$$

Here, Ψ_0 represents the electron wave function corresponding to the ground state of the molecule, and the summation is performed over all the electrons in the system. Gelius *et al.* [67] have demonstrated that the magnitude of the chemical shift $\Delta\sigma_A^d$ is given by the expression

$$\Delta\sigma_A^d = -\frac{\Delta E_A}{3mc^2} + \frac{e^2}{3mc^2}\Delta\sum_{B\neq A}\frac{Z_B}{R_{AB}}.$$

Therefore, if the chemical shifts ΔE_A and the relevant structural data are known, it is possible to determine the diamagnetic screening constant.

For the paramagnetic screening constant the following formula is obtained:

$$\sigma_A^p = -\frac{4}{3}\mu_B^2 \sum_{n=1}^{\infty} \left\langle \Psi_0 \middle| \sum_i L_i \middle| \Psi_n \right\rangle \times$$
$$\times \left\langle \Psi_n \middle| \sum_i L_i r_{Ai}^{-3} \middle| \Psi_0 \right\rangle (E_n - E_0)^{-1},$$

where Ψ_n represents the wave function of the nth excited state of the nonperturbed Hamiltonian, and $\mu_B = eh/2mc$ is the Bohr magneton. Calculation of the magnitude of σ_A^p is difficult since it requires knowledge of the wave functions of the whole system in the excited state. Unfortunately, it is just σ_A^p that is the dominant term in the NMR chemical shift.

The total shift $\Delta\bar{\sigma}_A$ is in many cases two orders of magnitude greater than $\Delta\sigma_A^d$, and for this reason a correlation between the chemical shifts observed in X-ray photoelectron spectroscopy and those observed by NMR is hardly to be expected. Figure 50 shows the shift of the boron 1*s* line, as determined with X-ray photoelectron spectroscopy, and the NMR shift of boron nuclei (^{11}B), in a series of 28 specimens containing boron atoms in various chemical environments [67]. The dotted line indicates the correlation that would be expected if the term related to ΔE_A were dominant in the NMR shift. Large deviations from such a correlation are caused by variations in the paramagnetic screening constant. Therefore, though in general a linear correlation between the shifts corresponding to the compounds in Figure 50 can hardly be expected, such a correlation is not out of the question in certain cases [169, 170]. Figure 51 shows the chemical shifts determined by X-ray photoelectron spectroscopy and by nuclear magnetic resonance for some compounds of the CH_nX_{4-n} type, where X may be fluorine, chlorine, bromine, or iodine, and n takes values from 1 to 4. This figure demonstrates that molecules can exist for which a nearly linear correlation is observed between the two shifts. It is to be noted that for the bromine compounds in Figure 51 there is a marked deviation from linearity.

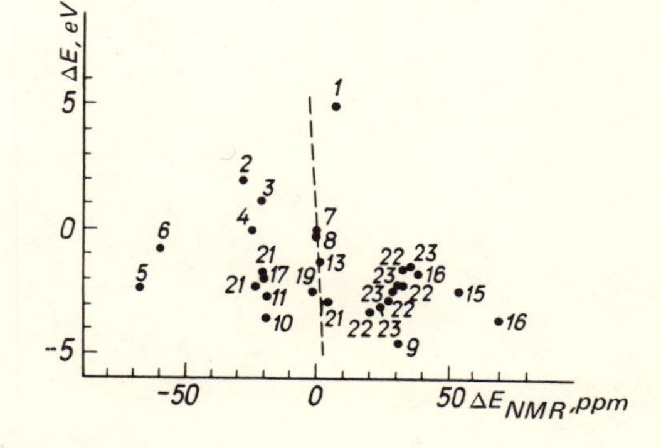

Figure 50. The chemical shift of the boron 1s level and the NMR shift on the ^{11}B nucleus for boron in various chemical environments, according to the data from [67]. The shifts have been measured with respect to boron in the compound $B(OCH_3)_3$.

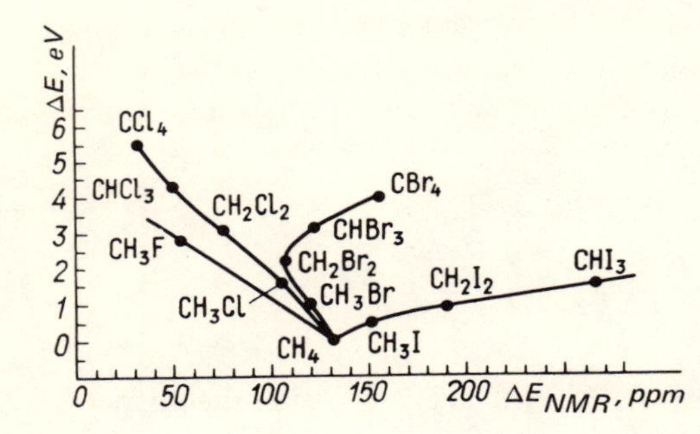

Figure 51. Chemical shift of the carbon 1s level and the NMR shift on the ^{31}C nucleus for compounds of the CH_nX_{4-n} type.

Lindberg [171] has discussed the correlation between the shifts determined by X-ray photoelectron spectroscopy and by nuclear magnetic resonance for a large series of compounds. A practically linear correlation has been found for methyl compounds of the CH_3X type (CH_3I, CH_3Br, CH_3Cl, CH_3F) between the chemical shift of the carbon 1s line and the NMR shift of the hydrogen nuclei, as well as for some phenyl compounds. In order to demonstrate this correlation, compounds should be chosen for which the NMR shift can occur as a result of the change of the effective charge and not as a result of magnetic anisotropy effects.

3

Metals and Alloys

The structure of the valence bands of many metals is characterized by a great complexity and contains features that can be observed experimentally only by using spectroscopic instruments with a resolving power of the order of some tenths of an electron volt. At present, however, since most electron spectrometers have a resolution of about 0.6–0.9 eV, it is not yet possible to determine the detailed structure of the valence bands of most transition metals from their electron spectra. Thus the resolution of available instruments is completely inadequate for a determination of the accurate valence band structure of transition metals in which the density of the *d* states is high and their bandwidth is not large. As the atomic number increases, however, the valence band broadens. For gold, for example, it is possible to determine experimentally the whole fine structure previously predicted by relativistic band calculations.

For alloys, the calculations are much more difficult and, therefore, in the study of the electronic nature of processes that accompany the formation of alloys, experiment will continue to play the predominant role.

Electron spectroscopy has already revealed a series of important characteristics of the structure of valence bands in alloys and has allowed the quantitative determination of the essential parameters in the development of the electronic theory of disordered alloys. Thus, the study of X-ray photoelectron spectra of alloys of transition and noble metals has revealed the existence of strongly localized *d* states. These experimental results have been explained by interpreting the electron structure of disordered alloys within the framework of the coherent potential method.

Density of States of Valence Electrons as Given by the X-Ray Photoelectron Spectra of Light Metals

The light metals have not been studied by X-ray photoelectron spectroscopy as extensively as the transition and noble metals. This is because their sample surface undergoes a rapid oxidation, and, consequently, it is necessary to perform the measurements in ultrahigh vacuum of the order of 10^{-10} torr. At present, experimental data exist for the X-ray photoelectron spectra of the valence bands of only two light metals, sodium and aluminum. Since their valence electrons may be treated in terms of the nearly free electron approximation, it is sufficient to use only a small number of orthogonal plane waves for the description of the band structure of these metals.

The X-ray photoelectron spectra of sodium have been studied by Kowalczyck *et al.* [172] and by Citrin [173]. In the work of Kowalczyck *et al.* [172], the sodium sample was prepared by evaporation on an aluminum substrate. The electron spectra were recorded at a pressure of 8×10^{-11} torr. The intensity of oxygen and carbon lines was insignificant, and, therefore, it can be considered that the presence of such minute amounts of oxygen and carbon on the sample surface does not result in a modification of the shape of the valence electron spectra. In the work of Citrin [173], the sample was prepared in a similar way, but since the pressure of residual gases in the spectrometer was somewhat higher (of the order of 10^{-9} torr), a number of evaporations of sodium onto the substrate were performed before its valence and core-electron spectra were recorded. Since the photoionization cross section of the valence electrons in sodium is low, the valence band spectra were recorded continuously for eight hours. An HP-5950A electron spectrometer was used. The long duration of the spectra recording might explain the relatively high fluctuations in the valence band photoelectron spectra of the sodium sample illustrated in Figure 52. A parabola may be drawn through the experimental points, which is consistent with the free-electron representation of the valence electrons of sodium. The width of the valence band of sodium determined from these data is equal to 3.2 ± 0.1 eV. However, if the linear background of secondary electrons is subtracted from the data in Figure 52, and a new parabola is drawn through the resulting points, a lower value for the width of the valence band is obtained, namely, 2.8 ± 0.1 eV. Since these approximations give the upper and the lower values, respectively, one may take the mean value 3.0 ± 0.2 eV as the more plausible value of the valence bandwidth. Studies of ultrasoft X-ray spectroscopy

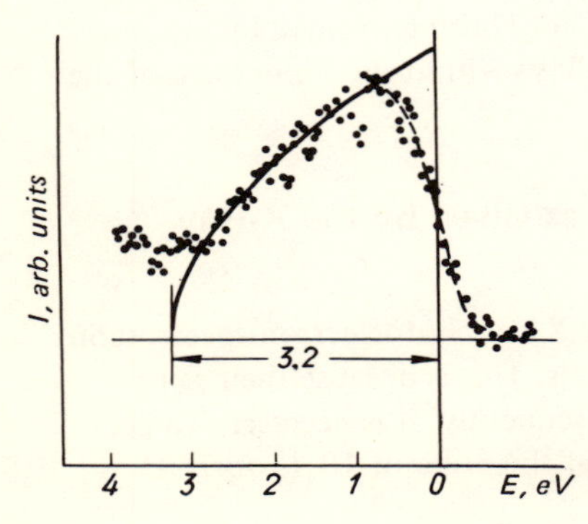

Figure 52. Photoelectron spectra of sodium valence electrons.

by Crisp and Williams [174] have yielded a value of 2.6 eV for the valence electron bandwidth.

Calculation of the bandwidth within the framework of the free-electron model, or by using the pseudopotential method [175] or the method of Green's functions [176], gives values of 3.2 eV, 3.1 eV, and 3.3 eV, respectively. It should be noted that the shape of the X-ray photoelectron spectra in the region of the Fermi energy is determined by the experimental resolution function of the instrument, which amounts to 0.5–0.6 eV. Temperature broadening is insignificant (of the order of some hundredths of an electron volt).

As is shown by Figure 53, in the X-ray photoelectron spectra of $1s$, $2s$, and $2p$ core electron levels there also exists a background of secondary electrons. This appears as a result of inelastic scattering processes involving photoelectrons on their way out of the sample. In these spectra, peaks are also observed, that correspond to the electrons that have lost part of their energy in the excitation of plasmon oscillations in the bulk of the sample (bulk plasmons) or on its surface (surface plasmons). In the spectrum of $1s$ electrons, besides the four bulk plasmons (v), peaks corresponding to the excitation of surface plasmons (s) and of combined surface and bulk plasmons ($v + s$) are also observed. The energy of the first bulk plasmon was found to be $\hbar\omega_p = 5.83 \pm 0.6$ eV, which is consistent with the results obtained by other authors: 5.85 ± 0.1 eV [177], 5.71 ± 0.1 eV [178]. The existence of surface plasmons has been predicted theoretically by Ritchie [179] and by Stern and Ferrel [180]. The experimental data on the energy of surface plasmons agree well with those predicted theoretically.

In the compound NaOH the $3s$ electrons of sodium take part in the formation of the chemical bond and cannot be considered as being free. This is the reason why in the X-ray photoelectron spectra of this compound, plasmons are not observed (Figure 54). The experimental values of the binding energies of sodium core electrons in the pure metal and in NaOH compound are given in Table 7. The accuracy in the determination of the binding energies is dependent on the uncertainty in the determination of peak positions by the method of least squares (which is of the order of 0.02 eV), the uncertainty in the determination of the Fermi level (approximately 0.06 eV), and the operating instability of the spectrometer itself. The instrument instability can be estimated by measuring the position of the same line over a long period of time. It also amounts to several hundredths of an electron volt. In spite of the fact that the binding energy values may be subject to an error of the order of 0.08 eV, the relative position of the lines can be determined with a greater precision. In the case of sodium in NaOH, the binding energy values were determined relative to the Fermi level of the spectrometer. It should be mentioned that the work function of sodium, as determined by van Oirschot *et al.* [181], is equal to 2.36 ± 0.02 eV.

Baer and Busch [56] have studied the X-ray photoelectron spectra of the valence band of aluminum. The method of sample preparation and the ex-

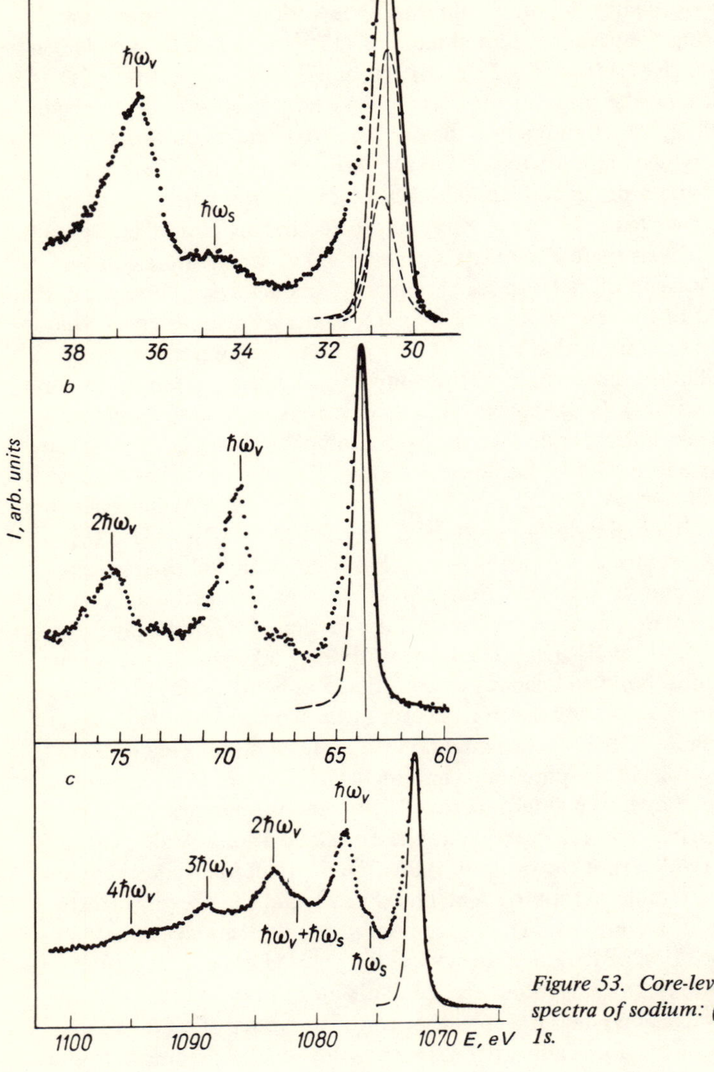

Figure 53. Core-level photoelectron spectra of sodium: (a) 2p; (b) 2s; (c) 1s.

perimental conditions have been mentioned in Chapter 1. The features of the density of states of valence electrons of aluminum predicted in the theoretical calculations of Rooke [182] (Figure 55a) were also found in the experimental X-ray photoelectron spectra (shown in Figure 55b). These features were also observed by Dimond [183] in the $L_{2,3}$ X-ray emission spectra of aluminum

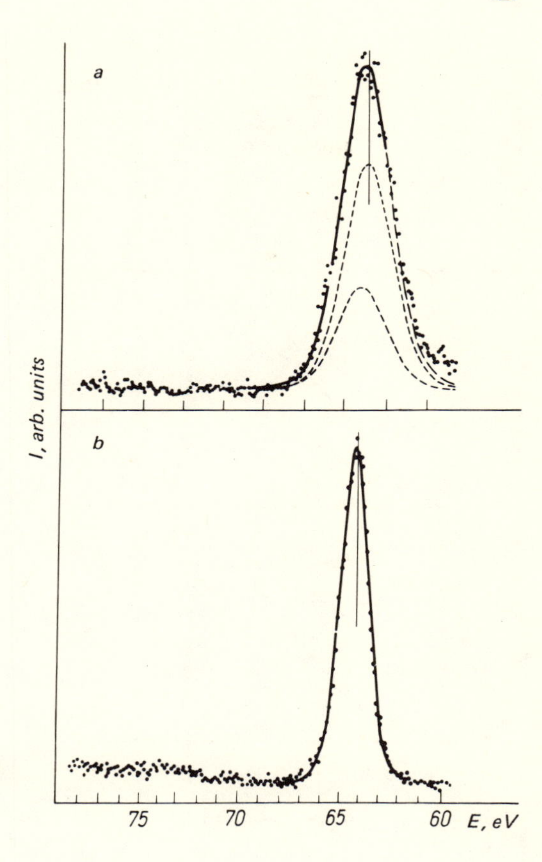

Figure 54. Core-level photoelectron spectra of sodium in NaOH: (a) 2p; (b) 2s.

(Figure 55d). In the K band of aluminum [184], owing to the influence of the transition probability, only the states connected to the singular points X_4 and K_1 of the Brillouin zone appear (Figure 55c).

The calculations of Smrčka [185] successfully reproduce the intensity distribution of the radiation in X-ray emission bands. Figure 56 shows the X-ray photoelectron spectrum of aluminum $2p$ electrons. Besides the main line corresponding to $2p$ electrons, additional lines related to the excitation of volume

TABLE 7. Binding Energy of Na Core Electrons in Metallic Na and in NaOH

	State		
Sample	$1s$	$2s$	$2p$
Na Metal	1071.76 ± 0.03	63.57 ± 0.03	30.52 ± 0.04
NaOH	1072.59 ± 0.04	64.21 ± 0.04	31.39 ± 0.05

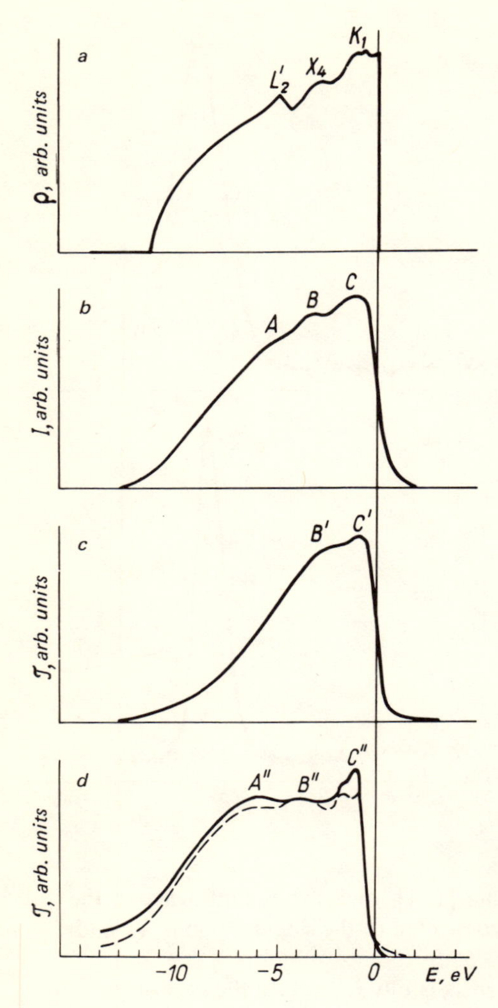

Figure 55. Electronic density of states of valence electrons, X-ray emission spectra, and X-ray photoelectron spectra of aluminum: (a) density of states; (b) photoelectron spectra; (c) Kβ emission band of aluminum; (d) experimental (——) and theoretical (– – –) $L_{2,3}$ band of aluminum.

and surface plasmons can be observed. The spectrum also exhibits plasmon satellites (structure a) arising from the excitation of $2p$ electrons by the $K\alpha_{3,4}$ satellite line of magnesium, with the subsequent energy loss $\hbar\omega_v$.

Studies of X-ray photoelectron spectra of sodium and aluminum indicate that in these cases the quasimomentum dependence of the electron excitation probability is not significant. This is because the isoenergetic surfaces of these metals in the **k** space are nearly spherical, and on each isoenergetic surface the probabilities corresponding to different values of the quasimomentum are about the same. This explains why the X-ray photoelectron spectra of valence electrons

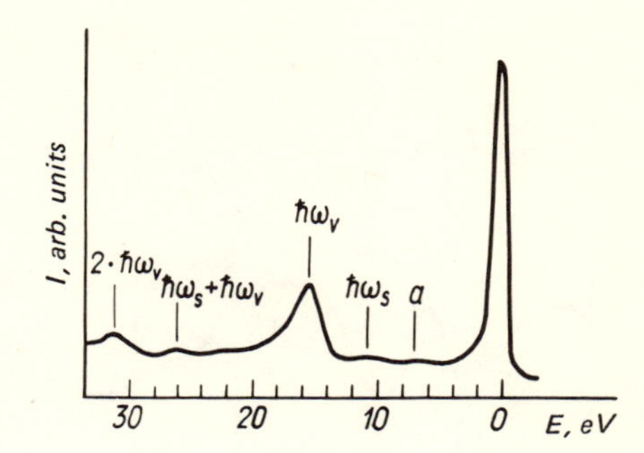

Figure 56. 2p photoelectron spectrum of aluminum.

in sodium and aluminum are so close to the distributions of the electron density of states.

Influence of Transition Probability on the Shape of X-Ray Photoelectron Spectra of Transition and Noble Metals

The electron structure of transition and noble metals is more complicated. This is because in these metals there exist two markedly different groups of s, p, and d electrons. The d electrons, unlike the s and p electrons, are localized, but the degree of their localization is lower than in free atoms. The structure of energy bands in the transition metals is determined essentially by the hybridization of s, p, and d states. The existence of localized states on the broader energy background of the relatively smooth distribution of s and p states has led to the concept of d resonance states [5]. The high localization of d electrons in solids may be understood if one realizes that, in the calculation of d-type radial wave functions, the term $l(l+1)/r^2$ that enters into the radial Schrödinger equation describes the centrifugal repulsion of the electron by the nucleus. In a bound state, for $l \geqslant 2$, the electron experiences both the centrifugal barrier and the atomic potential. By adding the term $l(l+1)/r^2$ to the crystal potential in the vicinity of the atom and taking into account the particular structure of the crystal potential, the effective potential with barrier that prevents the electron from being transferred to the neighboring atom is obtained. There exists, however, a quantum-mechanical probability of transition from bound states with $l = 2$ through the barrier, and therefore the resonance state is extended over a rather large energy region.

The electron structure of transition metals has been treated theoretically in a large number of investigations. In many cases, rather good agreement has been obtained between the calculated and experimental values of the parameters that describe the electron structure of these metals: the width of the valence band, the density of states at the Fermi level, and the energy position of van Hove singularities. When transition metals are studied by the method of electron

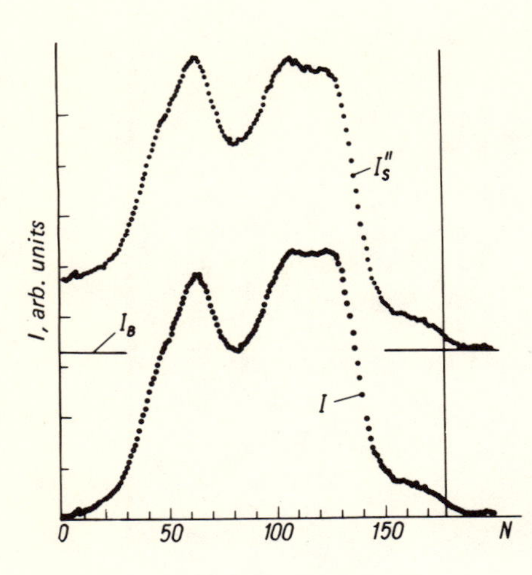

Figure 57. Photoelectron spectrum of gold valence electrons before (I''_S) and after (I) subtraction of the background of inelastic electron scattering.

spectroscopy, it is important to find out whether the density of states of the valence band electrons is reflected in the X-ray photoelectron spectra. In the first studies in this field [186–190], the electron spectra of transition metals were measured with a relatively low resolution, of the order of 1.2 eV. Recently, however, studies of some transition and noble metals have been made with a better resolution (0.5–0.6 eV) [191–193]. In these experiments, unlike the earlier ones [186–190], some fine structure could be observed in the X-ray photoelectron spectra. One of the most extensively studied metals is gold. Shirley [194] measured the valence band electron spectra of three gold samples. The samples were cut from a monocrystal in such a way that the three specimen surfaces were oriented parallel to the planes (100), (110), and (111), respectively. The samples were polished, chemically etched, and heated. The photoelectron spectra were produced by using the monochromatized $K\alpha_{1,2}$ aluminum line as excitation radiation. The spectra of the three specimens were very similar to each other, and therefore in Figure 57 only the spectrum corresponding to the (110) surface is shown. It has been smoothed by use of the formula:

$$I''_s(N) = \frac{1}{4}(I'(N-1) + 2I'(N) + I'(N+1)),$$

where $I'(N)$ represents the intensity in the Nth channel, before smoothing. From the quantity $I_S(N) = I''_S(N) - I_B$ (see Figure 57), the contribution of the background of inelastic electron scattering should also be subtracted. This correction is proportional to the area bounded by the curve of the intensity distribution and the abscissa axis, in an energy interval between the given energy value and the Fermi level. Therefore,

$$I(N) = I_s(N) - I_s(0)\left(\frac{\sum_{N'>N} I_s(N')}{\sum_{N'>0} I_s(N')}\right)$$

Shirley has performed a careful analysis of the valence band photoelectron spectra of the gold (110) surface by comparing them with the calculated spectra [195-199] and taking into account the relativistic effects in the density of states of valence electrons (Figures 58, 59). Figure 58 shows the results of the theoretical calculations performed by Ramchandani [195] and by Smith and Traum [196]. The poor agreement with the experimental data may be attributed to the fact that in these theoretical calculations some inadequate approximations were made. Ramchandani [195], for example, used only a low number of functions in the basis, whereas Smith and Traum [196] used an approximate scheme for the inclusion of relativistic effects, namely, the spin-orbit interaction was introduced in a band structure obtained by nonrelativistic calculations. In the work mentioned above [195], Ramchandani also performed calculations of the density of states by choosing for the parameter α of the exchange potential (V_{ex}^{α}) the values $\alpha = \frac{5}{6}$ and $\alpha = \frac{2}{3}$. The results obtained using these values of the parameter α show an even worse agreement with the experiment than those in Figure 58, which were obtained with $\alpha = 1$. In the calculations of Connoly and Johnson [197], based on the method of Green's functions, an exchange poten-

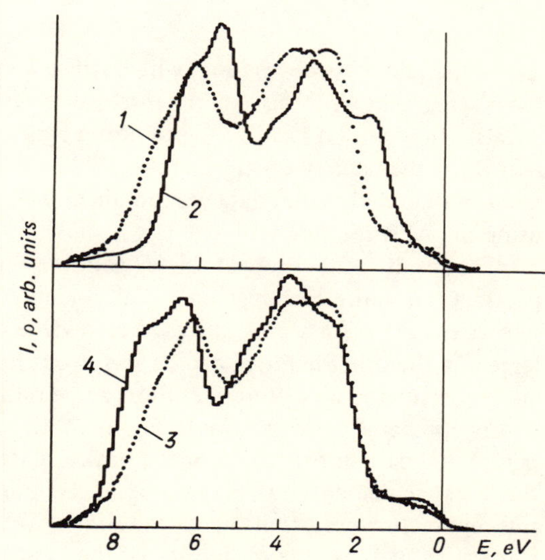

Figure 58. Photoelectron spectra of gold valence electrons as compared to the density-of-states curves: (1, 3) photoelectron spectra from [194]; (2, 4) density-of-states histograms calculated in [195] and [196], respectively.

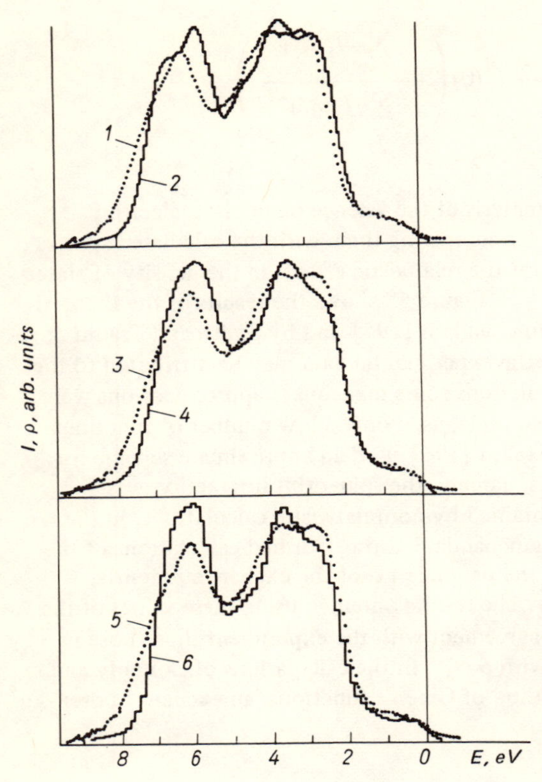

Figure 59. Photoelectron spectra of gold valence electrons as compared to the density of states curves: (1, 3, 5) photoelectron spectra from [194]; (2, 4, 6) density of states histograms calculated in [197], [198], and [199], respectively.

tial of the Slater type was used, which led to a value of the valence bandwidth in good agreement with the experimental value. Kupratakuln [198] used the APW method with $\alpha = 1, \frac{5}{6}, \frac{2}{3}$, but better agreement with the experiment is obtained for the density-of-states curves calculated with $\alpha = \frac{19}{24}$. The best agreement with experimental data was obtained by Christensen and Seraphin [199] using the APW method with $\alpha = 1$.

The results of theoretical calculations are very sensitive to the choice of the parameter α in the exchange potential. Sommers and Amar [200], for example, have shown that for $\alpha = \frac{2}{3}$, the predicted value of the valence bandwidth is too large. On the other hand, as can be seen from Figure 59, the density-of-states curves calculated by different authors agree rather well with each other. All the density-of-states curves shown in Figure 59 show the best agreement with experimental data in the region of the peak situated close to the Fermi level, and the worst agreement in the region of greater binding energy values. It is in just this region, however, that the partial density of s states is increasing. Therefore, if it is assumed that the magnitude of the ionization cross section is lower for s states than for d states, it becomes clear why the values of the density of states are higher than the intensity of the energy distribution of photoelectrons. We should mention that the scale for the density-of-states curves in Figure 59 was chosen so that the height of the maximum in the density of states at 3 eV fits the intensity at this point in the experimental spectrum. This type of analysis shows that experimental data provided by X-ray photoelectron spectroscopy may be used as a criterion to assess the validity of theoretical calculations of the band structure of gold.

So far, we have discussed only calculations in which relativistic effects have been taken into account. In all the work mentioned above, with the exception of the work of Smith and Traum [196], the relativistic Dirac equation was solved. The calculations performed by Smith and Traum [196] without taking into account the relativistic effects led to a distribution of electron density of states which exhibits only one maximum situated in the energy region in which the experimental curve exhibits a minimum. Therefore, the occurrence of two maxima in the intensity distribution of the X-ray photoelectron spectra originates entirely from the relativistic effects.

In order to give an idea of the magnitude of the spin–orbit interaction in gold, it can be noted that the energy distance between the states Γ_8^+ and Γ_7^+ resulting from the splitting of the state Γ_{25}' is equal to 1.3 eV, which is significantly lower than the experimental value (2.8 eV) of the distance between the maxima.

Eastman and Grobman [201] and Freeouf *et al.* [202] have studied the valence band photoelectron spectra of gold in the energy region 10.2–90 eV, by using synchrotron radiation as the excitation source. The samples were prepared by evaporation of a gold layer of 1000–5000 Å thickness onto the (111) surface of a silicon monocrystal as substrate. During operation, the vacuum in the spectrometer chamber was about 3×10^{-10} torr, while the evaporation was performed at about 1×10^{-10} torr. As is shown in Figure 60, the intensity in the energy distribution of photoelectrons, in the region 10.2–50 eV, depends strongly on the excitation energy, while this dependence is much weaker at other energy values. Even though at photon energies greater than 50 eV the overall structure of the photoelectron spectra does not undergo significant changes, their fine structure exhibits observable variations. These variations are probably due to the quasi-momentum dependence of the matrix elements of transitions from the valence band to the free states, as well as to Bragg-diffraction effects at the exit of electrons from the sample. However, as can be seen from Figure 61, the photoelectron spectrum of gold obtained at 80 eV energy differs from the X-ray photoelectron spectrum in the energy region in which both the first and the second peak are situated. It may be presumed that here the dependence of the transition matrix elements on the excitation energy is making itself felt.

It should be mentioned that, in the work of Eastman and Grobman [201] and Freeouf *et al.* [202], polarized radiation was used, and the geometry of the experiments was different in the two cases. Figure 61 also shows the $N_{6,7}$ X-ray emission spectrum of gold (corresponding to transitions from $5d$ to $4f$ states) as

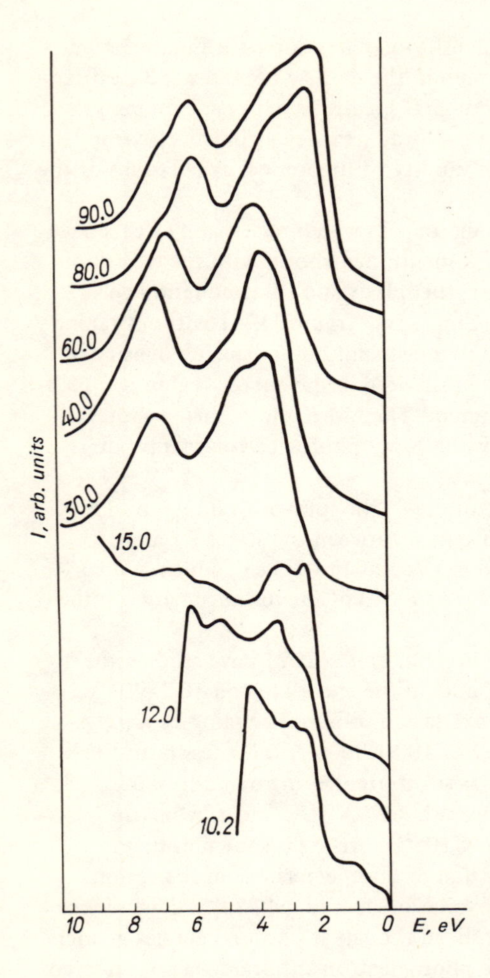

Figure 60. Photoelectron spectra of gold valence electrons, obtained at various excitation energies.

measured by McAlister *et al.* [203]. In their experiment, the resolution (0.5 eV) was close to the resolution in the X-ray photoelectron spectrum of gold (0.6–0.8 eV). Since the maxima in the valence-band density of states of gold are observed at about 3.4 eV and 6.2 eV, while the binding energy values of the $4f_{7/2}$ and $4f_{5/2}$ electrons are equal to 84.0 eV and 87.7 eV, respectively, it is to be expected that the maxima in the $N_{6,7}$ X-ray emission spectrum of gold will be situated at 77.8, 81.0, 81.5, and 84.7 eV. Figure 61 shows that this prediction is in fact correct, the energies of the three maxima a, b, d and the bump c agreeing rather well with the above values of transition energies.

Shirley has shown that the X-ray photoelectron spectrum of the valence electrons of gold provides rather reliable information about the density of electron states in the valence band. The observed discrepancy may be attributed to the

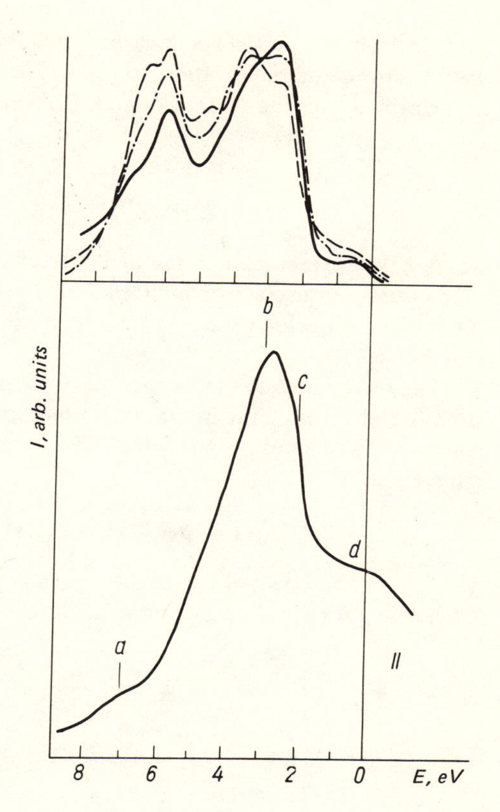

Figure 61. Photoelectron and X-ray emission $N_{6,7}$ spectra of gold valence electrons as compared to the density-of-states curves: I(——)–photoelectron spectrum obtained at 80 eV excitation energy; (–·–·–·)–photoelectron spectrum obtained at 1253.6 eV excitation energy; (– – –) –electron density of states; II–X-ray emission $N_{6,7}$ spectrum.

influence of transition probabilities on the shape of photoelectron spectrum. The problem of the influence of transition probabilities on the X-ray photoelectron spectra of a few transition and noble metals, namely, copper and silver, has been investigated in detail by Nemoshkalenko *et al.* [204, 205].

The energy distribution of electrons emitted from the sample under the action of incident X rays with frequency ω is given by formula (67), which, in the scheme of reduced bands, may be written as follows:

$$I(\omega, E) \approx \sum_{n,\mathbf{k},\mathbf{k}',\mathbf{Q}} |\langle \Psi_{n\mathbf{k}'} | W | \Psi_{\mathbf{k}+\mathbf{Q}} \rangle|^2, \tag{102}$$

where $\Psi_{n\mathbf{k}'}(\mathbf{r})$ and $\Psi_{\mathbf{k}+\mathbf{Q}}(\mathbf{r})$ represent the wave functions of electrons in the valence band and in the conduction band, respectively; W represents the operator describing the interaction of the emitted electron with the external electromagnetic field:

$$W = -\frac{e}{mc} \exp i\mathbf{q}\cdot\mathbf{r}\mathbf{A}\cdot\mathbf{p}.$$

The electromagnetic field is described by a linearly polarized plane wave with wave vector **q** and polarization vector **A**. In formula (102), the summation is performed over the vectors **k** and **k′**, which satisfy the conditions

$$\left.\begin{aligned} E(\mathbf{k}+\mathbf{Q}) - E_n(\mathbf{k}') &= \hbar\omega, \\ E = E(\mathbf{k}+\mathbf{Q}) = |\mathbf{k}+\mathbf{Q}|^2 &= \mathbf{P}^2, \end{aligned}\right\} \tag{103}$$

where $E(\mathbf{k}+\mathbf{Q})$ represents the final-state electron energy, and **P** represents the momentum of the electron emitted from the sample. In the interpretation of X-ray photoemission processes, the electron work function φ of the specimen may be neglected.

The wave functions of valence electrons of copper, silver, and palladium have been found by using the interpolation method of Hodges and Ehrenreich [146]. The wave function of valence electrons is given in this case by the following expression:

$$\Psi_{n\mathbf{k}}(\mathbf{r}) = \sum_{\mu} C_{\mu}^{n}(\mathbf{k})\, \varphi_{\mathbf{k},\mu}(\mathbf{r}) + \sum_{\mathbf{K}_i} C_{\mathbf{K}_i} \varphi_{\mathbf{k},\mathbf{K}_i}(\mathbf{r}), \tag{104}$$

where $\varphi_{\mathbf{k},\mu}$ for $\mu = 1, 2, 3$ represents the Bloch combination of t_{2g} orbitals:

$$\varphi_1 = \left(\frac{15}{4\pi}\right)^{1/2} R(r)\frac{xy}{r^2}, \; \varphi_2 = \left(\frac{15}{4\pi}\right)^{1/2} R(r)\frac{yz}{r^2}, \; \varphi_3 = \left(\frac{15}{4\pi}\right)^{1/2} R(r)\frac{zx}{r^2},$$

while for $\mu = 4$ and $\mu = 5$ it represents the Bloch combination of e_g orbitals:

$$\varphi_4 = \left(\frac{15}{16\pi}\right)^{1/2} (x^2 - y^2)\,\frac{R(r)}{r^2}, \qquad \varphi_5 = \left(\frac{5}{16\pi}\right)^{1/2} \frac{(3z^2 - r^2)}{r^2} R(r);$$

$$\varphi_{\mathbf{k},\mathbf{K}_i}(\mathbf{r}) = \frac{\exp i(\mathbf{k} + \mathbf{K}_i)\cdot\mathbf{r}}{\sqrt{N\Omega_0}}.$$

$\varphi_{\mathbf{k},\mathbf{K}_i}$ are plane waves characterized by the vectors of the reciprocal lattice:

$$K_1 = \frac{2\pi}{a}(0, 0, 0), \qquad K_2 = \frac{2\pi}{a}(0, \bar{2}, 0),$$

$$K_3 = \frac{2\pi}{a}(\bar{1}, \bar{1}, \bar{1}), \qquad K_4 = \frac{2\pi}{a}(\bar{1}, \bar{1}, 1).$$

Thus, in the method of Hodges and Ehrenreich, the basis contains nine functions, and therefore the matrix of the Hamiltonian operator

$$\left(\begin{array}{c|c} \text{PW-PW} & \text{PW-LCAO} \\ \hline \text{PW-LCAO} & \text{LCAO-LCAO} \end{array}\right) \tag{105}$$

has the dimension (9×9).

The most complicated expression has the block (LCAO-LCAO):

$$H_{\mu\mu'}(\mathbf{k}) = \langle \varphi_{\mathbf{k},\mu} | -\nabla^2 + V(\mathbf{r}) | \varphi_{\mathbf{k},\mu'} \rangle.$$

By taking into acount that

$$V(\mathbf{r}) = \sum_{\nu} v(\mathbf{r} - \mathbf{R}_\nu),$$

one obtains

$$H_{\mu\mu'} = [E_0 + \Delta(\delta_{\mu 4} + \delta_{\mu 5})]\,\delta_{\mu\mu'} + \\ + \sum_{\nu\neq 0} \exp(-i\mathbf{k}\cdot\mathbf{R}_\nu) \int a_\mu^*(\mathbf{r} - \mathbf{R}_\nu)(V - v)\, a_{\mu'}(\mathbf{r})\, dV, \qquad (106)$$

where

$$\int a_\mu^*(\mathbf{r})[-\nabla^2 + v(\mathbf{r})]\, a_\mu(\mathbf{r})\, dV$$

is equal to E_0 for the t_{2g} orbitals, and to $E_0 + \Delta$ for the e_g orbitals. If only the interaction of the nearest neighbors is taken into account, the matrix elements $H_{\mu\mu'}$ may be written as follows:

$$H_{11} = E_0 - 4A_1 \cos\xi \cos\eta + 4A_2 \cos\zeta(\cos\xi + \cos\eta);$$
$$H_{22} = E_0 - 4A_1 \cos\eta \cos\zeta + 4A_2 \cos\xi(\cos\eta + \cos\zeta);$$
$$H_{33} = E_0 - 4A_1 \cos\xi \cos\zeta + 4A_2 \cos\eta(\cos\xi + \cos\zeta);$$
$$H_{44} = E_0 + \Delta + 4A_4 \cos\xi \cos\eta - 4A_5 \cos\zeta(\cos\xi + \cos\eta);$$
$$H_{55} = E_0 + \Delta - \frac{4}{3}(A_4 + 4A_5)\cos\xi\cos\eta + \\ + \frac{4}{3}(2A_4 - A_5)\cos\zeta(\cos\xi + \cos\eta);$$
$$H_{12} = H_{21} = -4A_3 \sin\xi \sin\zeta;$$
$$H_{13} = H_{31} = -4A_3 \sin\eta \sin\zeta;$$
$$H_{23} = H_{32} = -4A_3 \sin\xi \sin\eta;$$
$$H_{14} = H_{41} = 0;$$
$$H_{24} = H_{42} = -4A_6 \sin\eta \sin\zeta;$$
$$H_{34} = H_{43} = 4A_6 \sin\xi \sin\zeta;$$
$$H_{15} = H_{51} = -(8/\sqrt{3})\, A_6 \sin\xi \sin\eta;$$
$$H_{25} = H_{52} = (4/\sqrt{3})\, A_6 \sin\eta \sin\zeta;$$
$$H_{35} = H_{53} = (4/\sqrt{3})\, A_6 \sin\xi \sin\zeta;$$
$$H_{45} = H_{54} = (4/\sqrt{3})(A_4 + A_5)\cos\zeta(\cos\eta - \cos\xi),$$

where $\xi = \pi k_x$, $\eta = \pi k_y$, and $\zeta = \pi k_z$.

The block of matrices (105), corresponding to the matrix element (PW–PW), has the following form:

$$\begin{pmatrix} \beta+\alpha\,|\mathbf{k}|^2 & V(4)F_2 & V(3)F_3 & V(3)F_4 \\ V(4)F_2 & \beta+\alpha\,|\mathbf{k}+\mathbf{K}_2|^2 & V(3)F_2F_3 & V(3)F_2F_4 \\ V(3)F_3 & V(3)F_2F_3 & \beta+\alpha\,|\mathbf{k}+\mathbf{K}_3|^2 & V(4)F_3F_4 \\ V(3)F_4 & V(3)F_2F_4 & V(4)F_3F_4 & \beta+\alpha\,|\mathbf{k}+\mathbf{K}_4|^2 \end{pmatrix}. \tag{107}$$

Thus, the matrix elements entering into this block depend on two Fourier components of the potential $V(3)$ and $V(4)$, and also on the quantities α and β. The occurrence of the symmetrization factors F_2, F_3, and F_4 is related to the fact that the bases that contain plane waves and symmetrized plane waves are essentially different from each other. At the points of high symmetry these factors are equal to zero or to unity, depending on whether or not the given plane wave enters into the symmetrized combination of plane waves having the lowest energy at the given point. If the factors F_2, F_3, and F_4 had not been taken into account then, in the calculated energy bands, undesirable shifts and splittings would have appeared. The values of symmetrization factors for the points of high symmetry in the Brillouin zone are given in Table 8. For the points inside the Brillouin zone, the following possible values of the symmetrization factors may be used:

$$F_2 = \frac{k_y - k_x}{k_y - k_x + 0.001\cos(\pi k_y)}\,(2k_y - k_y^2),$$

$$F_3 = \frac{k_x}{k_x + 0.001\cos(\pi k_x)\cos\left(\frac{2}{3}\pi k_x\right)}\left[\frac{12(k_x+k_y+k_z) - 4(k_x+k_y+k_z)^2}{9}\right],$$

$$F_4 = \frac{k_x - k_z}{k_x + k_z + 0.001\cos(\pi k_x)\cos\left(\frac{2}{3}\pi k_x\right)}\left[\frac{12(k_x+k_y) - 4(k_x+k_y)^2}{9}\right].$$

TABLE 8. Values of the Symmetrization Factors in the High-Symmetry Points of the Brillouin Zone

Symmetry point	$F_2(k)$	$F_3(k)$	$F_4(k)$
Γ (0, 0, 0)	0	0	0
X (0, 2, 0)	1	0	0
L (4, 4, 4)	0	1	0
K (3/4, 3/4, 0)	0	1	1
W (1/2, 1, 0)	1	1	1
U (1/4, 1, 1/4)	1	1	0

The matrix elements of the block (PW–LCAO) may be written as

$$H_{\mathbf{K},s} = B_2 j_2 (|\mathbf{k}+\mathbf{K}|B_1)\left[\frac{(\mathbf{k}+\mathbf{K})_\mu (\mathbf{k}+\mathbf{K})_\nu}{|\mathbf{k}+\mathbf{K}|^2}\right] F_{\mathbf{K}}(\mathbf{k}), \tag{108}$$

where s, μ, ν take, respectively, the values $(1, x, y)$, $(2, y, z)$, $(3, z, x)$ for the t_{2g} orbitals, and

$$\left.\begin{aligned} H_{\mathbf{K},4} &= B_3 j_2 (|\mathbf{k}+\mathbf{K}|B_1)\left[\frac{(\mathbf{k}+\mathbf{K})_x^2 - (\mathbf{k}+\mathbf{K})_y^2}{2|\mathbf{k}+\mathbf{K}|^2}\right] F_{\mathbf{K}}(\mathbf{k}), \\ H_{\mathbf{K},5} &= B_3 j_2 (|\mathbf{k}+\mathbf{K}|B_1)\left[\frac{3(\mathbf{k}+\mathbf{K})_z^2}{|\mathbf{k}+\mathbf{K}|^2} - 1\right] F_{\mathbf{K}}(\mathbf{k}) \end{aligned}\right\} \tag{109}$$

for the e_g orbitals. The matrix elements (106)–(109) contain 15 parameters. The values of these parameters are chosen by taking into account the results obtained in calculations of the energy bands of the metal, by applying the APW method and the method of Green's functions. The origin on the energy scale may be chosen so that the parameter β is equal to zero.

The interpolation method allows reproduction of the energy-band structure of the crystal with a mean square error of the order of several thousandths of a Rydberg. The wave functions obtained by Hodges and Ehrenreich [146] provide a good approximation to the density of electronic states in the valence band. The parameters that describe the energy-band structure of copper and silver were taken from the works of Stocks *et al.* [206, 207]. In the first article [206] on copper, these parameters were determined from the calculations of the copper band structure by Burdick [208], using the APW method with the potential proposed by Chodorow [209]. By comparing the results of calculations performed with different approximations [210], it can be concluded that the Chodorow potential leads to energy spectra that show the best agreement with experiment. For silver, the energy band structure has been calculated by the method of Green's functions, with a model of the crystal potential proposed by Mattheiss [211] and a Slater-type exchange. For copper, it is not yet necessary to take into account the spin–orbit splitting, since the experimental spectra were recorded with a resolution of 0.6–0.9 eV, but the spin–orbit splitting amounts to only 0.2 eV. For silver, however, it is desirable to take into account the spin-orbit interaction, since the spin–orbit splitting may reach values of the order of 0.5–0.6 eV.

Since the *d*-wave functions of valence band electrons in these crystals are localized mainly in the region 1.4–1.5 a.u., it follows that $qr \approx 0.6 < 1$, and therefore the dipole approximation may be used. This statement is also supported by the great similarity between the X-ray photoelectron spectra obtained at excitation energies 100–1486.6 eV. By using the plane waves

$$\varphi_{\mathbf{k}+\mathbf{Q}}(\mathbf{r}) = \frac{\exp i(\mathbf{k}+\mathbf{Q})\cdot\mathbf{r}}{\sqrt{N\Omega_0}}$$

as the wave functions of the electrons in the conduction band, formula (102) becomes for the dipole approximation:

$$I(\omega, E) \approx \sum_{n,\mathbf{k},\mathbf{Q}} \sum_{\mu;\mu'} C_{\mu}^{n*}(\mathbf{k})\, C_{\mu'}^{n}(\mathbf{k})\, \tilde{Y}_{2\mu}(\Theta_{\mathbf{k}+\mathbf{Q}}, \varphi_{\mathbf{k}+\mathbf{Q}}) \times$$

$$\times \tilde{Y}_{2\mu'}(\Theta_{\mathbf{k}+\mathbf{Q}}, \varphi_{\mathbf{k}+\mathbf{Q}})\,(A, (\mathbf{k}+\mathbf{Q}))^2\, B_d^2(|\mathbf{k}+\mathbf{Q}|), \qquad (110)$$

where

$$\tilde{Y}_{2\mu}(\Theta_{\mathbf{k}+\mathbf{Q}}, \varphi_{\mathbf{k}+\mathbf{Q}})$$

represents the linear combinations of spherical harmonics corresponding to the angular dependences of e_g and t_{2g} orbitals:

$$B_d(|\mathbf{k}+\mathbf{Q}|) = \int_0^\infty dr r^2 j_2(|\mathbf{k}+\mathbf{Q}|r|)\, R_{nd}(r).$$

In using the functions given by the expression (104), because $K_i << Q$, the contribution of s and p electrons to the energy distribution of electrons expressed by equation (110) is equal to zero. This fact, however, is not obviously reflected in the calculated results, because the contribution of the partial densities of s and p electrons to the total density of electronic states in the valence band of copper, silver, and palladium is low [206, 207].

At an acceptance angle of photoelectrons in the electron spectrometer equal to 8°, the summation in formula (110) is performed over all the states (n, k) in the Brillouin zone compatible with the law of conservation of energy. In this case, the law of conservation of electron momentum is satisfied for all the points k in the Brillouin zone. Expression (110) can be significantly simplified if the summation over the whole Brillouin zone is replaced by the summation over its irreducible $\frac{1}{48}$ part. If the symmetry properties of electron wave functions are taken into account, it can be shown that the cross terms in formula (110) compensate each other, so that the intensity of the photoelectron spectrum can be written as follows:

$$I(\omega, E, P) \sim \sum_{\mathbf{k},n} [2C_{t_{2g},n}(\mathbf{k})\,(\tilde{Y}_1^2 + \tilde{Y}_2^2 + \tilde{Y}_3^2)$$

$$+ 3C_{e_g,n}(\mathbf{k})\,(\tilde{Y}_4^2 + \tilde{Y}_5^2)]\ \sigma_d(E), \qquad (111)$$

where

$$C_{t_{2g},n}(\mathbf{k}) = C_{1,n}^2(\mathbf{k}) + C_{2,n}^2(\mathbf{k}) + C_{3,n}^2(\mathbf{k}),$$

$$C_{e_g,n}(\mathbf{k}) = C_{4,n}^2(\mathbf{k}) + C_{5,n}^2(\mathbf{k}),$$

$$\sigma_d(E) = B_d^2(|P|) = B_d^2(E).$$

In formula (111), the summation is performed only over the electron quasi-momenta in $\frac{1}{48}$ part of the Brillouin zone. Therefore, the energy dependence of the angular distribution of photoelectrons is determined by the t_{2g} and e_g components of the density of states:

$$\rho_{t_{2g}} = \frac{1}{\Delta E} \sum_{n,k} C_{t_{2g},n}(\mathbf{k}),$$

$$\rho_{e_g} = \frac{1}{\Delta E} \sum_{n,k} C_{e_g,n}(\mathbf{k}).$$

The variation of the photoionization cross section of d electrons, σ_d, over the whole range of the occupied part of the valence band, at a photon energy of the order of 1500 eV, is not greater than 1%. At photoelectron emission along the (001) direction, the functions $\widetilde{Y}_1$, $\widetilde{Y}_2$, and $\widetilde{Y}_3$ become equal to zero for $\theta_p = \varphi_p = 0$, while $\widetilde{Y}_4$ and $\widetilde{Y}_5$ remain different from zero. In the case of photoemission along the (111) direction, and for $\theta_p = 54.7°$, $\varphi_p = 45°$, the functions $\widetilde{Y}_1 = \widetilde{Y}_2 = \widetilde{Y}_3$ are different from zero, while $\widetilde{Y}_4$ and $\widetilde{Y}_5$ are equal to zero. Therefore, in the first case, the photoelectron spectrum must be determined by the e_g component of the density of states, and in the second case by its t_{2g} component. One should notice that, for polycrystalline materials, the intensity of photoelectron spectra can be represented as follows:

$$I(\omega, E) \sim \sigma_d \rho_d(E).$$

For polycrystalline copper and silver, calculation of the density of states of d electrons has been done by using the interpolation method of Hodges and Ehrenreich. As Figures 62 and 63 show, the main features of the electron density of states of copper and silver are seen in the experimental spectra obtained by Hüffner *et al.* [192]. Their experiment was performed using an HP-5950A electron spectrometer with a resolution of 0.6 eV and a vacuum of $\sim 10^{-9}$ torr. However, the shape of the experimental and theoretical curves are significantly different. In the valence band photoelectron spectrum of copper, the intensity of the maximum at ~2.5 eV is greater than that of the maximum at ~3.5 eV, while in the density-of-states curve, the greatest maximum is that

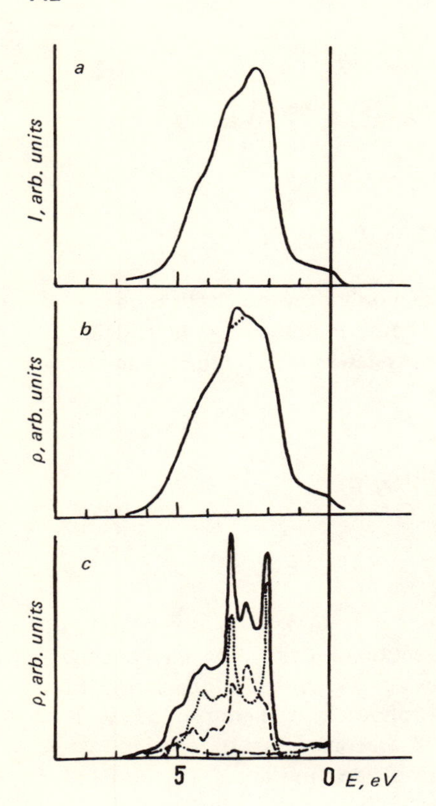

Figure 62. The X-ray photoelectron spectrum and the density of states of valence electrons in copper: (a) the photoelectron spectrum; (b) the density of states calculated: (——) without taking into account the spin-orbit interaction; (· · · · ·) by taking into account the spin-orbit interaction (the theoretical curves were smoothed by a dispersion curve with a half-width of 0.25 eV); (c) the density of states calculated without taking into account the spin-orbit interaction: (· · · · ·) t_{2g} states; (– – –) e_g states; (—·—·—) s states; (——) total.

situated nearer the Fermi level. Consideration of the spin-orbit splitting does not change significantly the density of states of valence electrons of copper. One possible explanation is that the discrepancy between the photoelectron spectra and the partial density of states of d electrons is due to relaxation effects, which might be different for electrons of t_{2g}- and e_g-symmetry type. One can admit that, in copper, the contribution of electrons of e_g-symmetry type is greater in the photoelectron spectrum than in the density of states. An increase in the density of e_g states near the Fermi level can be caused by a small energy shift of the position of e_g states with respect to t_{2g} states. Therefore, the t_{2g} and e_g components of the density of states should show, separately, a rather good consistency with the photoelectron spectra corresponding to photoemission along (111) and (001) directions, respectively. In the case of silver, as it results from Figure 63, it is necessary to consider also the relativistic effects. The density of states calculated by taking into account the relativistic effects [192] has a minimum that is absent in the density-of-states curve calculated without inclusion of the relativistic effects.

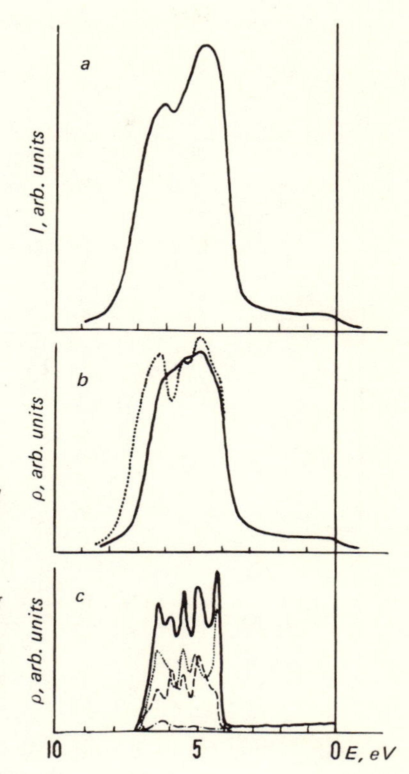

Figure 63. The X-ray photoelectron spectrum and the density of states of valence electrons in silver: (a) the photoelectron spectrum; (b) the density of states calculated: (——) without taking into account the spin-orbit interaction; (· · · · ·) by taking into account the spin-orbit interaction (the theoretical curves were smoothed by a dispersion curve with a half-width of 0.25 eV); (c) the density of states calculated without taking into account the spin-orbit interaction: (· · · · ·) t_{2g} states; (– – –) e_g states; (—·—·—) s states; (——) total.

The discrepancy between the photoelectron spectrum and the density of states of valence electrons in silver consists in a different intensity ratio of the two main maxima. In order to illustrate the energy distribution of s, t_{2g}, and e_g electron states throughout the valence band of copper and silver, the partial density of these states is also shown in Figures 62 and 63. Figures 64 and 65 show the theoretical curves and the experimental data for photoelectron emission along (001) and (111) directions in copper, and along (100) and (111) directions in silver [211a, 211b]. One should mention that, for cubic crystals, the (001) and (100) directions are equivalent. The agreement between the theoretical and experimental data for copper and silver is quite good. One should notice that in copper the maximum in the photoelectron spectrum corresponding to e_g states is shifted toward the Fermi level, as compared to the corresponding maximum in the partial density of e_g states.

Dobbyn *et al.* [213] and Liefeld [214] have studied the M and L X-ray emission spectra of copper. As can be seen from Figure 66, the L band is narrower than the M band. This is explained by the authors [213] as being due

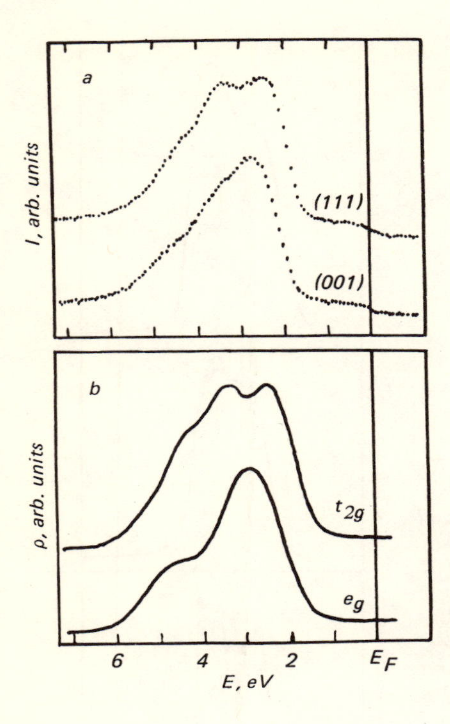

Figure 64. (a) X-ray photoelectron spectra of copper along the (111) and (001) directions. (b) Calculated partial density of states of electrons with t_{2g} and e_g symmetry. The theoretical curves were smoothed by a Gaussian curve with a half-width of 0.4 eV.

to a different energy dependence of the radial wave functions for 3*p* and 2*p* electrons of copper. However, the work of Hüfner *et al.* [193] indicates that the width of the copper 3*p* level is equal to 1.4 eV, and that of the 2*p* level is equal to 0.8 eV. This result is essentially different from that obtained by Dobbyn *et al.* [213]. Therefore, although the transition probability has an influence on the width of the X-ray emission band, if in this case the calculations of Hüfner *et al.* [193] are correct, it follows that the difference in the width of the *L* and *M* bands is mainly related to the different widths of the 3*p* and 2*p* levels.

Wehner *et al.* [214a] have studied the valence band photoelectron spectra of silver using synchrotron radiation in the energy range 32–250 eV. This offered the possibility to observe the intensity change of photoelectron spectra caused by a specific atomic effect due to the radial singularity of the 4*d* electron wave function of silver, since in this case there exists an energy range in which the photoionization cross section has a strong energy variation. The valence band electron spectra of silver, obtained by using synchrotron radiation, are consistent with those obtained by using the K_α radiation of aluminum. In the region

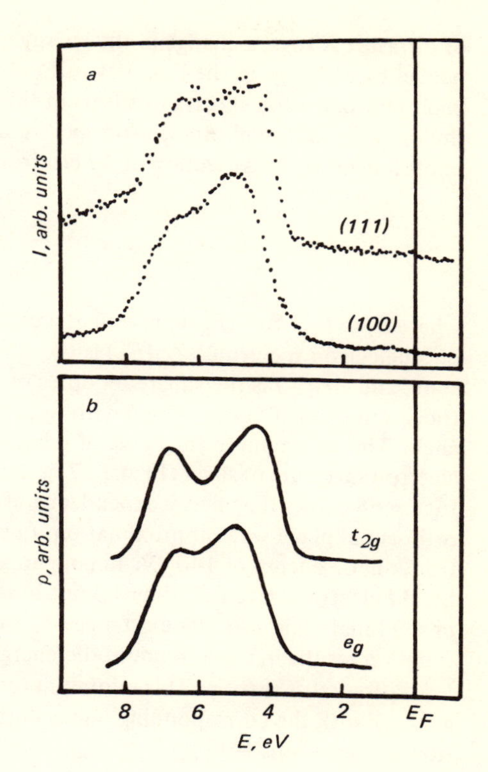

Figure 65. (a) The X-ray photoelectron spectra of silver along the (111) and (001) directions. (b) Calculated partial density of states of electrons with t_{2g} and e_g symmetry. The theoretical curves were smoothed by Gaussian curves with a half-width of 0.4 eV in the case of t_{2g} states, and 0.5 eV in the case of e_g states.

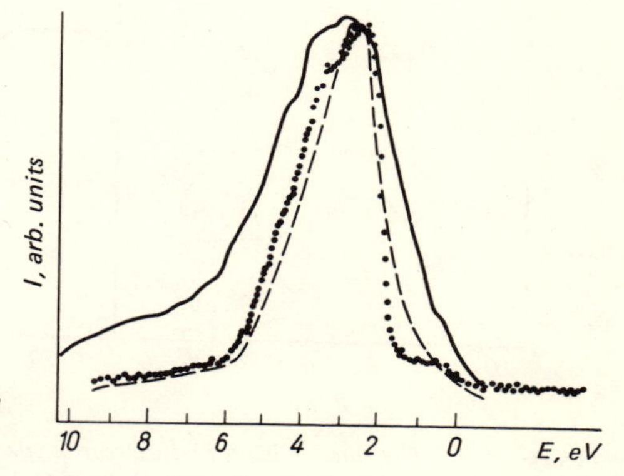

Figure 66. $M_{2,3}$ (——) and $L_{2,3}$(– – –) X-ray emission spectra, and valence band photoelectron spectrum (· · ·) of copper.

of ultrasoft X rays, a change in the energy of the exciting photons is accompanied by a change in the intensity ratio of the two main peaks in the valence band photoelectron spectra of silver. In this energy region of exciting radiation, the intensity of the photoelectron spectrum of silver is proportional to the photoionization cross section of 4*d* electrons:

$$I(\omega) \sim \sigma(\omega, E) \sim \frac{1}{\omega} \langle |t_{fi}(E_f)|^2 \rangle N_f(E_f),$$

where $E_f = \overline{E}_j + \hbar\omega$ ($\overline{E}_j$ represents the center of gravity of the valence band photoelectron spectrum), $N_f(E_f)$ is the electron density of states in the final state, and $\langle |t_{fi}|^2 \rangle$ is the square modulus of the matrix element of the electron transition probability, averaged with respect to the photoelectron emission angle. The minimum in the curve of relative intensity of the valence band photoelectron spectrum of silver (Figure 67) is due to the energy dependence of $\langle |t_{fi}|^2 \rangle$. Figure 68 shows the energy dependence of $\langle |t_{fi}|^2 \rangle$ for silver and copper in the orthogonal plane wave approximation. In silver, the minimum occurs at an electron kinetic energy of 140 eV. In copper, such a minimum does not exist, since the 3*d* electron wave functions do not have radial singularities. In conclusion, in solid materials also, the existence of a radial singularity of the electron wave function results in a minimum in the energy dependence of the electron photoionization cross section. This minimum can be named "Cooper minimum," in analogy with the corresponding one occurring in the photoionization cross section of free atoms [91].

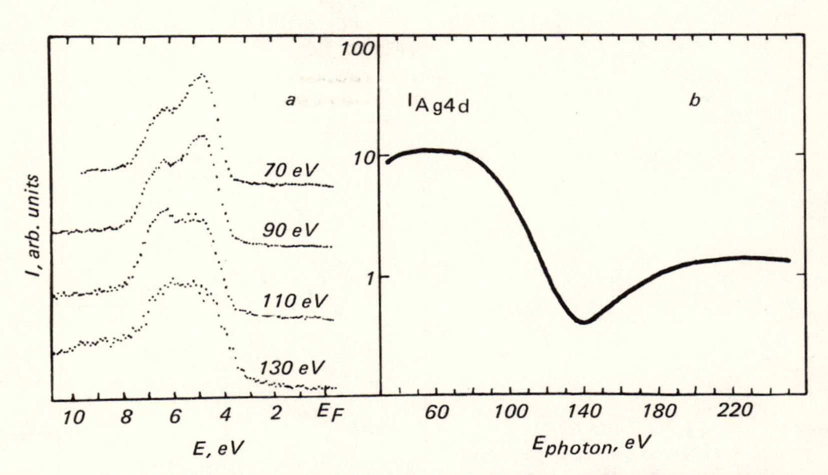

Figure 67. (a) The valence band photoelectron spectra of silver obtained at an exciting photon energy of 70 eV, 90 eV, 110 eV, 130 eV. (b) The relative intensity of the 4d peaks in the photoelectron spectra of silver.

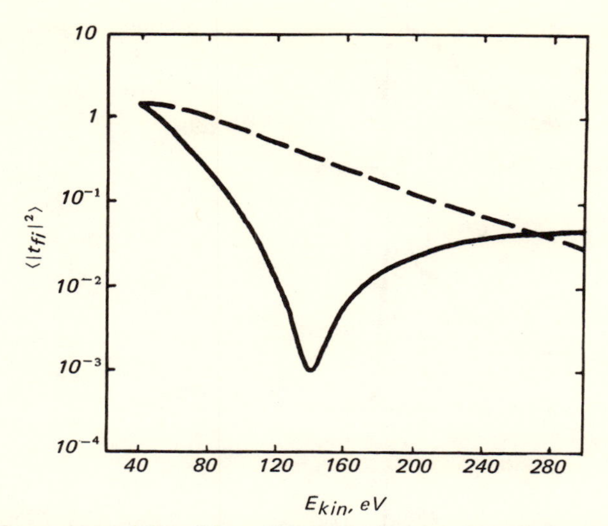

Figure 68. The square modulus of the radial matrix element of the electron transition probability, as a function of the electron kinetic energy for silver (——) and copper (– – –).

Hüfner and Wertheim [215] have studied the valence band photoelectron spectra of a series of transition metals from manganese to copper, using an HP-5950A electron spectrometer with a resolution of 0.5 eV. The measurements were performed with a vacuum of 10^{-9} torr. For all of the metals investigated, with the exception of manganese, the detected oxygen and carbon signals were low and did not increase significantly even after 24 hours of continuous operation of the spectrometer. In the case of manganese, an oxide layer had already formed on the sample surface after 2 hours. This limited the duration of spectra recording and consequently led to statistically poorer results. Figure 69 shows the experimental valence band photoelectron spectra of manganese, iron, cobalt, nickel, and copper. As has already been pointed out, the fine structure of the X-ray photoelectron spectrum of copper successfully reproduces the energy position of the singular points in the density of states of copper. For ferromagnetic nickel, the calculations of Zornberg [216] have predicted that certain particular features should appear in the X-ray photoelectron spectrum at 0.4, 1.6, and 3.2 eV, and in fact these predictions agree with the experimental data. For ferromagnetic cobalt, Wong *et al.* [217] have predicted peaks situated at 0.5, 1.8, and 2.5 eV, and these are also found in the experiment. For manganese, the density of states of valence electrons calculated by Yamashita *et al.* [218] does not show any significant structure, which again is confirmed by experiment. The broad peak at 6 eV in nickel is possibly due to plasma oscillations [186]. In approximately the same energy region, weak structures have also been

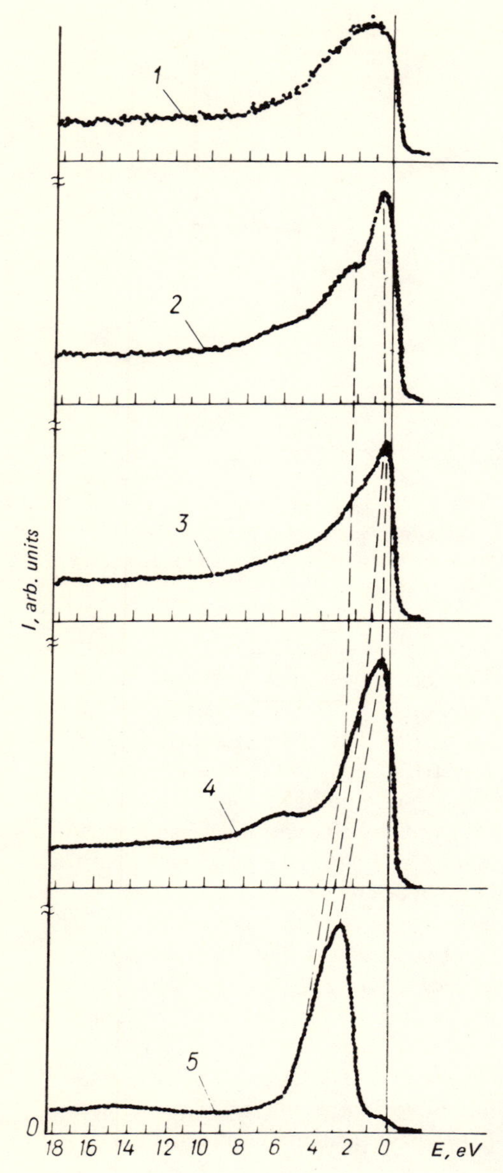

Figure 69. Valence band photoelectron spectra of (1) manganese; (2) iron; (3) cobalt; (4) nickel; (5) copper.

observed in the X-ray photoelectron spectra of cobalt and iron. The dotted line in Figure 69 connects the corresponding features observed in the spectra of copper, nickel, cobalt, and iron.

Hüfner *et al.* [219] have measured the valence band photoelectron spectra of rhodium, palladium, silver, and gold (Figure 70) using an HP-5950A electron spectrometer with a resolution of 0.5 eV, and the $K\alpha_{1,2}$ line of aluminum as exciting radiation. The specimens were prepared as evaporated thin films, or as foils cleaned by argon ion bombardment. The spectra of silver and gold differ only slightly from the spectra shown in Figures 63 and 57. For rhodium, the band structure calculated by Christensen [220] is completely consistent with the experimental data. Also shown in Figure 70 are the spectra of iridium and

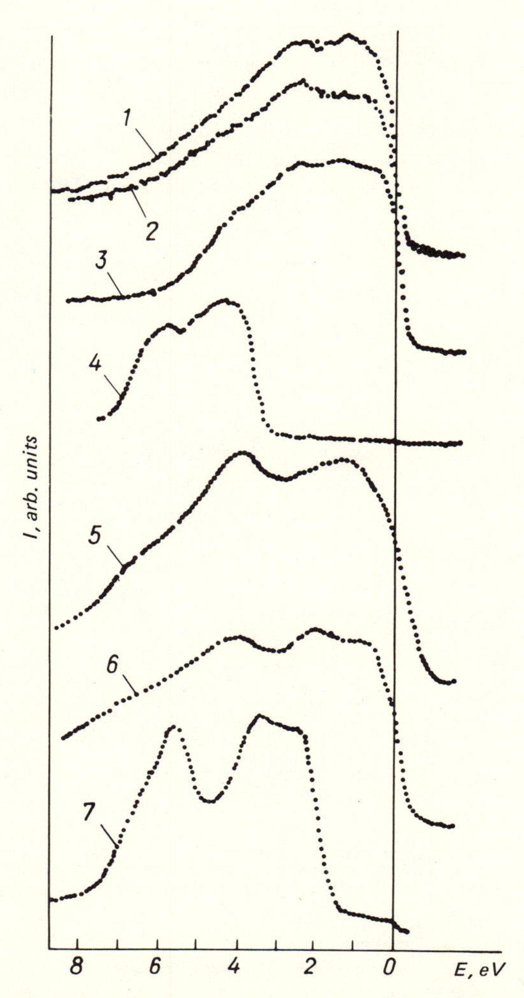

Figure 70. Valence band photoelectron spectra of (1) rhodium (foil, cleaned by argon ion bombardment); (2) rhodium (evaporated onto a substrate); (3) palladium; (4) silver; (5) iridium; (6) platinum; (7) gold.

palladium measured by Kowalczyk *et al.* [221] on high-purity monocrystalline samples. Before starting the experiment, the specimens were mechanically polished and then electropolished. Inside the spectrometer, the cleaning operation was continued by argon ion bombardment. During the measurements the vacuum level in the spectrometer chamber was 8×10^{-9} torr. From Figure 70 it can be seen that the spectra of iridium and platinum are rather similar to the spectrum of gold, which demonstrates the great similarity between the energy band structures of these metals.

The same group of authors have also studied the X-ray photoelectron spectra of the series of elements from palladium to xenon, in the energy region in which the 4*d*, 5*s*, and 5*p* electrons of these elements are localized. The X-ray photoelectron spectra of all of these elements, excepting iodine, were measured using an HP-5950A electron spectrometer provided with a monochromatized source of $K\alpha_{1,2}$ radiation of aluminum. All the samples were monocrystals of high purity except antimony and iodine, which were polycrystalline. The method of sample preparation was similar to that used in the work mentioned previously [221].

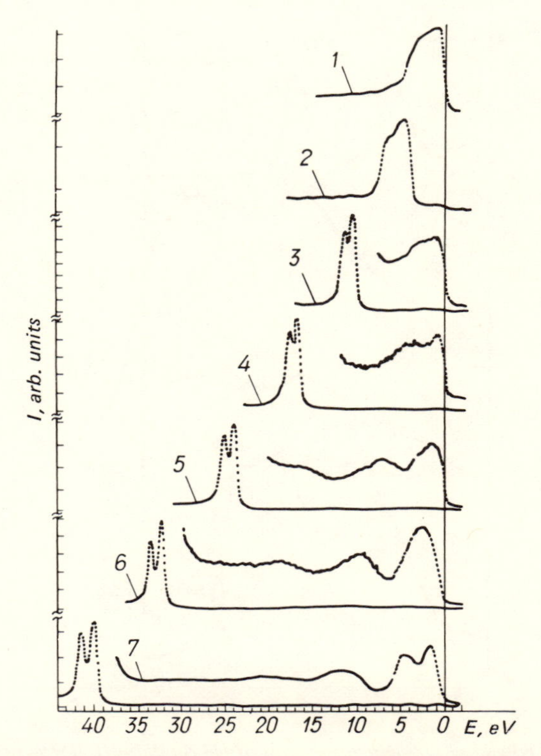

Figure 71. Valence band photoelectron spectra of (1) palladium; (2) silver; (3) cadmium; (4) indium; (5) tin; (6) antimony; (7) tellurium.

The experimental resolution in all these cases, except for iodine, amounted to 0.55 eV. Iodine was studied by using an iron-free electron spectrometer constructed at Berkeley. The doublet structure of the $4d$ levels of these elements is clearly seen in Figure 71. Figure 72 shows the energy distance between the doublet components as well as the theoretical values obtained from the data of Carlson *et al.* for free atoms [223]. The values in Figure 72 are matched so that for antimony the experimental and theoretical values of the splitting coincide.

Such a matching procedure yields rather good values of the spin-orbit splitting for the investigated crystals. This has been confirmed, for xenon, for example, by the good agreement between the magnitude of spin-orbit splitting, determined by extrapolating the theoretical curve and its value determined experimentally for xenon gas [2]. This normalization lowers the value of the splitting for free atoms by 0.1 eV. The intensity ratio of the two components of the doublet is of the order of 1.3, which is consistent with the value 1.5 resulting from the multiplicity of $4d_{3/2}$ and $4d_{5/2}$ states.

A lack of agreement between the theoretical and the experimental values of the splitting has been observed for cadmium and silver. This indicates how important it is for these metals to account for the effects related to the structure of energy bands. Figure 72 also shows the variation of the full-width-at-half-maximum of the $4d$ line, for the series of elements investigated. This quantity increases for elements situated after indium, which is explained by the increase of the magnitude of spin-orbit interaction in these elements. For cadmium, the linewidth is greater than for indium, which indicates the presence of band-structure effects.

The fine structure of $5s$ and $5p$ energy bands has been studied for cadmium, indium, tin, antimony, and tellurium. In the case of tin and antimony, these

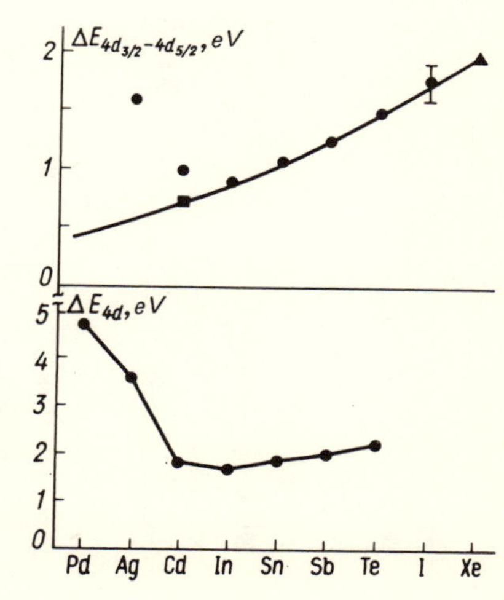

Figure 72. Splitting of $4d_{3/2}$ and $4d_{5/2}$ photoelectron lines and their full width in the series of elements from palladium to xenon. The value for cadmium, indicated by ■, was obtained for cadmium atoms evaporated on the (111) surface of gold.

TABLE 9. Binding Energies of 4*d*, 5*s*, and 5*p* Electrons for Elements from Silver to Xenon

Element	State				Energy splitting of 4*d* states	Ratio of the areas for the 5*s* and 5*p* states	Ratio of the occupation numbers n_s/n_p
	$4d_{3/2}$	$4d_{5/2}$	5*s*	5*p*			
Silver	–	–	–	–	1.6	–	–
Cadmium	11.5	10.5	2.2	–	1.0	–	–
Indium	17.6	16.7	4.1	0.7	0.9	1.8	2.0
Tin	24.8	23.7	7.0	1.2	1.1	0.7	1.0
Antimony	33.4	32.1	9.1	2.3	1.2	0.6	0.7
Tellurium	41.8	40.3	11.5	4.0 1.1	1.5	0.5	0.5
Iodine	–	–	–	–	1.8	–	–
Xenon	–	–	–	–	2.0	–	–

energy bands contain two peaks, while in the case of tellurium they exhibit three peaks. In indium, tin, antimony, and tellurium, the band that is closest to the Fermi level is the 5*p* band, and therefore the peak in the spectrum situated at the greatest binding energy corresponds mainly to the 5*s* states.

Figure 71 shows how the atomization of 5*s* electrons occurs as one progresses in the series of elements. From the data in Table 9, it follows that the ratio of the areas under the characteristic 5*s* and 5*p* lines is the same as the ratio of the occupation numbers for the free atoms with configuration $5s^n 5p^m$. Figure 73 shows the position relative to the Fermi level and the width at half-height of the 5*s* and 5*p* lines, for the series of elements investigated. For elements from cadmium to tellurium, all the energy bands are significantly broader than the corresponding levels in xenon. The splitting of 5*p* states in tellurium is due to the formation of energy bands, since the magnitude of the splitting (2.9 eV) is much greater than the spin-orbit splitting in xenon (1.3 eV).

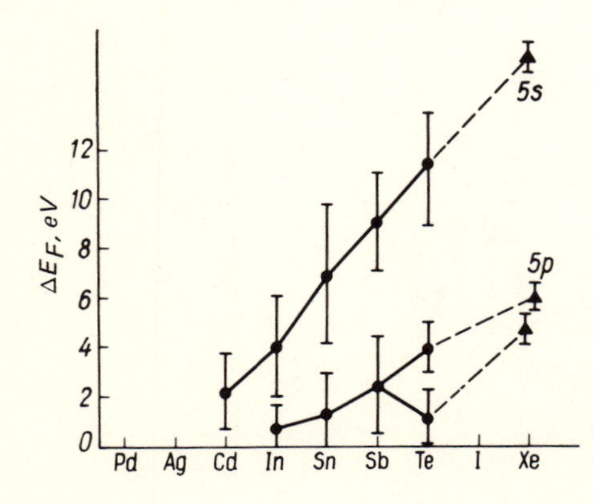

Figure 73. Position of the 5s and 5p bands relative to the Fermi level. The vertical bars represent the bandwidth at half-maximum.

For the elements in the sixth period of the periodic system—thallium, lead, and bismuth—in contrast to lighter elements, the spin-orbit splitting of valence energy bands is of the same order of magnitude as the splitting due to the crystal field. The X-ray photoelectron spectra of the valence electrons in these crystals have been measured by Ley *et al.* [224] using an HP-5950A electron spectrometer and the monochromatized $K\alpha_{1,2}$ radiation of aluminum. Cleaning of the sample surfaces was performed by argon ion bombardment. The top of the valence band is mainly occupied by $6p$ states, while at the bottom of the valence band, mainly $6s$ states are localized. Figure 74 shows that, when passing from thallium to bismuth, the atomization of $6s$ states takes place, while the $6p$ states undergo splitting. The spin-orbit splitting at a number of high-symmetry points of the Brillouin zone has been calculated for these metals by Soven [225], Loucks [226], and Ferriera [227]. Typical values for this splitting are 0.3 eV for thallium, 1.4 eV for lead, and 2.0 eV for bismuth. As seen in Table 10, the calculated values are consistent with the experimental ones. This indicates that the spin-orbit interactions in the valence band lead to greater splitting than the crystal field. Table 10 also shows that the value of spin-orbit splitting of $6p$ states, calculated by Lu *et al.* [228] for the free atom of bismuth, also agrees well with the experimental data obtained for crystalline bismuth.

For the elements indicated in Table 10, besides the spin-orbit splitting, account should also be taken of other relativistic effects, i.e., the dependence of mass on the velocity of light, and the Darwin correction.

Comparison of binding energy values for the s bands in thallium, lead, and

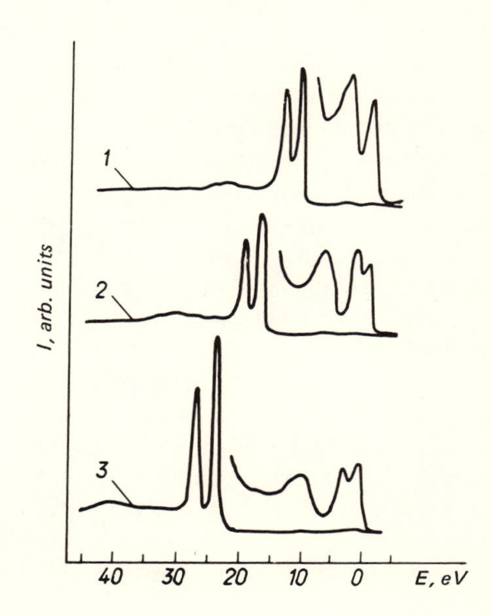

Figure 74. Valence band photoelectron spectra of (1) thallium, (2) lead, and (3) bismuth.

TABLE 10. Binding Energies of Valence Electrons of Tl, Pb, and Bi

Parameter of the electron structure	Thallium	Lead	Bismuth
Binding energy of $5d_{3/2}$ electrons	14.5	20.3	26.9
Binding energy of $5d_{5/2}$ electrons	12.3	17.7	23.9
Energy splitting of $5d$ states	2.2	2.6	3.0
Energy splitting of $5d$ states in free atoms	2.4	2.8	3.3
Binding energy of $6s$ electrons	4.9	7.7	9.9
Binding energy of $6p$ electrons	0.8	2.3	3.3
		0.5	1.2
Energy splitting of $6p$ states	–	1.8	2.2
Energy splitting of $6p$ states in free atoms	–	–	2.2

bismuth with those in indium, tin, and antimony indicates that in heavy elements the binding energy of the s electrons increases. This also is related to relativistic effects.

Characteristics of the Structure of X-Ray Photoelectron Spectra of the Rare-Earth Metals

The physical and chemical properties of the rare-earth elements can be explained mainly by the fact that they have an unfilled $4f$ shell. Energetically, the $4f$ electrons are situated in the valence band of the crystal, but because of the high orbital quantum number l, the wave functions of the f electrons are rather localized. The basic configuration for the atoms of the rare-earth group in crystalline form is $4f^n 5d^1 6s^2$, with the exception of europium and ytterbium whose configurations are $4f^7 6s^2$ and $4f^{14} 6s^2$, respectively. It should be mentioned here that numerous attempts have been made to study the structure of the $4f$ states of these elements by optical methods, as well as by ultraviolet spectroscopy. These attempts have not, however, been successful because of the low value of the probability of transition from the valence band to the conduction band. Hagström and co-workers have studied the valence band photoelectron spectra of ytterbium [229], neodymium, samarium, dysprosium, erbium [230], europium, gadolinium, lutetium, and holmium [231]. The photoelectron spectra of the rare-earth metals were measured in a vacuum of 10^{-7} torr, using an electron spectrometer similar to that described by Siegbahn *et al.* [1]. The samples were prepared by evaporation *in situ*. As excitation radiation, the $K\alpha_{1,2}$ line of aluminum was used. Formation of oxide layers on the sample surface was already detected 5–10 min after evaporation. Therefore, during the experiment, a series of repeated evaporations was performed. Because of these circumstances, the data obtained from the analysis of the above-mentioned experiments [229–231] offer only a general representation of the energy localization of $4f$ electrons in the valence band of rare-earth metals.

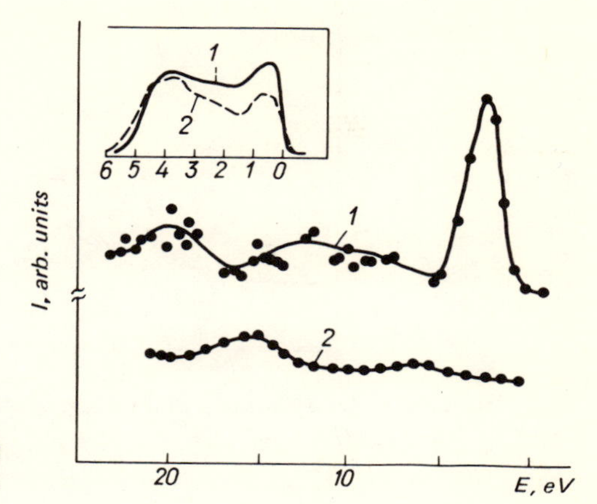

Figure 75. Valence band photoelectron spectra of (1) europium and (2) barium.

Brodén *et al.* [232] have shown that the valence band photoelectron spectra of barium and europium at an excitation energy of 7.7 eV are very similar to each other (see Figure 75), which means that at this value of the excitation energy, the $4f$ electrons have a low photoionization cross section. The photoionization cross section of f, d, and s electrons have a different dependence on the energy of the exciting photons. In the case of atoms of the rare-earth elements, the photoionization cross section for different subshells increases with increasing aximuthal quantum number l [232]. Therefore, the principal features in the structure of X-ray photoelectron spectra of rare-earth metals are due to the excitation of $4f$ electrons. Figure 75 shows also the valence band photoelectron spectrum of barium, in which the outer subshell of the free atom is characterized by the configuration $6s^2$, and also the spectrum of europium, this element being characterized by the configuration $4f^7 6s^2$. The X-ray photoelectron spectrum of barium up to about 15 eV does not exhibit any clear structure related to the s states. The maximum at about 15 eV corresponds to the excitation of $5p$ electrons. In the spectrum of europium, a clear maximum is observed at a binding energy of about 2.1 eV, corresponding to the $4f$ electrons. The X-ray photoelectron spectra of the metals with completely filled or half-filled $4f$ shells, such as europium, gadolinium, ytterbium, and lutetium, have a rather simple structure.

A number of rare-earth metals have been studied under ultrahigh vacuum (5×10^{-12} torr) using the AEI-100 photoelectron spectrometer. The samples were prepared by evaporation onto a quartz substrate, at a pressure of the order of 10^{-10} torr. The valence band photoelectron spectra of terbium, holmium, thullium, and ytterbium were measured with a resolution of about 0.8 eV, using the $K\alpha_{1,2}$ line of magnesium as excitation source. In the case of samarium, dysprosium, and erbium, a resolution of 0.3 eV was achieved using

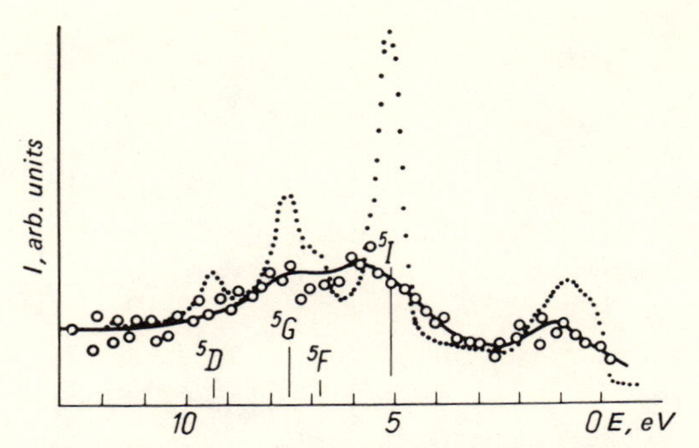

Figure 76. Valence band photoelectron spectra of samarium, according to the data from [230] (···) and [233] (——).

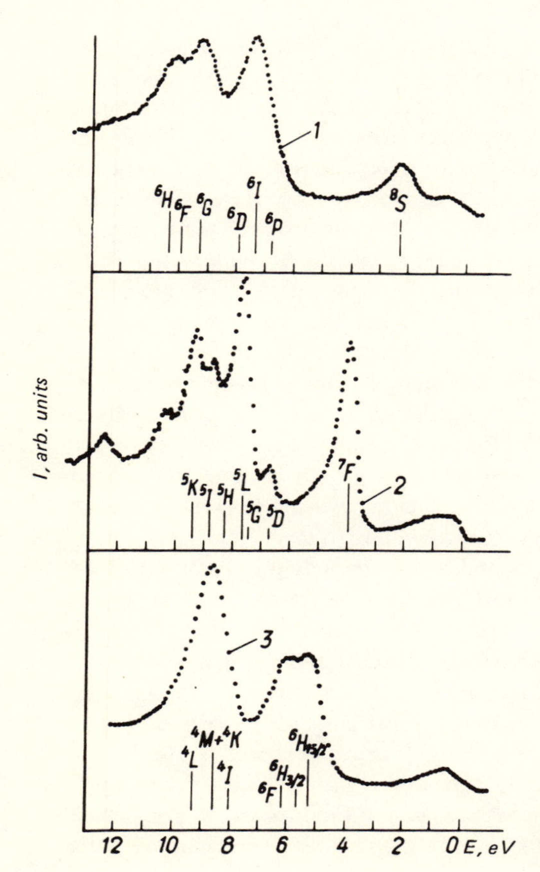

Figure 77. Valence band photoelectron spectra of (1) terbium; (2) dysprosium; (3) holmium.

the $K\alpha_{1,2}$ line of aluminum, a rotating anode, and a monochromator with a spherically bent quartz crystal. The photoelectron spectra of samarium measured by Hedén *et al.* [230] and by Baer and Busch [233], illustrated in Figure 76, are chosen as an example to show how drastically the photoelectron spectra of rare-earth metals are affected by the resolution of electron spectrometers and by the quality of vacuum. The spectrum measured in better experimental conditions is characterized by a fine structure that reflects the multiplet structure caused by electron emission from the incompletely filled $4f^n$ shell.

Figures 77 and 78 show the valence band photoelectron spectra of the rare-earth metals measured by Baer and Busch [233]. Cox *et al.* [235] have calculated the intensity of photoelectrons emitted from the $4f^n$ shell, for different final states of the $4f^{n-1}$ shell. In these calculations it was assumed that the intensity of photoelectron lines is proportional to the square of the fractional parentage coefficient. The calculated data reproduce well the observed values of the intensities of electron transitions from the $4f^n$ shell to the free states, and

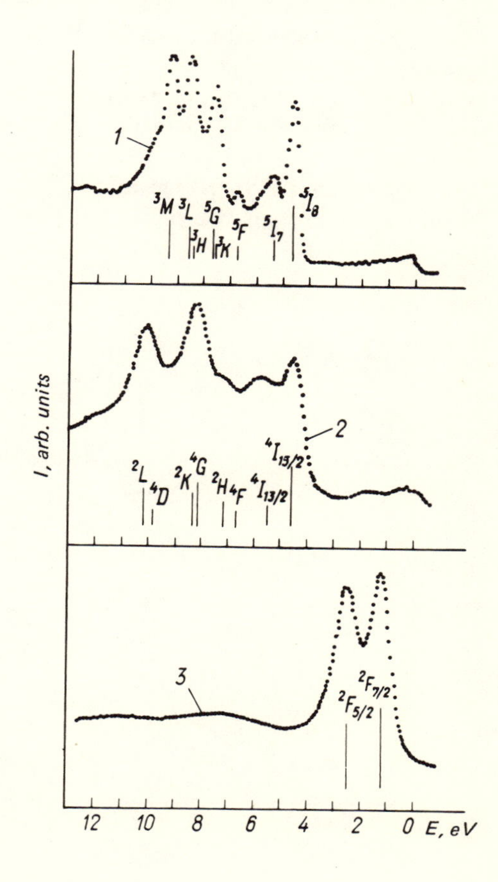

Figure 78: Valence band photoelectron spectra of (a) erbium; (2) thulium; (3) ytterbium.

also the splitting energy of the multiplet states. McFeely *et al.* [236] have shown that the energy split of final states in the far-ultraviolet absorption spectra is 8–13% lower than the value observed in X-ray photoelectron spectroscopy. This discrepancy is explained as being caused by a stronger intra-atomic potential due to the additional positive charge felt by the electrons in the photoemission final state.

The valence band in rare-earth metals is made up of the hybridized 6*s* and 5*d* states. In the vicinity of the Fermi level, the partial density of *d* states is significantly higher than the partial density of *s* states. In the study of valence band photoelectron spectra, it is important to use monochromatized radiation. Otherwise, the high-energy satellites corresponding to transitions from the $4f^n$ shell will hinder appreciably the observation of the structure of electron spectra near the Fermi level. Figure 79 illustrates the electron density of states near the Fermi level, calculated by Keeton and Loucks [237] taking into account the distortion introduced by the instrument. The figure also shows the corresponding X-ray photoelectron spectra. Correct determination of the energy position of the 6*s* and 5*d* valence band is difficult because of the background of electron inelastic scattering and because of the tails of 4*f* photoelectron lines. As can be seen from Figure 79, the photoelectron spectra reproduce only the most significant changes in the electron density of states on passing from gadolinium to erbium.

Fuggle *et al.* [238] have measured the valence band photoelectron spectra of uranium and thorium by using an ESCA-3 electron spectrometer equipped with

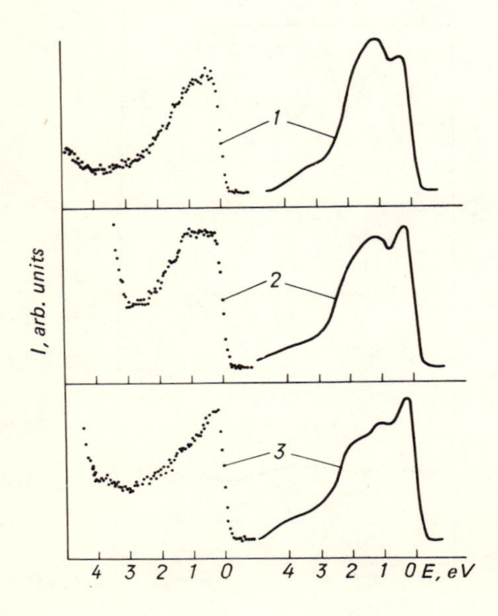

Figure 79. Valence band photoelectron spectra (· · ·) and the density of states (——) of (1) gadolinium, (2) dysprosium, (3) erbium.

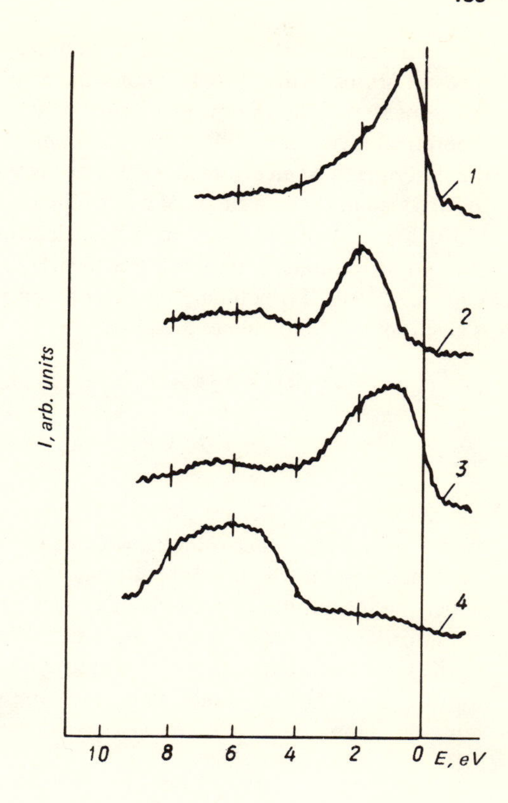

Figure 80. Valence band photoelectron spectra of (1) uranium; (2) uranium oxide; (3) thorium; (4) thorium oxide.

a magnesium $K\alpha_{1,2}$ radiation source. The specimens were prepared by evaporation inside a special sample preparation chamber, at a pressure of 10^{-8} torr. During the measurements, the pressure in the analyzer was lower than 5×10^{-9} torr. It should be pointed out that the uranium and thorium specimens had an exceptionally high oxidation reactivity. Figure 80 shows the spectra of uranium and thorium in pure metallic state and in oxidized state. The oxidation process was performed at a pressure of 10^{-3} torr. The spectrum of thorium exhibits a maximum on the side of higher binding energies, which is due to plasma energy losses.

Influence of Transition Probability on the Shape of X-Ray Photoelectron Spectra of Alloys of Transition Metals

Electron spectroscopy has made it possible to show experimentally that, in the disordered binary alloys of the transition and noble metals, the *d* states of both of the components are strongly localized. This circumstance can result in different values of the excitation probability for the d electrons of the different types of atoms in the alloy and consequently to an energy distribution of photoelectrons that differs from the density-of-states distribution. Unfortunately, the theoretical investigation of this problem involves great difficulties with regard to the description of electron states in disordered alloys. This is due principally to the absence in these systems of a periodic potential. Nemoshkalenko and Aleshin [239] have calculated the X-ray photoelectron spectra of binary alloys by using the two-band model proposed by Schwartz *et al.* [240] and by Brouers and Vedyaev [241]. For a binary alloy with composition $A_x B_{1-x}$, the Hamiltonian is given by the following expression:

$$H = \sum_{\mathbf{k}} E_s(\mathbf{k}) \,|\, \mathbf{k}_s\rangle \langle \mathbf{k}_s \,| + \sum_{k} \varepsilon_d(\mathbf{k}) \,|\, \mathbf{k}_d\rangle \langle \mathbf{k}_d \,| + \sum_{n} \varepsilon_d^n \,|\, n_d\rangle \langle n_d \,| +$$
$$+ \sum_{\mathbf{k}} \gamma \,[|\, \mathbf{k}_s\rangle \langle \mathbf{k}_d \,| + |\, \mathbf{k}_d\rangle \langle \mathbf{k}_s \,|\rangle],$$

where $E_s(\mathrm{k})$ and $\epsilon_d(\mathrm{k})$ are the kinetic energies of s and d electrons; ϵ_d^n is the energy of the resonance d levels, which can take one of the two values ϵ_d^A and ϵ_d^B depending on which atom is situated at the node n; γ is a constant that determines the magnitude of s–d hybridization.

In order to describe the energy distribution of photoelectrons, use is made of the Green's function $\bar{G} = \langle G \rangle$ averaged over all the atomic configurations in the alloy. This Green's function has the following expression in the $|k_s\rangle$, $|k_d\rangle$ representation:

$$\bar{G} = \begin{pmatrix} z - E_s(\mathbf{k}) & -\gamma \\ -\gamma & z - \varepsilon_d(\mathbf{k}) - \sum_d(z) \end{pmatrix}^{-1}$$

Here, $\Sigma_d(z)$ is the operator of the eigenenergy. A self-consistent choice of this operator may be made by using the method of coherent potential. In this case, according to Brouers and Vedyaev [241], the eigenenergy operator satisfies the following equation:

$$\sum_d(z) = \bar{\varepsilon}_d - (\varepsilon_d^A - \sum_d(z))\, F_{dd}(z, \sum_d(z))\, (\varepsilon_d^B - \sum_d(z)), \qquad (112)$$

where

$$\bar{\varepsilon}_d = x\varepsilon_d^A + (1 - x)\,\varepsilon_d^B,$$
$$F_{dd}(z, \sum_d(z)) = \frac{\Omega_0}{(2\pi)^3} \int d^3k \,\{z - \sum_d(z) - \varepsilon_d(\mathbf{k}) - \gamma^2 [z - E_s(\mathbf{k})]^{-1}\}^{-1}.$$

It can be shown that the relation between the energy distribution of photoelectrons and the Green's function $\bar{G}$ is analogous to the relation between the

Green's function and the intensity distribution of radiation in X-ray emission spectra [242]:

$$I(\omega, E) \sim \sum_{\mathbf{k},\mathbf{Q}} \operatorname{Im} \langle \mathbf{k}+\mathbf{Q} \,|\, W^{+}\overline{G}W \,|\mathbf{k}+\mathbf{Q}\rangle. \tag{113}$$

In this expression, W represents the operator describing the interaction of the electron with the electromagnetic radiation:

$$W = i\frac{e}{mc}A\cdot\nabla; \quad \overline{G} = \overline{G}(\omega - E_{\mathbf{k}+\mathbf{Q}} + i\delta),$$

where $E_{\mathbf{k}+\mathbf{Q}}$ is the energy of an electron having momentum $\mathbf{k}+\mathbf{Q}$ situated in the conduction band of the alloy. In the study of X-ray photoemission processes, it is convenient to assume that the states of the excited photoelectrons can be described by plane waves:

$$\langle \mathbf{r} \,|\, \mathbf{k}+\mathbf{Q}\rangle = \frac{1}{\sqrt{N\Omega_0}} \exp i(\mathbf{k}+\mathbf{Q})\cdot\mathbf{r}.$$

The summation in formula (113) is performed over all the vectors $\mathbf{k}$ and $\mathbf{Q}$ that give rise to the same value of the energy of the emitted electron:

$$E = |\mathbf{k}+\mathbf{Q}|^2 = P^2. \tag{114}$$

We will not consider here the processes that accompany the exit of electrons from the specimen. We will assume that these processes do not appreciably affect the energy distribution of photoelectrons. Use of the dipole approximation for describing the processes of electron excitation by X-ray quanta is justified up to high values of photon excitation energies, of the order of 1500 eV, since the wave functions of electrons in crystals (and especially the wave functions of d electrons) have a great amplitude only in a limited region of the elementary cell of the crystal, namely, the region in which $\mathbf{q}\cdot\mathbf{r} < 1$ ($\mathbf{q}$ is the wave vector of the X rays).

Let us transform formula (113) by introducing the Wannier functions $\varphi_{\mathbf{R}l}$ ($\mathbf{R}$ is the radius vector of the node in which the Wannier function is centered, and l denotes the type of state). From the strong localization of the Wannier functions in the elementary cell of the crystal, the following is obtained:

$$I(\omega, E) \sim \operatorname{Im} \sum_{\mathbf{k},\mathbf{Q}} \sum_{\mathbf{R}_1,\mathbf{R}_2,l_1,l_2} \langle \mathbf{k}+\mathbf{Q} \,|\, \mathbf{R}_1 l_1\rangle \langle \mathbf{R}_1 l_1 \,|\, \overline{G} \,|\, \mathbf{R}_2 l_2\rangle \times$$
$$\times \langle \mathbf{R}_2 l_2 \,|\, \mathbf{k}+\mathbf{Q}\rangle (A, \ (\mathbf{k}+\mathbf{Q}))^2. \tag{115}$$

We will now transform this expression further, by assuming that the Wannier functions $\varphi_{\mathbf{R}l}$ are given approximately by the expression $\varphi_{\mathbf{R}l} \approx \varphi_l(\mathbf{r}-\mathbf{R})$. In this case, the following relation is valid:

$$\langle \mathbf{k}+\mathbf{Q} \mid \mathbf{R}l \rangle = \exp\left[-i(\mathbf{k}+\mathbf{Q})\cdot\mathbf{R}\right] \langle \mathbf{k}+\mathbf{Q} \mid 0l \rangle. \tag{116}$$

By using the Fourier expansion of the Green's function

$$\langle \mathbf{R}_1 l_1 \mid G \mid \mathbf{R}_2 l_2 \rangle = \frac{1}{N} \sum_{\mathbf{k}} G_{l_1 l_2}(\mathbf{k}) \exp i\mathbf{k}\cdot(\mathbf{R}_1 - \mathbf{R}_2)$$

and the expression (116), the following expression for $I(\omega, E)$ is obtained:

$$I(\omega, E) \sim \sum_{\mathbf{k},\mathbf{Q}} \operatorname{Im} \sum_{l_1 l_2} \langle 0l_2 \mid \mathbf{k}+Q \rangle \langle \mathbf{k}+\mathbf{Q})\, 0l_1 \rangle\, G_{l_1 l_2}(\mathbf{k})\, (A, (\mathbf{k}+\mathbf{Q}))^2.$$

Because of the high energy of the X-ray quanta, the summation in this expression may be performed over all the vectors $\mathbf{k}$ in the Brillouin zone, since it is always possible to find vectors $\mathbf{Q}$ of the reciprocal lattice that satisfy the relation (114). The number of these vectors is proportional to $\mathbf{Q}^2$. By taking into account that

$$\frac{1}{N} \sum_{\mathbf{k}} G_{l_1 l_2}(\mathbf{k}) = \langle 0l_1 \mid \bar{G} \mid 0l_2 \rangle,$$

one obtains

$$I(\omega, E) \sim E^{3/2} \sum_{l_1 l_2} \langle Ol_2 \mid \mathbf{P} \rangle \langle \mathbf{P} \mid Ol_1 \rangle \langle Ol_1 \mid \bar{G} \mid Ol_2 \rangle \cos^2\theta,$$

where θ is the angle between the vectors $\mathbf{A}$ and $\mathbf{P}$. After averaging over these angles, taking into account that the crystal has one s band and one degenerate d band (its degree of degeneracy is 5), and assuming that the angular dependence of Wannier functions is the same as for atomic functions, then for crystal samples the following expression is obtained:

$$I(\omega, E) \sim E^{3/2}\left[P_d(E)\,\rho_d(E) + P_s(E)\,\rho_s(E) + P_{sd}(E)\,\rho_{sd}(E)\right]. \tag{117}$$

Here

$$\rho_d(E) = -\frac{5}{\pi} \operatorname{Im} \langle Od \mid \bar{G} \mid Od \rangle$$

and

$$\rho_s(E) = -\frac{1}{\pi}\,\mathrm{Im}\,\langle Os\,|\,\bar{G}\,|\,Os\rangle$$

represent the d and s components of the density of states of valence electrons;

$$\rho_{sd}(E) = -\frac{1}{\pi}\,\mathrm{Im}\,\langle Os\,|\,\bar{G}\,|\,Od\rangle,$$

and the radial factors of the transition probability have the following expressions:

$$P_d(E) = \frac{\pi}{3}\left[\int R_d(r)\,j_2(\sqrt{E}r)\,r^2dr\right]^2,$$

$$P_s(E) = \frac{\pi}{3}\left[\int R_s(r)\,j_0\sqrt{E}r)\,r^2dr\right]^2,$$

$$P_{sd}(E) = \frac{4\pi}{3\sqrt{5}}\left[\int R_d j_2 r^2dr\right]\left[\int R_s j_0 r^2dr\right],$$

where R_s and R_d represent the radial parts of the Wannier functions, and j_2 and j_0 represent the spherical Bessel functions.

The density of states ρ_d and ρ_s and the magnitude of ρ_{sd} may be determined by the following formulas:

$$\left.\begin{aligned}
&\rho_d(z = E + iO) = \\
&= -\frac{5}{\pi}\frac{\Omega_0}{(2\pi)^3}\,\mathrm{Im}\int d^3k\,\frac{z - E_s(k)}{(z - E_s(\mathbf{k}))\,(z - \varepsilon_d(\mathbf{k}) - \sum_d) - \gamma^2},\\
&\rho_s(z = E + iO) = \\
&= -\frac{1}{\pi}\frac{\Omega_0}{(2\pi)^3}\,\mathrm{Im}\int d^3k\,\frac{z - \varepsilon_d(\mathbf{k}) - \sum_d}{(z - E_s(\mathbf{k}))\,(z - \varepsilon_d(\mathbf{k}) - \sum_d) - \gamma^2},\\
&\rho_{sd}(z = E + iO) = \\
&= -\frac{1}{\pi}\frac{\Omega_0}{(2\pi)^3}\,\mathrm{Im}\int d^3k\,\frac{\gamma}{(z - E_s(\mathbf{k}))\,(z - \varepsilon_d(\mathbf{k}) - \sum_d) - \gamma^2}.
\end{aligned}\right\}\quad (118)$$

Nemoshkalenko and Aleshin [239] have studied a model equiatomic alloy $A_{0.5}B_{0.5}$ characterized by the following parameters: the width of the s band $2w_s = 1.11$ Ry, the hybridization constant $\gamma = 0.0571$, $E_s(\mathbf{k}) = \epsilon_s(\mathbf{k}) - 0.278$, $\epsilon_d(\mathbf{k}) = 0.167\epsilon_s(\mathbf{k})$, the distance δ between the d resonances undergoing scattering, $\delta = 0.14$ Ry. The eigenenergy $\Sigma_d(z)$ was determined self-consistently from equation (112). The obtained $\Sigma_d(z)$ value was used for the determination of $\rho_d(E)$, $\rho_s(E)$, and $\rho_{sd}(E)$ from formulas (118). Because of the model character of the problem, it was meaningless to perform an accurate calculation of $P_s(E)$ and $P_d(E)$. Since at high excitation energies these functions exhibit a weak energy dependence, we have assumed that the ratio $\int R_d j_2 r^2\,dr / \int R_s j_0 r^2\,dr$ is equal to 1. If this ratio were to differ from 1 by a factor of several units, then an unrealistically high or low ratio between the contributions of d and s electrons in the X-ray photoelectron spectrum would be obtained. The results of these calculations are shown in Figure 81 for the two values $\gamma_1 = 3\gamma$ and $\gamma_2 = \gamma$. This figure shows clearly that the magnitude of the hybridization constant determines in a significant degree the shape of X-ray photoelectron spectrum. The third term in formula (117) is of comparable magnitude to the second term, which causes some deformation of the shape of the band. Particularly large deformations occur for the alloy with hybridization constant γ_1. Therefore, in the study of $A_x B_{1-x}$ alloy systems, it cannot in general be considered that the magnitude of the ordinate values, taken at the Fermi energy, is proportional to the density of the corresponding electron states. This conclusion is also valid for pure metals, since in this case, in formula (117), $\epsilon_d^A = \epsilon_d^B$. The model described above is of limited validity because one of the most important assumptions was the equality of Wannier functions centered at different nodes in the alloy. Thus the model is more likely to be valid for alloys in which the atoms A and B have similar scattering properties.

Consideration of the difference in excitation probability for atoms of different types in disordered alloys is a difficult problem. However, some informa-

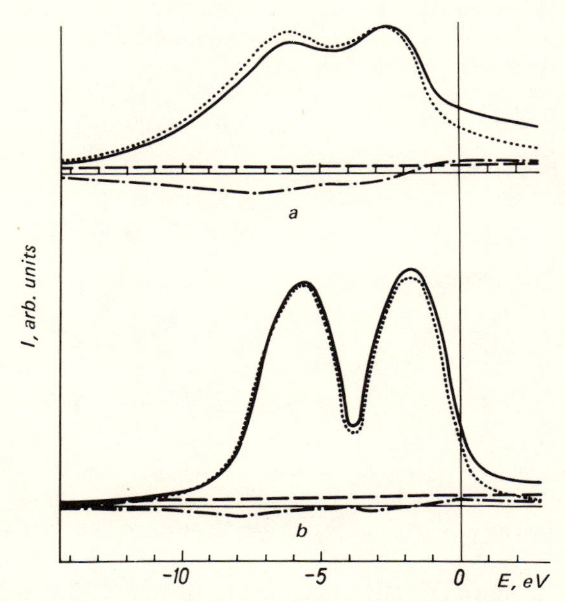

Figure 81. Structure of the valence band photoelectron spectrum of the alloy $A_{0.5}B_{0.5}$, with hybridization constants: $\gamma_1 = 3\gamma(a)$ and $\gamma_2 = \gamma(b)$: (——) photoelectron spectrum of valence electrons; (· · · ·) d-electron contribution to the photoelectron spectrum; (– – –) s electron contribution to the photoelectron spectrum; (—·—·—) s-d term contribution to the photoelectron spectrum.

tion about the influence of transition probability on the shape of X-ray photoelectron spectra of alloys may be obtained by comparing experimental data with the theoretically calculated electron density of states. Stocks *et al.* [206, 207] and Nemoshkalenko *et al.* [243-245] have calculated the density of electronic states in the valence bands of the disordered alloys nickel–copper, palladium–silver, copper–palladium, and manganese–copper by using the two-band model of alloys proposed by Kirkpatrick [246]. This model allows for the existence in the valence band of alloys of states of t_{2g}- and e_g-symmetry type. It also takes into account a more correct hybridization of states of the s-, p-, and d-symmetry type, than that considered by Schwartz *et al.* [240] and by Brouers and Vedyaev [241].

The Hamiltonian of the alloy may be written in matrix form as

$$H = W + D = \begin{pmatrix} W_{ss} & \begin{matrix} W_{sd} & \end{matrix} \\ W_{ds} & \begin{matrix} W_{EE} & W_{ET} \\ W_{TE} & W_{TT} \end{matrix} \end{pmatrix} + \begin{pmatrix} 0 & \begin{matrix} 0 & \end{matrix} \\ 0 & \begin{matrix} D_{EE} & 0 \\ 0 & D_{TT} \end{matrix} \end{pmatrix}. \tag{119}$$

The different blocks in the Hamiltonian are labeled by the index s for the s states, d for the d states, and E and T for the states of the e_g- and t_{2g}-symmetry type. The basis functions used in the representation of the Hamiltonian by formula (119) are not invariant under translation operations, since the orbitals centered at different sites of the lattice may be different from each other. However, approximately, it may be considered that the matrix elements of the operator W are translation invariant and configuration independent. The block D of the Hamiltonian (119) is also not translation invariant. The total dd block of the Hamiltonian may be written as follows:

$$H_{dd} = \sum_{n,\mu} | \mu n\rangle \varepsilon_{\mu n} \langle \mu n | + \sum_{n \neq n', \mu, \mu'} | \mu n\rangle t_{\mu n, \mu' n'} \langle \mu' n' |, \tag{120}$$

where $|\mu n\rangle$ represents the d orbital centered at the nth site. For $\mu = 1, 2, 3$, the orbital at the nth site is characterized by a symmetry of the t_{2g} type, whereas for $\mu = 4, 5$, it is characterized by a symmetry of the e_g type. If the nth site is occupied by an atom A, then the orbital $|\mu n\rangle$ is chosen as the orbital of the atom A. If instead the nth site is occupied by an atom B, then the orbital $|\mu n\rangle$ is taken as the orbital of the atom B. The nondiagonal elements $t_{\mu n, \mu' n'}$ in formula (120) are translation invariant. The configuration-averaged Green's function $\langle G\rangle$ may be written as

$$\langle G\rangle = \langle (z - W - D)^{-1}\rangle = \frac{1}{z - W - \sum}, \tag{121}$$

and

$$\rho(E) = -\frac{1}{\pi N} \operatorname{Im} Sp\langle G(E + i\delta)\rangle.$$

By making use of the projection operators for the space of functions with s-, t_{2g}-, and e_g-symmetry type, it follows that the density of states $\rho(E)$ may be expressed as a sum:

$$\rho = \rho_s + \rho_{t_{2g}} + \rho_{e_g},$$

where

$$\rho_s = -\frac{1}{\pi N} \operatorname{Im}[Sp(1 - P)\langle G\rangle],$$

$$\rho_{t_{2g}} = -\frac{1}{\pi N} \operatorname{Im}[SpP_T\langle G\rangle],$$

$$\rho_{e_g} = -\frac{1}{\pi N} \operatorname{Im}[SpP_E\langle G\rangle].$$

ρ_{e_g} and $\rho_{t_{2g}}$ satisfy the sum rule

$$\int dE\rho_{e_g}(E) = 2,$$

$$\int dE\rho_{t_{2g}}(E) = 3.$$

The eigenenergy Σ entering into formula (121) may be written as

$$\Sigma = \sum_n P_n \Sigma_n P_n,$$

where $P_n = P_{nE} + P_{nT}$ represents the operator of projection to the space of the states which characterize the node n. Then

$$\Sigma_n = \Sigma_E P_{nE} + \Sigma_T P_{nT}, \quad P_n\langle G\rangle P_n = F_E P_{nE} + F_T P_{nT},$$

where Σ_{ET}, F_E, and F_T represent scalars determined by the equations

$$\Sigma_E = \bar{\varepsilon}_E - (\varepsilon_E^A - \Sigma_E) F_E (\varepsilon_E^B - \Sigma_E),$$

$$\Sigma_T = \bar{\varepsilon}_T - (\varepsilon_T^A - \Sigma_T) F_T (\varepsilon_T^B - \Sigma_T),$$

in which F_E and F_T are functions of Σ_E and Σ_T. Making the approximation

$$\Delta_E = \sum_E - \varepsilon_E^A \approx 0, \quad \Delta_T = \sum_T - \varepsilon_T^A \approx 0,$$

then for the alloy with high content of the A component the following is obtained:

$$F_E(z) = F_E^A(z - \sum_E + \varepsilon_E^A),$$
$$F_T(z) = F_T^A(z - \sum_T + \varepsilon_T^A),$$

where F_E^A and F_T^A are characteristic for the pure A component. In integral representation, these functions are given by

$$F_E^A = \frac{1}{2}\int \frac{\rho_E^A(\zeta)}{z-\zeta}\,d\zeta, \quad F_T^A = \frac{1}{3}\int \frac{\rho_T^A(\zeta)}{z-\zeta}\,d\zeta.$$

Then

$$\rho_E(E) = -\frac{2}{\pi}\,\mathrm{Im}\,F_E(E + i0),$$

$$\rho_T(E) = -\frac{3}{\pi}\,\mathrm{Im}\,F_T(E + i0).$$

The local density of states ρ_E^A, ρ_T^A, ρ_E^B, ρ_T^B with e_g- and t_{2g}-symmetry type may also be determined. Thus

$$\rho_E^A = -\frac{1}{\pi N}\,\mathrm{Im}\,SpP_E^A G(E + i0).$$

ρ^A and ρ^B may be expressed as follows:

$$\rho^A = \rho_E^A + \rho_T^A, \quad \rho^B = \rho_E^B + \rho_T^B,$$
$$\rho_E = \rho_E^A + \rho_E^B, \quad \rho_T = \rho_T^A + \rho_T^B.$$

The main parameter that determines the density of states $\rho(E)$ of the alloy is $\delta_\mu = \epsilon_\mu^A - \epsilon_\mu^B$. It gives the energy distance between the centers of gravity of the d states in the alloy. For this reason, the study of alloys with different δ_μ is of particular interest.

Binary Alloys of Neighboring Elements of the Same Period

Nickel-Copper Alloys. As can be seen from Figure 82, copper is characterized by a smaller density of states at the Fermi level than nickel. In nickel-

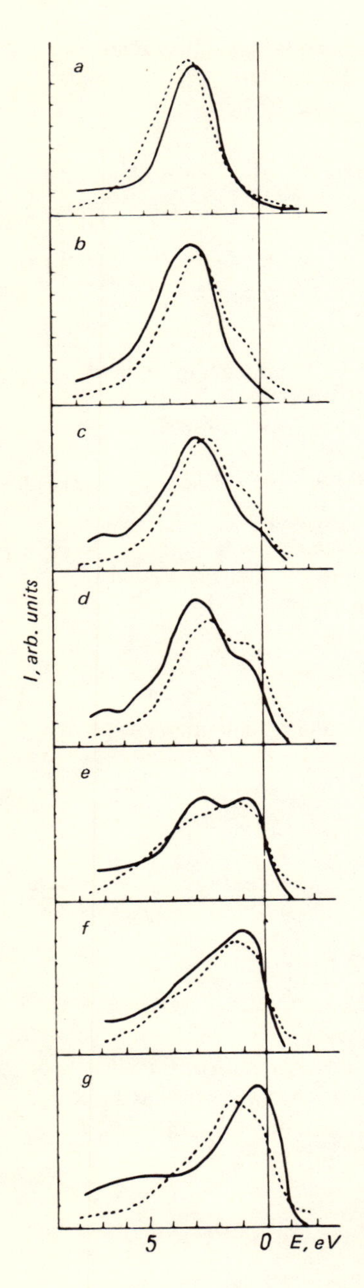

Figure 82. Valence band photoelectron spectra of nickel, copper and their alloys [247]: (a) copper; (b) $Ni_{13}Cu_{87}$; (c) $Ni_{23}Cu_{77}$; (d) $Ni_{38}Cu_{62}$; (e) $Ni_{61}Cu_{39}$; (f) $Ni_{89}Cu_{11}$; (g) nickel; (· · · ·) theoretical curves [206].

copper alloys, the density of states at the Fermi level increases when the concentration of nickel in the alloy increases [247]. In these alloy systems, the position of the *d* states of nickel and copper remains practically unchanged upon the formation of the alloy. Studies of the X-ray emission *M* spectra of nickel and copper in nickel–copper alloys [248] have revealed that these spectra also preserve the shape characteristic of the pure elements. As in the case of X-ray photoelectron spectra, the *M* emission spectra mainly reflect the energy distribution of *d* electrons, although they are the result of transitions from *s*- and *d*- to 3*p*-type states. A similar distribution for electron of *s*-type symmetry cannot be revealed for a number of reasons, namely: their low number, the broad energy extension of their distribution, the high-energy satellites of the *M* bands and also the very high density of electrons of *d*-type symmetry. The X-ray photoelectron spectra of nickel–copper alloys have also been studied by Hüfner *et al.* [249], but at a different concentration of the components.

The X-ray photoelectron spectra of nickel–copper alloys have been measured at those nickel concentrations for which Stocks *et al.* [206] calculated the density of valence band electron states using the coherent potential method. In these calculations, the following assumptions were made: (a) the two-band model was assumed to be valid, allowing for the presence of only *s* and *d* electrons; (b) a common *s* band was supposed to exist in the crystal; (c) translation invariance was assumed; and (d) the nondiagonal elements of the Hamiltonian were assumed to be independent of concentration. The calculated density-of-states curves of the alloys, corrected by taking account of instrumental distortion, have been compared with the X-ray photoelectron spectra (see Figure 82). The theoretical results illustrated in this figure are consistent with the experimental data with respect to the shape and the width of bands and also the number of observed structures. The salient feature of the spectra obtained for the alloys is the occurrence of *d* resonances characteristic for the pure elements.

Hüfner *et al.* [249] have suggested that the X-ray photoelectron spectra of valence bands in alloys, because of their simplicity, may be obtained simply by superposition of the spectra corresponding to the pure components. The data of Stocks *et al.* [206] show, however, that this procedure is not justified. The authors of the work mentioned above also calculated the local density of states of nickel and copper. It was found that it is significantly different from the total density of electronic states in the valence band of the alloy. It can be stated that the main contribution to the density of electronic states of these alloys in the region situated at about 2-eV distance from the Fermi level is given by the *d* states of nickel, whereas at a greater distance, the total density of electronic states is mainly determined by the local density of the *d* states of copper.

It should be noted that the local density of *d* states in nickel and copper in alloys is essentially different from that of pure nickel and pure copper. The

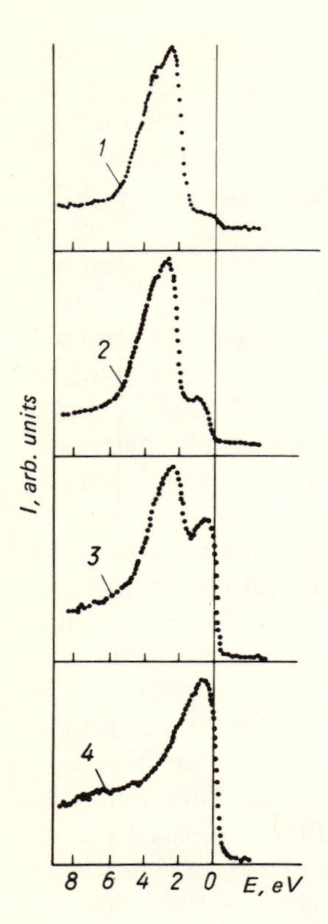

Figure 83. Valence band photoelectron spectra of nickel, copper and their alloys [193]: (1) copper; (2) $Ni_{10}Cu_{90}$; (3) $Ni_{53}Cu_{47}$; (4) nickel.

increased resolving power of X-ray photoelectron spectroscopy makes it possible to distinguish the difference between the valence band density of states in nickel–copper alloys and the simple superposition of the density of states corresponding to the individual components. Thus Hüfner *et al.* [193] have used an instrument with high resolution (0.6 eV) to measure the X-ray photoelectron spectra of the valence bands in pure nickel, pure copper and a number of nickel-copper alloys (Figure 83). The X-ray photoelectron spectrum of the $Ni_{10}Cu_{90}$ alloy clearly shows a maximum corresponding to the *d* resonances of copper, but in the spectrum of the $Ni_{13}Cu_{87}$ alloy no maximum is observed in this energy region (see Figure 82b). In Figure 83 it can be seen that the detailed structure characteristic of pure copper is less evident in the spectrum of alloys and becomes already unobservable in the spectrum of $Ni_{53}Cu_{47}$ alloy. Figure 84 shows the energy position of the maxima corresponding to the *d* states of nickel and copper in nickel–copper alloys, according to the data of Hüfner *et al.* [193]. These authors also showed that the band structure in the region where the resonance *d* states of nickel in the $Ni_{10}Cu_{90}$ alloy are localized may be described by

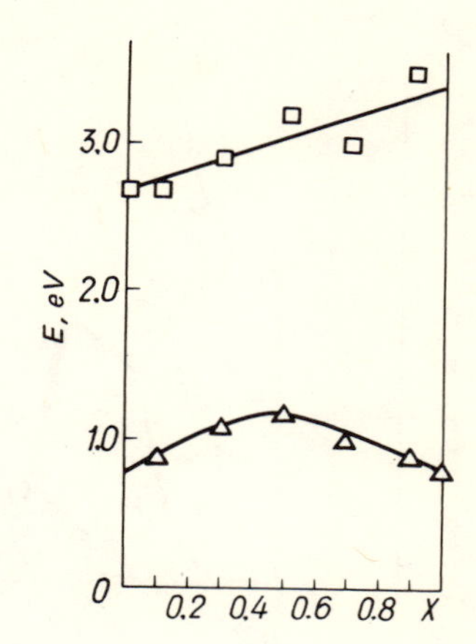

Figure 84. Energy position of the maxima in the valence band photoelectron spectra of nickel-copper ($Cu_{1-X}Ni_X$) alloys: □*–copper,* Δ*–nickel.*

the formula

$$I(E) = \frac{g}{\pi} \frac{\Delta}{(E - E_d)^2 + \Delta^2},$$

where $\Delta = 0.35$ eV and $E_d = 0.8$ eV.

For nickel-copper alloy systems, the binding energies of the core electrons have also been measured. It was found [247] that the binding energies of the $2p_{3/2}$, $3s_{1/2}$, and $3p_{1/2}$ electrons of nickel, and the $2p_{1/2}$, $3s_{1/2}$, and $3p_{1/2}$ electrons of copper are practically the same in the alloys as in the pure elements. The observed shifts do not, as a rule, exceed 0.3 eV, which means that the corresponding charge transfer cannot be large. The absence of a significant charge transfer between the components of nickel-copper alloys is consistent with the results obtained from the measurement of Mössbauer isomer shifts by Love *et al.* [250]. These data indicate that the electron density at the site of the nickel atoms does not depend on the concentration of copper in the alloys. The calculations of Stocks *et al.* [206] have also shown that, in nickel-copper alloys, the charge density of the d electrons of nickel and copper atoms in the alloy does not undergo any change. It is to be noted that in the coherent-potential method, the parameter δ is used, which, on the assumption that the alloy has a common s band, determines the energy distance between the scattering d resonances. Stocks *et al.* [206] used in their calculations the value $\delta = 1.8$ eV, while the value determined experimentally is $\delta = 2.2$ eV. Use of this experimental value in the calculations would result in an improvement of the agreement between theory and experiment.

Palladium-Silver Alloys. The X-ray photoelectron spectra of valence electrons in palladium-silver alloys and of the corresponding pure elements, as mea-

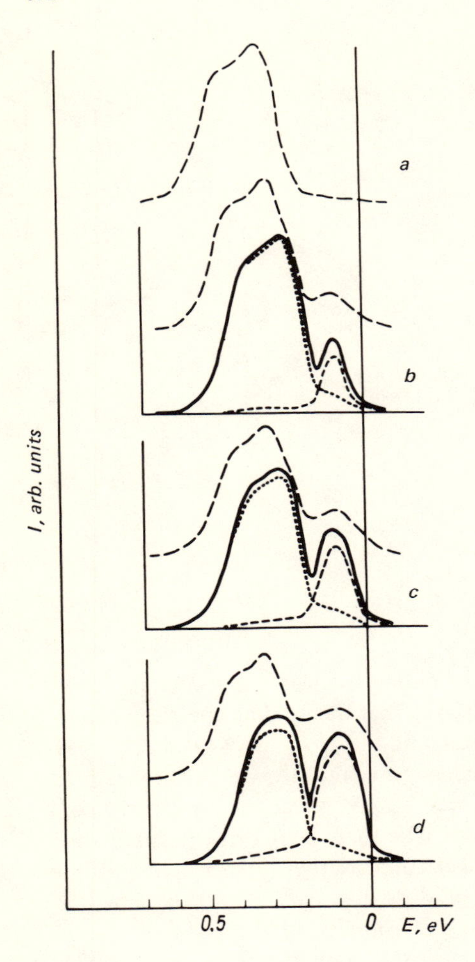

Figure 85. Valence band photoelectron spectra and the density of states in silver-rich palladium-silver alloys [212]: (– – –) photoelectron spectra; (——) density of states; (. . .) local density of states of silver; (– – –) local density of states of palladium; a–silver; b–$Ag_{86.2}Pd_{13.8}$; c–$Ag_{75.4}Pd_{24.6}$; d–$Ag_{59.8}Pd_{40.2}$.

sured by Nemoshkalenko *et al.* [212], are shown in Figures 85 and 86.† The top of the *d* band of silver is situated 4 eV under the Fermi level, whereas palladium exhibits a high density of states at the Fermi level and the position of its Fermi level does not correspond to the midpoint of the low-energy branch of the spectrum. It should be pointed out that in some experimental works [186, 251], the calibration of X-ray photoelectron spectra has been made on the assumption that the Fermi level of palladium corresponds exactly to the midpoint of the low-energy part of the spectrum. This assumption is not justified either by ex-

†*Editors' note:* This alloy system, as well as others discussed below, was first investigated with UV-photoelectron spectroscopy. For Ag–Pd, see, e.g., C. Norris and P. O. Nilsson, *Solid State Commun.* **6**, 649 (1968) and C. Norris and H. P. Myers, *J. Phys. F* **1**, 62 (1971).

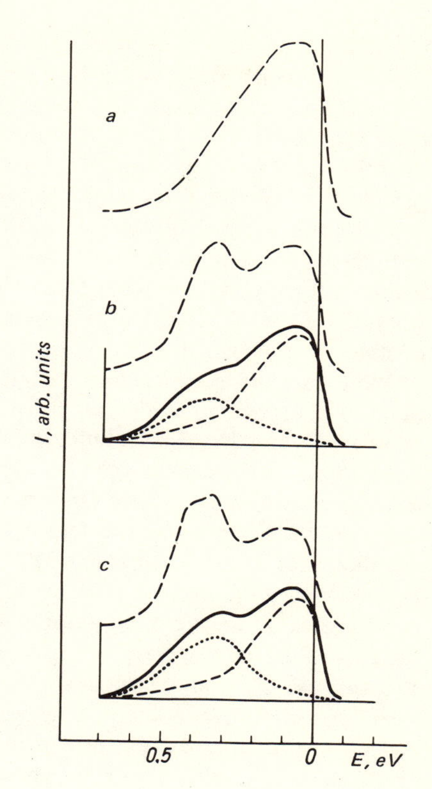

Figure 86. Valence band photoelectron spectra and the density of states in palladium-rich palladium–silver alloys [212]: (–––) photoelectron spectra; (——) density of states; (· · · ·) local density of states of silver; (– – –) local density of states of palladium; (a) palladium; (b) $Pd_{70}Ag_{30}$; (c) $Pd_{57.4}Ag_{42.6}$.

periment or by the existing theoretical calculations of the density of states in the valence band of palladium made by Mueller *et al.* [252]. These calculations were performed by the relativistic augmented plane-waves method, with 10^6 **k** points in the Brillouin zone.

Nemoshkalenko *et al.* [244] have calculated the distribution of *d* states in the valence band of palladium–silver alloys. As a measure of the density of the *d* states of the pure components, the intensity distributions of the X-ray photoelectron spectra of palladium and silver, measured earlier by the same research group, were taken [212]. It has been pointed out earlier that the intensity distributions in the X-ray photoelectron spectra of both palladium and silver differ from the corresponding distributions of the density of electronic states. However, to a first approximation, it can be considered that these differences are not too great.

The palladium–silver alloy system, like the system nickel–copper, is suited to a calculation of the density of electronic states by the coherent-potential method since these elements are neighbors in the periodic system. Therefore, the nondiagonal matrix elements of the block *d–d* of the Hamiltonian can approximately be considered as being translation invariant. Some contribution to the valence band density of states of pure palladium and silver comes also from the density of *s* electrons, but the magnitude of this density is much lower than the density of *d* electrons. In the determination of the density of states of palladium–silver alloys, the experimental value $\delta = 0.18$ Ry was used. The calculated density of states of valence band electrons in these alloys are also included in Figures 85 and 86. In the calculation of the density of states for alloys with a high-percentage silver content, the density of states of pure silver was used, whereas for the alloys with high palladium content, use was made of the density of states of pure palladium.

As Figures 85 and 86 show, the calculated density of states allows a description of effects related to alloying. First, it is to be noted that in the alloys containing 13.8 and 24.6% palladium, the valence states corresponding to palladium appear in the density-of-states curve of the alloys as a structureless enhancement situated between the maximum corresponding to the *d* states of silver and the Fermi level. The density of states of the palladium component in the alloys has a Lorentzian shape. The density of states of pure silver exhibits two maxima; with increasing concentration of palladium, the structure of this part becomes more and more smeared out. At high palladium concentrations, the part of the band corresponding to the valence states of silver becomes structureless. The Fermi level in alloys is determined on the assumption that the valence band of silver contains 10 *d* electrons, and that of palladium contains 9 *d* electrons. An alternative choice of the occupation numbers is also possible, for example, 9.5 *d* electrons for both silver and palladium, that would not result in a significant change of the position of the Fermi level in alloys.

The density of states of palladium–silver alloys calculated by Stocks *et al.* [207] using the coherent-potential method is consistent with the data in Figures 85 and 86. These figures show a qualitative agreement between the structure of the calculated density of states and the shape of the energy distribution of photoelectrons. With increasing palladium concentration there occurs a significant change of the energy distance between the maxima in that part of the spectrum corresponding to pure silver. Thus, in pure silver, the energy separation between these maxima is 1.6 eV, but in the alloy containing 70% palladium, it is 0.9 eV. The magnitude of spin–orbit splitting in pure silver is relatively low, and therefore the two maxima in the density of states of pure silver can be attributed mainly to the particularities of the structure of the *d* bands rather than to the spin–orbit splitting of the *d* states. A large change of the energy separation between the maxima characteristic for pure silver has also been observed by Fuggle *et al.* [57] in the study of silver–aluminum alloys. Our experiments [212] have also confirmed the presence of a flat maximum in the energy region characteristic for pure palladium in alloys with 70% and 57.4% palladium content and its absence at lower palladium concentrations. However, the intensity ratio of the maxima associated with silver and palladium obtained from the density-of-states curves of the alloy differs from that measured on energy distribution of valence band photoelectrons. Comparison of the experimental curves of the valence electron distribution with the theoretical curves of the valence electron density of states indicates that the photoionization cross section of the d electrons of silver in alloys is approximately 1.5 times greater than that of the d electrons of palladium. Comparison of the electron density of states of alloys with the local density of states confirms that the structure of the density of states in the energy region characteristic for pure silver is determined mainly by the local density of states of silver, and similarly, the structure of the density of states in the energy region characteristic for palladium is determined by the local density of states of palladium. This explains why the shape of the energy distribution of photoelectrons in different regions of the spectrum is determined mainly by the local density of electronic states of silver and palladium.

Table 11 shows the values of the core-level binding energies of pure atomic palladium and silver and of palladium–silver alloys. It can be seen from this table that the binding energies of the core electrons of palladium in alloys coincide with those of pure palladium, to within 0.1 eV, but for silver, the values of core-level binding energies differ by up to 1.0 eV. That the core-level binding energies of silver in alloys are lower than in pure silver indicates that, in the alloys, silver is an electron acceptor. This is consistent with the enhancement of electron density at the site of silver atoms. At first sight this contradicts the generally accepted representation of the charge transfer, i.e., from the silver atoms to the palladium atoms, resulting in the occupation of 0.5 holes in the d shell of palladium. It should be realized, however, that the concept of charge transfer between the atoms of the components of the alloy is valid only if the entire electron density is localized at the atomic sites of the components of the alloy. It is well known, however, that part of the electron density is concentrated in the crystal lattice between the atoms of the components of the alloy,

TABLE 11. Values of Core-Electron Binding Energies of Pd and Ag Atoms in Pd–Ag Alloys

Sample	Level					
	$3p_{1/2}$	$3d_{3/2}$	$3d_{5/2}$	$3p_{1/2}$	$3d_{3/2}$	$3d_{5/2}$
Pd	559.7	340.2	334.9			
$Pd_{70}Ag_{30}$	559.7	340.1	334.8	602.9	373.1	367.1
$Pd_{57.4}Ag_{42.6}$	559.8	340.1	334.8	603.2	373.5	367.5
$Pd_{40.2}Ag_{59.8}$	559.5	340.1	334.8	603.2	373.6	367.6
$Pd_{24.6}Ag_{75.4}$	560.0	340.1	334.8	603.5	373.7	367.7
$Pd_{13.8}Ag_{86.2}$	–	340.1	334.8	603.6	373.8	367.8
Ag				603.8	374.1	368.1

and a change of this charge density does not directly affect the magnitude of the shifts of atomic core levels. Consequently, the measured shifts may only indicate an enhancement of the electron density in the vicinity of silver atoms and its stability near palladium atoms. The case when the observed shifts indicate an increase of the electron density in the vicinity of the atoms of both components of the alloy is also possible. In such alloys, the electron density in the intermediate region between the atoms will probably decrease. As we will see later on, such a situation is encountered in the case of copper–gold alloys.

The study of the X-ray emission spectra of the components of palladium–silver alloys [4] has shown that the mechanism of the change of their electronic structure is much more complicated than for a simple donor–acceptor interaction. On one hand, this agrees well with the X-ray photoelectron spectra and the shifts of core levels discussed in the present monograph. On the other hand, it explains completely satisfactorily the magnetic behavior of these alloys.

The X-ray photoelectron spectra of palladium-silver alloys have also been studied by Hedman *et al.* (the system $Pd_{29}Ag_{71}$) [251] and by Hüfner *et al.* [253, 193] using an HP-5950A electron spectrometer. Figures 87 and 88 illus-

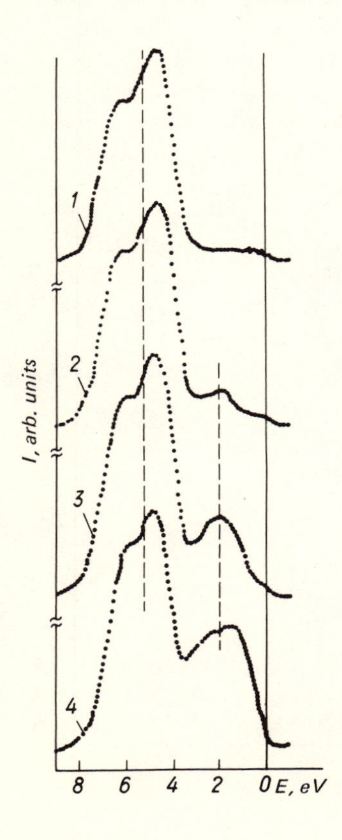

Figure 87. Valence band photoelectron spectra and the density of states in silver-rich palladium-silver alloys [187]: (a) silver; (2) Pd_5Ag_{95}; (3) $Pd_{25}Ag_{75}$; (4) $Pd_{35}Ag_{65}$.

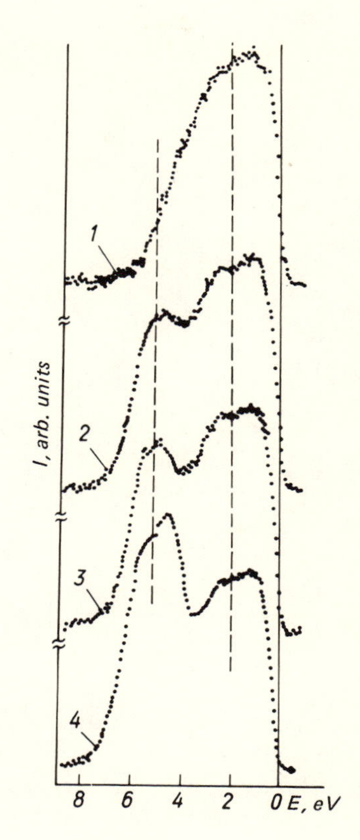

Figure 88. Valence band photoelectron spectra and the density of states in palladium-rich palladium–silver alloys: (1) palladium; (2) $Pd_{90}Ag_{10}$; (3) $Pd_{75}Ag_{25}$; (4) $Pd_{60}Ag_{40}$.

trate more clearly than Figures 85 and 86 the change of the fine-structure characteristic for silver and palladium in the alloy. The energy position of the maxima in the spectra, corresponding to the *d* electron states of palladium and silver in the alloys, is practically constant, while their width changes appreciably with composition. Figure 89 illustrates the change of width of the characteristic peaks of palladium and silver as a function of the concentration in the alloys.

Platinum–Gold Alloys. The X-ray photoelectron spectra of platinum–gold alloys are shown in Figure 90. The valence bands of pure platinum and pure gold exhibit a strong overlap in energy. In the calculation of the valence band density of states of both gold and platinum, it is very important to take into account the relativistic effects. To illustrate the significance of the relativistic effects, it is enough to point out that if these effects are neglected, then the calculated density of states of gold will exhibit a maximum [196] in that energy region in which the X-ray photoelectron spectrum exhibits a minimum situated between two maxima. The necessity to include relativistic effects makes the calculation of the valence band density of states of alloys containing gold extremely difficult. Therefore, experiment is as yet the only source of information about the

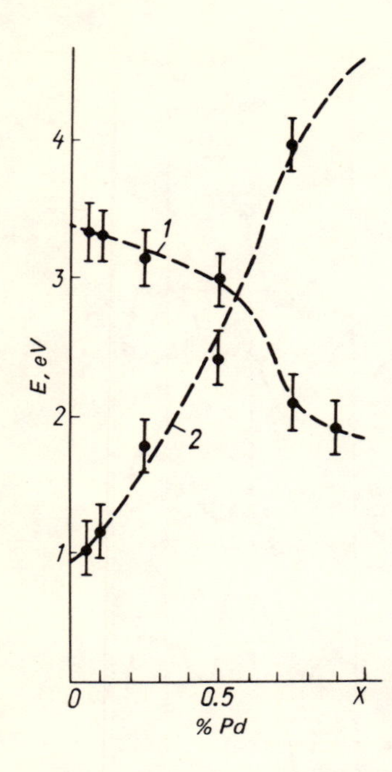

Figure 89. Energy width of the d bands of palladium and silver in palladium-silver ($Pd_X Ag_{1-X}$) alloys: (1) silver; (2) palladium.

density of electronic states in the valence bands of disordered binary alloys of heavy elements.

From the experimental data illustrated in Figure 90, it can be seen that in the valence bands of platinum-gold alloys the characteristics of gold appear more prominent. As in the case of palladium, the position of the Fermi level

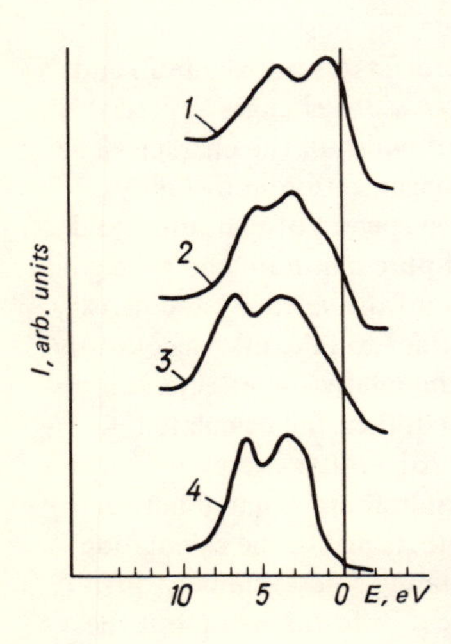

Figure 90. Valence band photoelectron spectra of platinum, gold, and their alloys: (1) platinum; (2) $Pt_{60}Au_{40}$; (3) $Pt_{20}Au_{80}$; (4) gold.

TABLE 12. Values of Core-Electron Binding Energies of Pt and Au Atoms in Pt–Au Alloys

Sample	Level					
	$4d_{5/2}$	$4f_{5/2}$	$4f_{7/2}$	$4d_{5/2}$	$4f_{5/2}$	$4f_{7/2}$
Pt	314.5	74.4	71.0			
$Pt_{60}Au_{40}$	314.1	74.0	70.6	334.5	87.2	83.5
$Pt_{20}Au_{80}$	314.0	74.0	70.6	335.0	87.7	84.0
Au				335.1	87.5	83.8

does not coincide with the position of the midpoint of the low-energy part of the X-ray photoelectron spectrum. With increasing gold content, the density of states at the Fermi level decreases, while the maximum in the density of *d* states of platinum, which is situated 1 eV below the Fermi level, is shifted approximately 2 eV toward the bottom of the valence band. In the alloy Pt_{60}/Au_{40}, a small enhancement is observed on the low-energy part of the electron spectrum, caused by the contribution of the *d* electrons of platinum. It should be noted that even in the alloy with 80% gold, the high-energy maximum does not coincide with the corresponding maximum in the electron spectrum of gold.

The values of core-level binding energies of the component elements of platinum–gold alloys are given in Table 12, based on the data from the work of Nemoshkalenko *et al.* [254]. The binding energies of the $4d_{5/2}$, $4f_{5/2}$, and $4f_{7/2}$ core electrons of platinum diminish somewhat (by up to 0.4 eV) upon the formation of disordered binary alloys. The analogous energies for gold remain almost unchanged, although the scatter of the data for the given alloys is somewhat higher than usual.

Alloys of Elements from Different Periods

Copper–Palladium Alloys. Nemoshkalenko *et al.* [243] studied the X-ray photoelectron spectra of the valence bands of copper–palladium alloys, using an HP-5950A electron spectrometer.

The samples were prepared by alloying palladium with copper with a high-frequency furnace in a helium atmosphere, followed by annealing and quenching. The X-ray photoelectron spectra were recorded with an instrumental resolution of 0.6–0.7 eV. Before measurements were made, the samples were cleaned by argon ion bombardment at a pressure of 10^{-4} torr. After cleaning, the intensity of the oxygen 2*s* line diminished by a factor of 10, while the carbon 1*s* line practically disappeared. The vacuum level in the spectrometer was about 10^{-9} torr.

In palladium–silver and nickel–copper alloys, the centers of gravity of the *d* bands of the pure components are relatively far apart, namely, at energy separations of 2.5 and 1.9 eV, respectively. Copper–palladium alloys are different since the *d* bands of pure copper and palladium overlap considerably. It is therefore interesting to investigate the changes in the fine structure of the valence band as a function of alloy composition to establish whether the localization of the *d* states of both components is preserved.

Nemoshkalenko *et al.* [255] have used the coherent-potential method to calculate the electron density of states in the valence band of copper–palladium alloys. The calculations were performed as follows. Energy values in the high-symmetry points of the Brillouin zone were determined by the APW method. The crystal potential included the Coulomb and the exchange contribution. A Slater-type exchange potential was chosen. The contribution to the potential of the neighbors situated on the three nearest coordination spheres was found by using the scheme proposed by Mattheiss [211]. From the values of the eigenenergy at the points of high symmetry of the Brillouin zone (Γ, X, L, W, K), the parameters of the Hodges–Ehrenreich interpolation scheme were determined for each concentration of the alloy. To do this, a system of 15 equations with 15 unknowns was solved (three of these equations were nonlinear). The parameters obtained allowed the calculation of the energy values at any point of the Brillouin zone. The density of states of pure copper and palladium were found from the energy values calculated for 5230 points in 1/48 of the Brillouin zone. The calculations also yielded the e_g, t_{2g}, and s components of the density of states.

The density of states of pure copper and palladium was used for calculating the density of states in the valence band of the alloys. For the alloy with high copper content, the valence band was calculated on the basis of the density of states of pure copper, and for the alloy with high palladium content, it was calculated on the basis of the density of states of pure palladium. The experimental and calculated results are shown in Figures 91 and 92. The X-ray photoelectron spectra of pure copper and palladium, also measured with an HP-5950A spectrometer, were taken from the work of Hüfner *et al.* [193]. Owing to the low partial density of the s states of the pure components, the density of s states of the alloy was taken as a superposition of the density of states of both components (taking into account their concentration). As Figures 91 and 92 show, the basic features of the density of electronic states are reflected in the spectra. In particular, the X-ray photoelectron spectra of the valence band in the alloys with high-percentage copper content, illustrated in Figure 91, change markedly with concentration: the energy extension of the spectra increases, the intensity ratio of the peaks changes, and the fine structure gradually disappears. In alloys with high-percentage copper content, the *d* states of palladium appear as a resonance peak that moves closer to the Fermi level with increasing palladium content from 10% to 40%. This result is confirmed by the calculations. As can be seen from Figure 92, in alloys with high-percentage palladium content, an addition of 10% copper results in a change of the fine structure characteristic for pure palladium. The maximum of the local density of states of copper in alloys with

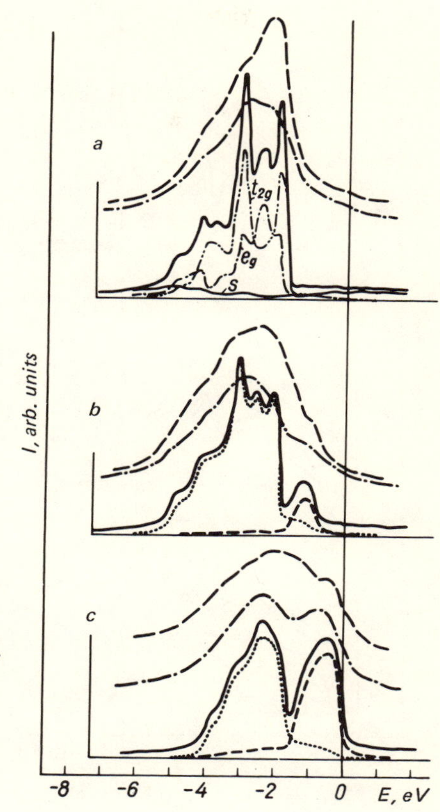

Figure 91. Valence band photoelectron spectra and the density of states in copper-rich copper–palladium alloys: (— —) photoelectron spectrum; (——) electron density of states; (—·—·—) smoothed density of states; (· · · ·) local electron density of states of copper; (– – –) local electron density of states of palladium; (a) copper; (b) $Cu_{90}Pd_{10}$; (c) $Cu_{60}Pd_{40}$.

high palladium content is situated at about 5 eV from the Fermi level, whereas in alloys with high copper content, this separation is about 3 eV. Figure 92 also shows that in the alloys with high palladium content investigated, the *d* resonance states of copper are not observed.

In conclusion, calculations performed using the coherent-potential method provide also, in the case of copper–palladium alloys, a general picture of the shape of the energy distribution of electrons. The discrepancies observed are due not only to the inadequacies of the theoretical model in the calculation of the valence band density of states in the alloys, but also to the fact that, in these calculations, the energy dependence of the photoionization cross sections of valence electrons has not been taken into account.

Table 13 gives the binding energies for a number of core electrons of copper and palladium in alloys and in the pure metals, as measured with an IEE-15 electron spectrometer [254]. For copper, the binding energy values are approx-

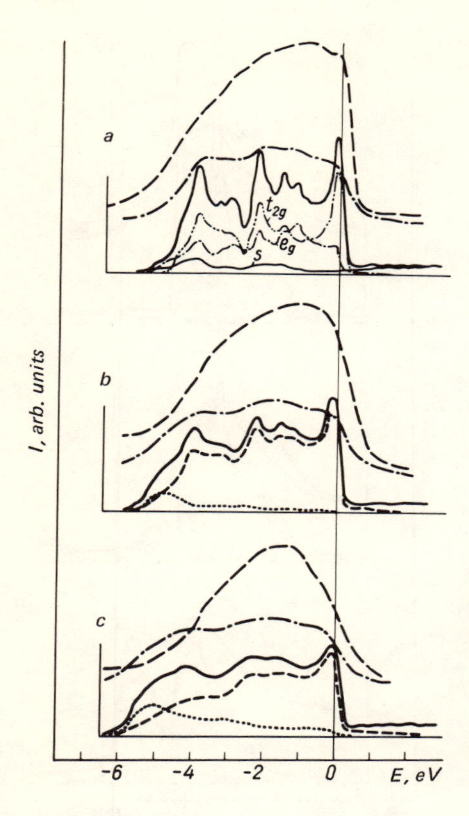

Figure 92. Valence band photoelectron spectra and the density of states in palladium-rich copper-palladium alloys: (– –) photoelectron spectrum; (——) electron density of states; (–·–·–) smoothed density of states; (· · ·) local electron density of states of copper; (– – –) local electronic density of states of palladium; (a) palladium; (b) $Pd_{90}Cu_{10}$; (c) $Pd_{75}Cu_{25}$.

imately 1 eV lower in alloys than in the pure metal. For palladium, the shifts in alloys as compared with the pure metal are lower, but the binding-energy values are higher in alloys than in the pure metal. Therefore, upon the formation of copper–palladium alloys, there occurs an increase in the electron density at the copper-atom sites and a decrease of it at the palladium-atom sites.

TABLE 13. Values of Core-Electron Binding Energies of Pd and Cu Atoms in Cu–Pd Alloys

	Level						
Sample	$2p_{1/2}$	$2p_{3/2}$	$3s_{1/2}$	$3p_{1/2}$ $3p_{3/2}$	$3p_{1/2}$	$3d_{3/2}$	$3d_{5/2}$
Pd					559.7	340.2	334.9
$Cu_{10}Pd_{90}$	951.2	931.3	121.8	74.5	559.8	340.3	335.0
$Cu_{60}Pd_{40}$	951.4	931.6	121.4	74.2	559.8	340.4	335.1
$Cu_{90}Pd_{10}$	951.7	932.0	121.8	74.5	559.7	340.7	335.4
Cu	952.1	932.1	122.2	75.0			

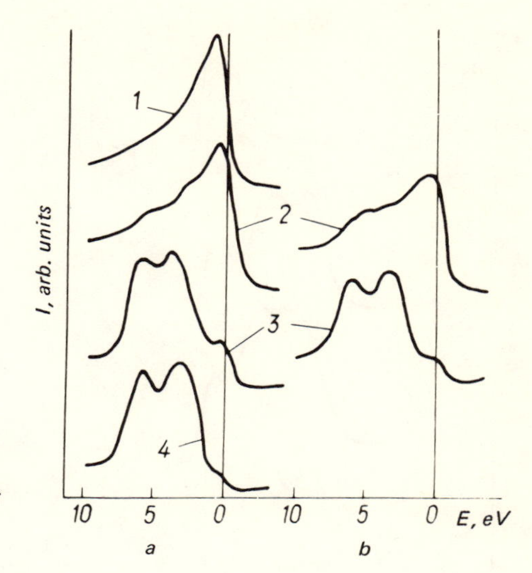

Figure 93. Valence band photoelectron spectra of iron, gold, and their alloys in quenched (a) and unquenched (b) states: (1) iron; (2) $Fe_{99.14}Au_{0.86}$; (3) $Fe_{47}Au_{53}$; (4) gold.

Iron–Gold Alloys. The iron–gold spectra studied by Nemoshkalenko and Aleshin [54] are interesting in that *d* subbands of the valence bands of the pure metals have very different widths. Furthermore, the maximum in the density of the iron *d* states is situated near the Fermi level, but for gold, the *s* and *p* states that have a low density are situated in this energy region.

One of the iron–gold alloys studied had a low gold content (0.86%) but, as is seen in Figure 93a, the distribution of photoelectrons in this alloy differs from that of pure iron on the high-binding-energy side of the main maximum. For the other alloy that contained 53% gold, a narrow maximum corresponding mainly to the local *d* states of iron appeared near the Fermi level.

Figure 93b shows the X-ray photoelectron spectra of the same alloys in the unquenched state. Since in this case the alloys consist of a mixture of two phases, it is easy to understand the smooth structure of the spectra; the spectrum of pure iron is superimposed over the X-ray photoelectron spectrum of the solid solution. Measurements of the binding energies of the $2p_{1/2}$ core electrons of iron and of the $4d_{3/2}$ and $4f_{5/2}$ electrons of gold have shown that in these alloys iron behaves as an electron acceptor, but the charge state of gold practically does not change.

Nickel–Gold Alloys. Nickel–gold alloys are similar to the iron–gold alloys discussed above. Figure 94 shows the X-ray photoelectron spectra of the valence electrons in nickel–gold alloys, as measured by Nemoshkalenko *et al.* [256]. It can be seen that in these alloys there is also a narrow peak corresponding to the *d* states of nickel reflected in the valence band of nickel–gold alloys. The binding energies of the core electrons of gold in the nickel–gold alloys are lower

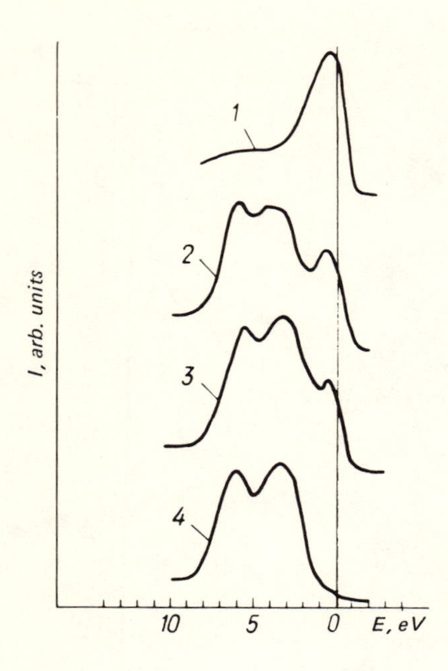

Figure 94. Valence band photoelectron spectra of nickel, gold, and their alloys: (1) nickel; (2) $Ni_{70}Au_{30}$; (3) $Ni_{30}Au_{70}$; (4) gold.

than in pure gold, which indicates that in alloys the electron density at the gold-atom sites is increased. For nickel, the shift of the core levels falls inside the experimental error bars (Table 14).

Palladium–Gold Alloys. Palladium–gold alloys are formed from elements characterized by broad d bands extended over energy ranges that exhibit considerable overlap. Figure 95 shows the X-ray photoelectron spectra of palladium–gold alloys produced from the data of Ref. 257. Also shown in Figure 95 are the corresponding spectra of pure palladium and gold taken from the works of Hüfner *et al.* [193] and Shirley [194]. It is seen that also in this system, the characteristic features of the spectra of pure palladium and gold appear in the spectrum of the alloys.

TABLE 14. Values of Core-Electron Binding Energies of Ni and Au Atoms in Ni–Au Alloys

Sample	Level							
	$2p_{1/2}$	$2p_{3/2}$	$3s_{1/2}$	$3p_{1/2,3/2}$	$4d_{3/2}$	$4d_{5/2}$	$4f_{5/2}$	$4f_{7/2}$
Ni	869.5	852.2	110.2	65.9				
$Ni_{70}Au_{30}$	869.4	852.1	110.1	65.9	352.7	334.7	87.3	83.6
$Ni_{30}Au_{70}$	869.5	852.2	110.0	66.0	352.9	334.8	87.3	83.6
Au					353.1	335.1	87.5	83.8

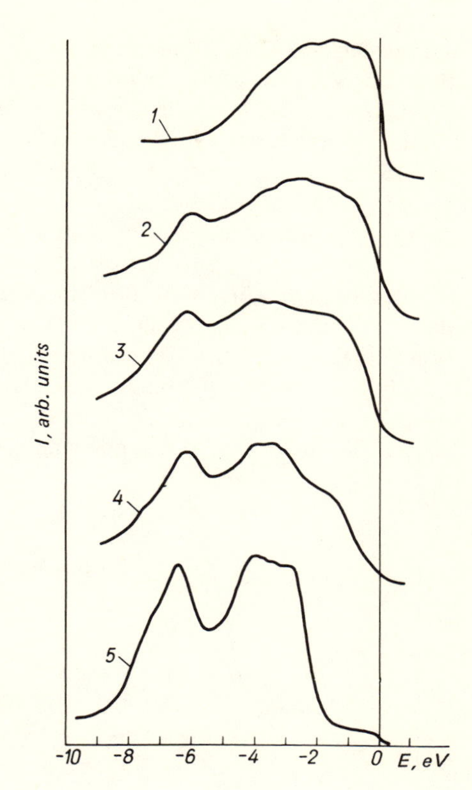

Figure 95. Valence band photoelectron spectra of palladium, gold, and their alloys: (1) palladium; (2) $Pd_{75}Au_{25}$; (3) $Pd_{50}Au_{50}$; (4) $Pd_{25}Au_{75}$; (5) gold.

Alloys of Isoelectronic Elements

Copper–Gold Alloys. Metals situated in the same subgroup of the periodic table may have essentially different valence band energy spectra even though they have the same electronic configuration in the free atomic state. This results in an appreciable change in their electronic properties in alloys [258]. Particularly marked is the change of the valence band density of states versus atomic number for isoelectronic metals at the end of the long periods. The magnitude of spin–doublet splitting actually increases sharply in the series Cu → Ag → Au, as also does the width of the valence band.

An interesting feature of copper–gold alloys is that they can be studied both in the disordered and ordered states. To obtain homogeneous solid solutions, the alloys $Cu_{75}Au_{25}$ and $Cu_{25}Au_{75}$ were quenched from a temperature of 650°C in a solution of 15% KOH.† To obtain ordered systems, the $Cu_{75}Au_{25}$ alloy was

†*Note added in proof:* The experiment on $Cu_{25}Au_{75}$ and $Cu_{75}Ag_{25}$ was performed by V. V. Nemoshkalenko, K. V. Chuistov, V. G. Aleshin, and A. I. Senkevich, *J. Electron Spec.* 9(2), 169 (1976).

annealed at 350°C for 12 hr, and the $Cu_{25}Au_{75}$ alloy at 230°C for the same time. The alloy $Cu_{75}Au_{25}$ exhibits a large degree of ordering of copper–gold alloys, but in $Cu_{25}Au_{75}$ the degree of ordering is less.

The X-ray photoelectron spectra of copper–gold alloys, measured with an IEE-15 electron spectrometer with the $K\alpha_{1,2}$ line of magnesium as excitation source, are shown in Figure 96 for both the disordered and ordered states. The background of inelastic electron scattering was corrected for.

The greatest difference between the photoelectron spectra of the two states is observed in the $Cu_{75}Au_{25}$ alloy, in which significant changes in the electron density of states occur. At the same time, the probability of photoelectron emission in the region of the high-energy maximum decreases.

The magnitude of the binding energies of the core-level electrons in pure copper and gold, and in copper–gold alloys, after the data from Ref. 254, are given in Table 15. For both copper and gold, it was found that the binding en-

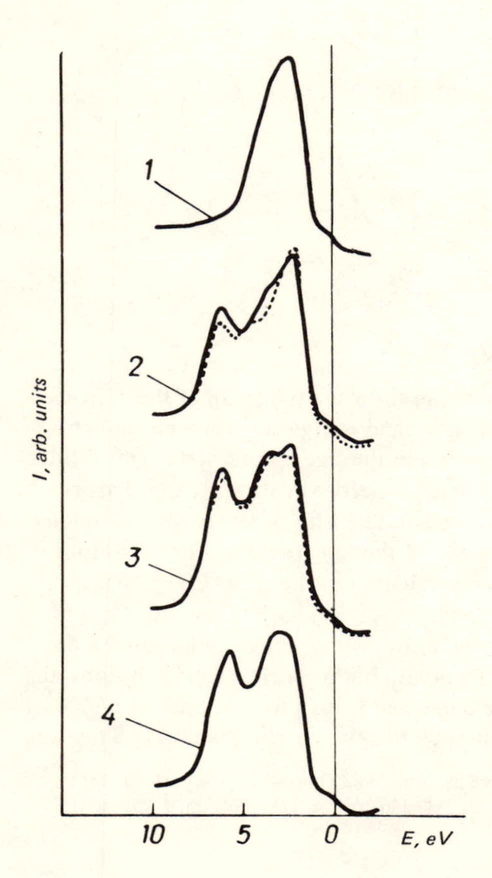

Figure 96. Valence band photoelectron spectra of copper, gold, and their compounds in the disordered (——) and ordered (· · ·) states, respectively: (1) copper; (2) Cu_3Au; (3) $CuAu_3$; (4) gold.

TABLE 15. Values of Core-Electron Binding Energies of Cu and Au Atoms in Cu–Au Compounds

Sample	Level					
	$2p_{1/2}$	$2p_{3/2}$	$4d_{3/2}$	$4d_{5/2}$	$4f_{5/2}$	$4f_{7/2}$
Cu	951.2	932.3				
Cu_3Au	951.6	931.8	352.8	334.8	87.4	83.7
$CuAu_3$	951.5	931.7	352.7	334.7	87.2	83.4
Au			353.1	335.1	87.5	83.8

ergies of the core electrons of the pure elements are lower than in the alloys. This means that in alloys, the electron density at the sites of both copper and gold atoms is increased relative to that in the pure metals. Upon ordering, no significant change of the core-level binding energies occurs.

Silver–Gold Alloys. The X-ray photoelectron spectra of silver–gold alloys are illustrated in Figure 97 [54]. As in the case of nickel–copper alloys, the width of the valence band in silver–gold alloys increases with the concentration of the component with larger valence bandwidth. Thus in silver the valence bandwidth is 8.9 eV, and in gold it is 9.4 eV. For both silver and gold, the density of states at the Fermi level is relatively low. It is in general determined by the *s*- and *p*-type states. The X-ray photoelectron spectra clearly show the max-

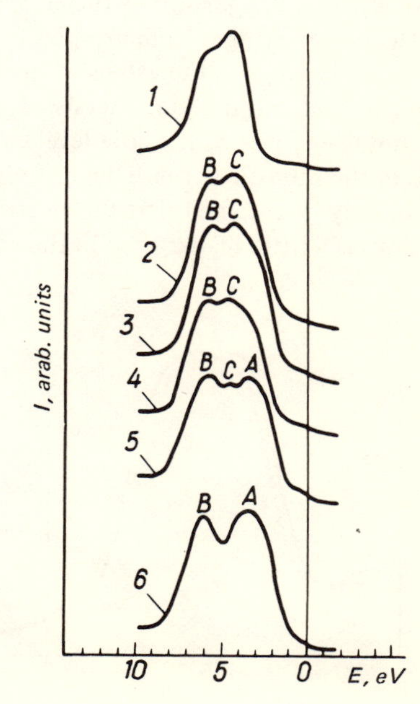

Figure 97. Valence band photoelectron spectra of silver, gold, and their alloys: (1) silver; (2) $Ag_{91}Au_9$; (3) $Ag_{75}Au_{25}$; (4) $Ag_{49}Au_{51}$; (5) $Ag_{11}Au_{89}$; (6) gold [54].

imum *B* corresponding to the high-energy maximum in the X-ray photoelectron spectra of pure gold and silver, as well as the peak *C* corresponding to the low-energy maximum in the spectrum of pure silver. Also apparent is the enhancement on the low-energy branch of the spectrum corresponding to the maximum *A* of pure gold.

Studies of silver–gold alloys have also been performed by Siegbahn *et al.* [1]. Even at a relatively low concentration of the second component, it was possible to observe changes in the structure of the valence band of the atoms. Thus a concentration of only 11% silver results in the appearance of the maximum *C*, while 9% gold causes changes in the band structure at the bottom and at the top of the valence band.

We have mentioned earlier that in a number of investigations of nickel–copper alloys, attempts were made to represent the valence band of the alloy as the superposition of the bands of the pure components. Since the resolution of X-ray photoelectron spectrometry is not very high (amounting to 1.1 eV in the work of Nemoshkalenko *et al.* [259], in which the X-ray photoelectron spectra of valence bands of silver–gold alloys were measured), and since the spectra of nickel–copper are quite narrow and exhibit practically no fine structure excepting the two maxima corresponding to the *d* electrons of nickel and copper, it follows that the validity of this approach is difficult to test for these alloys.

Figure 98 illustrates the X-ray photoelectron spectrum of the alloy $Ag_{49}Au_{51}$ and also the superposition of the spectra of pure gold and silver, weighted in proportion to the concentration of these elements in the alloy. The effect of alloying is evident.

Table 16 gives the values of the core-level binding energies of gold and silver in alloys and in the pure metals, after the data from Ref. 254. It is seen that the binding energies of the core-level electrons in the alloys are practically the same as in the pure elements. The differences do not exceed 0.2 eV. Therefore, it can only be assumed that the electron density in the vicinity of silver atoms in alloys is a little higher than in the pure metal. This assumption does not contra-

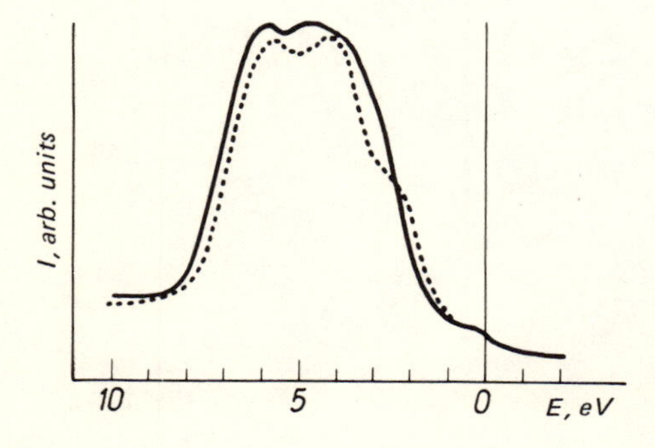

Figure 98. Valence band photoelectron spectra of the alloy $Ag_{49}Au_{51}$ (——) and the curve (· · ·) obtained by superposition of the photoelectron spectra of the pure components.

TABLE 16. Values of Core-Electrons Binding Energies of Ag and Au Atoms in Ag–Au Alloys

Sample	Level							
	$3p_{1/2}$	$3p_{3/2}$	$3d_{3/2}$	$3d_{5/2}$	$4d_{3/2}$	$4d_{5/2}$	$4f_{5/2}$	$4f_{7/2}$
Ag	603.8	572.8	374.1	368.1				
$Ag_{91}Au_9$	603.7	572.9	374.0	368.0	353.0	334.9	87.5	83.8
$Ag_{75}Au_{25}$	603.7	572.9	374.1	368.1	353.1	335.2	87.7	84.0
$Ag_{49}Au_{51}$	603.7	573.0	374.0	368.0	353.1	335.1	87.6	83.9
$Ag_{11}Au_{80}$	603.7	572.8	374.0	368.0	353.1	335.1	87.6	83.9
Au					353.1	335.1	87.5	83.8

dict the results of Levin and Ehrenreich [260], who, on the basis of the analysis of the parameters of optical spectra, have concluded that some transfer of charge occurs from gold to silver in silver–gold alloys.

All the alloys investigated by Nemoshkalenko and co-workers in the works mentioned above [54, 254, 256, 257, 259] were prepared from metals of high purity (99.99%), in vacuum or in a noble gas atmosphere. Since the studies were concerned with solid solutions, specimens were quenched from a temperature about 100°C above the transformation temperature for those alloys that normally become two-phased (Pt–Au, Cu–Pd, Cu–Au) or ordered (Pt–Au, Cu–Pd, Cu–Au) upon cooling. The effectiveness of quenching was checked by X-ray diffraction.

Alloys of Noble Metals with Aluminum and Magnesium

Fuggle *et al.* [57] studied the X-ray photoelectron spectra of the valence electrons of aluminum alloys with copper, silver, and gold. Their method of sample preparation and of spectra measurement have been discussed in Chapter 1. Figures 99–101 illustrate the X-ray photoelectron spectra of aluminum-copper, aluminum-silver, and aluminum-gold alloys. The spectra of the pure metals are consistent with the spectra obtained by Baer *et al.* [186]. In all the spectra shown in Figures 99–101 the Fermi level is well-defined. Most clearly appear in these spectra the *d* states. The splitting of the *d* states is absent in the copper alloys, but appears to be rather great in silver and gold alloys. In each alloy system, the energy distance of the *d* levels from the Fermi level increases with the aluminum content. In the X-ray photoelectron spectra of the valence electrons in Al_2Au measured by Hüfner *et al.* [261] using an IEE-15 electron spectrometer with the magnesium $K\alpha_{1,2}$ line as excitation source, the splitting of *d* states is less evident than in the spectra in Figure 99. This can possibly be explained by the fact that, in the experiment of Hüfner, Wernick, and West, the sample was cleaned by argon ion bombardment. According to the results of

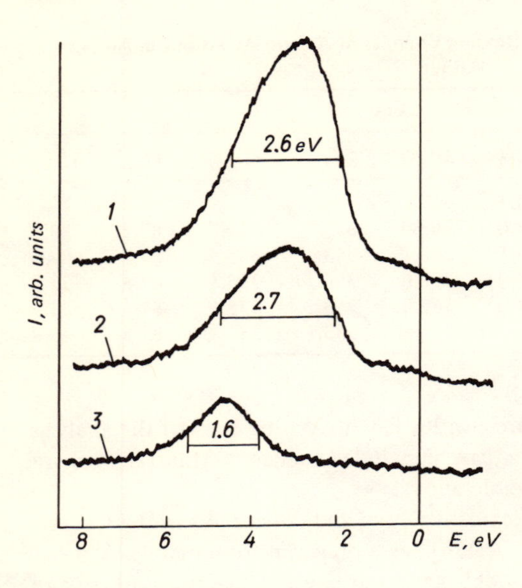

Figure 99. Valence band photoelectron spectra of aluminum-copper compounds: (1) copper; (2) Al_4Cu_9; (3) Al_2Cu.

Hüfner *et al*. [58], this treatment can result in the formation of an amorphous layer on the specimen surface. Particularly evident is the discrepancy between the spectra of palladium prepared by evaporation and palladium cleaned by argon ion bombardment.

One should notice that the alloys corresponding to the formulas Al_4Cu_9

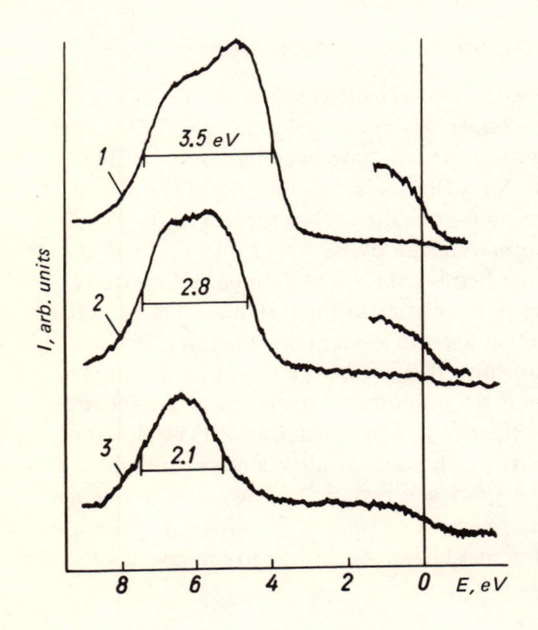

Figure 100. Valence band photoelectron spectra of aluminum–silver compounds: (1) silver; (2) $AlAg_2$; (3) $AlAg_p$.

TABLE 17. Energy Position of the Peaks in the Photoelectron Spectra of Valence Electrons of Compounds of Al with Cu, Ag, and Au

Sample	Peaks position	FWHM
Cu	2.7	2.6
Al_4Cu_9	2.9	2.7
Al_2Cu	4.3	1.6
Ag	4.5 6.2	3.5
$AlAg_2$	5.7 6.5	2.8
$AlAg_p$	6.2	2.1
Au	3.3 6.1	5.3
$AlAu_2$	4.5 7.0	4.6
AlAu	5.7 7.4	4.2
Al_2Au	6.1 7.4	3.2

and $AlAg_2$ are, in fact, not intermetallic compounds, since a whole series of possible alloys exists near the stoichiometric composition. $AlAg_p$ form a metastable solid solution of silver and aluminum (approx. 20 at. % Ag). The energy position of the maxima in the X-ray photoelectron spectra and their full width at half-maximum (FWHM) are shown in Table 17.

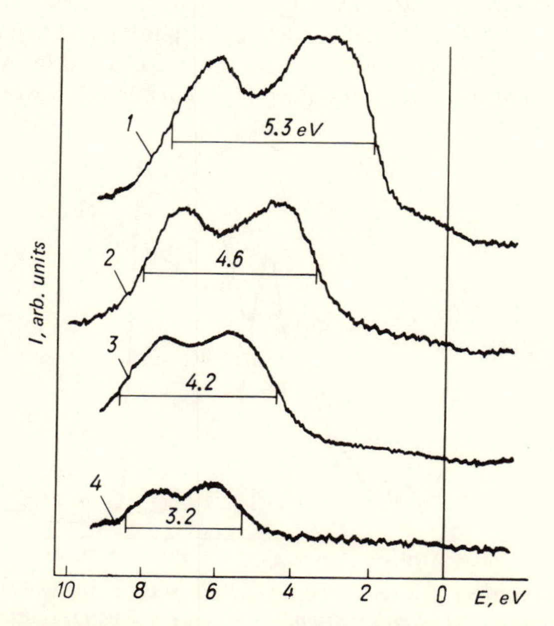

Figure 101. Valence band photoelectron spectra of aluminum–gold compounds: (1) gold; (2) $AlAu_2$; (3) AlAu; (4) Al_2Au.

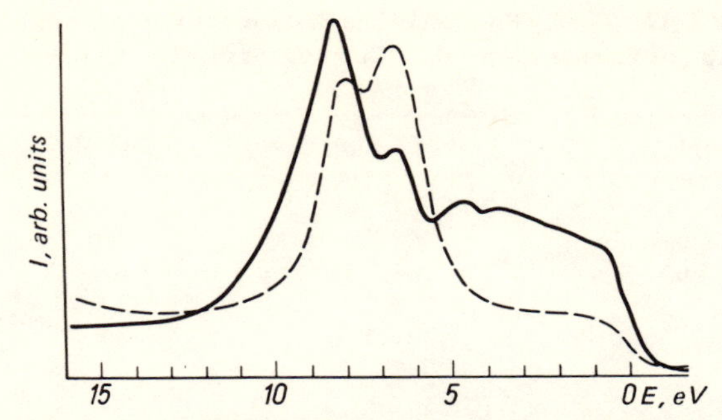

Figure 102. Photoelectron and X-ray emission spectra of the compound Al_2Au: (– – –) valence band photoelectron spectrum; (——) $L_{2,3}$ emission spectrum of aluminum.

Figures 102 and 103 show the X-ray photoelectron spectra of alloys Al_2Au, Mg_3Al and the $L_{2,3}$ emission line of aluminum, according to the data from Ref. 262. The two peaks corresponding to the splitting of the *d* states in the alloys may easily be correlated to the energy positions of the two peaks in the X-ray emission spectra. However, in the emission spectra there exists an additional structure which is absent in the X-ray photoelectron spectra. It reflects the distribution of *s* electrons in the higher part of the valence band of these alloys. In the X-ray photoelectron spectra this is suppressed by the *d* electrons, which have a much higher density of states and a much higher photoionization cross section.

Hüfner *et al.* [261] have also studied the alloys GaAu and In_2Au. The magnitude of *d* state splitting in these intermetallic compounds is lower than in pure

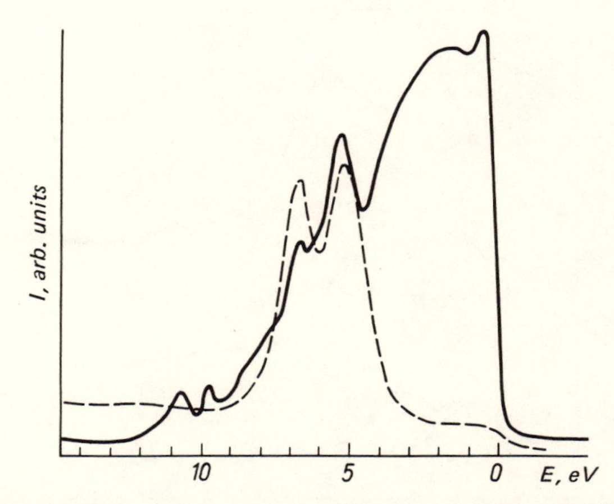

Figure 103. Photoelectron and X-ray emission spectrum of the compound Mg_3Au: (– – –) valence band photoelectron spectrum, (——) $L_{2,3}$ emission spectrum of magnesium.

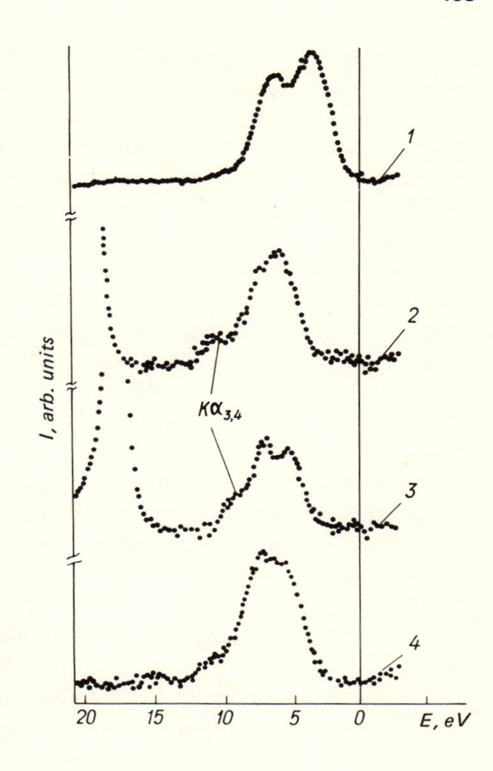

Figure 104. Valence band photoelectron spectra of pure gold and its compounds: (1) gold; (2) Ga_2Au; (3) In_2Au; (4) Al_2Au.

gold (Figure 104). The widths of the *d* bands determined experimentally and theoretically [263] are significantly different from each other.

Model Representations of the Density of States of Alloys

Large advances have been made in the study of the electron structure of many pure metals, but this cannot be said about disordered alloys. Use has largely been made of a number of very simplified theoretical models of the electron structure of alloys, for example, the model of rigid bands [264], the model of virtual bound states [265, 266], the model of virtual potential [267], and the two-band model [268].

In the model of rigid bands, the assumption is made that in the alloy a common valence band exists containing the valence electrons of both component atoms. The structure of this band may be obtained from the valence band structure of one of the components by simply shifting the Fermi level up to the energy value corresponding to the mean concentration of valence electrons for the given alloy. The model is most extensively used for calculation of the density of states in the valence band using measured values of the electronic specific thermal capacity and paramagnetic susceptibility [269]. Attempts to introduce restrictions into the model of rigid bands are well-known. Thus, analysis of the X-ray spectra of transition metals and of alloys of elements of groups III and VI led Nemnonov [270] to develop a modified rigid-band model, which he named the model of collective (generalized) d, s band. Unfortunately, many recent experimental studies [271–276] have proved that this model is not valid for the description of the properties of a series of alloys.

The model of virtual bound states has a limited area of validity, namely, for highly dilute solid solutions in which the valency of the impurity atoms is different from that of the atoms of the basic component. The excess charge formed at the impurity atom is effectively screened by the local deformation of the valence band. However, this model is not valid when the impurity concentration is sufficiently high to cause significant interaction between the impurity atoms.

In the model of virtual potential, the disordered alloy of components A and B, having the potentials v_A and v_B at the lattice nodes of the A and B atom sites, is replaced by an ordered system having the potential $cv_A + (1 - c)v_B$ (c is the concentration of A atoms). It is assumed that the dispersive potential $v_A - v_B$ is small. This assumption is not valid, however, for alloys of metals having dispersive resonances in the conduction band, that is, for the transition and noble metals.

Another special case of the rigid-band model was treated by Varley [268], and was named the two-band model. In this model, it is assumed that the valence electrons exist in two separate groups of energy states, each of them being related to the potential fields of the ionic components of the alloy. The models of virtual bound states and virtual potential are intermediate between the model of rigid bands and the two-band model.

The most important difficulties in the theoretical analysis of disordered systems arise from the lack of periodicity of the potential. Therefore, in the treatment of disordered alloys within the framework of the one-electron approximation, it is necessary to introduce simplifications, not only with respect to the choice of the one-electron potential for each of the alloy components, but also in the solution of the resulting one-electron Schrödinger equation. As we have already mentioned, a recently developed and successful method of solution of the one-electron Schrödinger equation for disordered systems is the coherent-potential approximation [158, 159]. It is more realistic than the virtual potential approximation and may be used for calculation of the density of states, in alloys of arbitrary concentration of the components. The coherent-potential approximation has led to the conclusion that the density of electron states localized in the vicinity of the atom of a given alloy component may differ significantly from the mean density of electronic states. This implies that the local density of electron states will be extremely sensitive to the local environment.

The band structure of binary alloys with a low concentration of the second component, obtained by using the coherent-potential method is intermediate between those obtained using the virtual bound states of Friedel and Anderson and the two-band model of Varley.

Stocks *et al*. [276] have compared the results of theoretical calculations using the rigid-band model, the virtual-potential model, and the coherent-potential model with the experimentally measured energy position of the midpoint of the uppermost subband. This midpoint characterizes the position relative to the Fermi level of the subband of the valence band constituted mainly of the copper states in nickel–copper alloys. The samples investigated were nickel–copper alloys of various compositions, up to 40% nickel.

The predictions of these theoretical models differ considerably among themselves. The rigid-band and virtual-potential models predict that an increase in nickel concentration in the alloy results in an almost linear approach of the subband to the Fermi level. The coherent-potential model predicts an insignificant change in its energy position, thus providing the best agreement with experiment.

We have in this chapter studied alloys formed from a variety of pure elements situated either closely to each other in the periodic system or in different periods.

It should be noted that in the alloys of elements in groups VIII and IB, the number of d electrons is much higher than that of p electrons, and therefore the X-ray photoelectron spectra mainly reveal the behavior of the d electron distribution in the valence band. The strong individuality of the valence d electrons of the components in alloys is to a great extent caused by their high centrifugal barrier, since this determines their energy and space localization.

The density of states in alloys is not a result of the simple superposition of the density of states of pure metals. In disordered alloys, the local density of electronic states, which is significantly different from the density of states in the pure components, is very important. Consequently, if the energy positions of the centers of weight of the d states of the pure components are separated from each other, then the total density of states in various parts of the energy spectra is mainly determined by the local density of states of one or other of the components.

Comparison of calculated valence band density of states in alloys, obtained with the coherent-potential method, with the corresponding X-ray photoelectron spectra indicates that this method provides a rather satisfactory theoretical description of the electron structure of disordered alloys.

4

Crystals with a Sphalerite-Type Lattice

Crystals with a sphalerite-type† lattice, characterized by a tetrahedral distribution of atoms, have a relatively simple crystal structure and are suitable for theoretical calculations. For a long time, these crystals have been the object of study in order to test both old and new theoretical and experimental methods of examining the band structure of solid state materials. In particular, their optical and ultraviolet photoemission properties have been actively investigated. This has made it possible to establish how well theoretical predictions about their band structure agree with experiment. The work of Cohen and Bergstrasser [138] using the pseudopotential approach has been of great importance to the theoretical and experimental investigations of the large group of crystals including diamond, silicon, germanium, A_3B_5, and A_2B_6 compounds. These authors have shown that in the interpretation of a great number of experimental results, the empirical pseudopotential method can be successfully used. It has been shown that, for the interpretation of the band structure of GaP, GaAs, GaSb, CdTe, ZnSe, InAs, InP, and InSb compounds, it is sufficient to use six parameters, three symmetrical and three antisymmetrical form factors for the pseudopotential. For diamond, silicon, and germanium, three parameters are sufficient, since in this case the antisymmetric form factors of the potential are equal to zero.

In the work of Cohen and Bergstrasser [138], the parameters were determined mainly from the optical reflection spectra. Studies of the optical properties of crystals are in general an important factor in the determination of their band structure. However, a detailed interpretation of the optical reflection spectra is seldom possible without a semiquantitative calculation of the energy-band structure, since the optical excitations fall into the energy region for which the valence band states and the states of the conduction band make a significant

†*Editors' note:* The more common name for sphalerite in English is zinc blende.

contribution to the observed structure of the spectra. In their turn, the reality and the accuracy of these calculations depend very much on the correctness of interpretation of the experimental results. This process for the determination of the band structure is, undoubtedly, a method of trial and error, but it often leads to a noncontradictory, quantitative, and to a high degree detailed picture of the band structure of semiconductors in a limited energy region in the vicinity of the forbidden band gap. However, the band structure of a great number of semiconductors is not as yet satisfactorily known, in spite of the fact that for these materials reliable experimental data exist.

Much progress has also been made with regard to the determination of the initial and final states of electrons participating in transitions, by studying the photoemission in the far-ultraviolet region. This information is extracted from the photoemission spectra by analyzing the dependence of their shape on the excitation energy. However, it is also necessary to take into account preliminary theoretical conclusions regarding the energy structure of the valence and conduction bands. The shape of the optical and photoemission spectra is strongly affected by the structure of the conduction band. Only when the energy of the exciting photons is increased to 50–100 eV, will the density of final states become a smooth function.

Such a condition is satisfied for exciting energies of the order of 50–100 eV. However, even in this case, the photoemission spectra do not allow a direct determination of the density of states of valence electrons, since the magnitude of the photoionization cross section for electrons of different symmetries varies as a function of the excitation energy. It has, however, been established that at low values of photon energies this variation has a rather small influence on the shape of the spectra, in comparison with that due to the structure of the final states. Therefore, in dielectrics, as in metals, the density of valence states can be determined more accurately from photoemission spectra measured at high excitation energies, than from spectra measured at low excitation energies or from the optical spectra.

The energy bands calculated using the values of the parameters determined by Cohen and Bergstrasser [138] are consistent with the experimental data in the energy region close to the top of the valence band and the bottom of the conduction band. This is just that energy region in which are found the main optical transitions that have been used for the determination of the form factors of the pseudopotential. The method of vacuum ultraviolet spectroscopy does not allow a study of energy states over the whole width of the valence band, which may be relatively large (of the order of 12–15 eV) in the majority of crystals.[†]

[†]*Editors' note:* Most UV photoemission experiments today are carried out either with resonance lamps, which give photon energies up to 40.8 eV, or with synchrotron radiation, which in this connection is used up to several hundred electron volts.

The accuracy in the description of states situated at the bottom of the valence band may be assessed by comparing the calculated results with the experimental data provided by the X-ray emission spectroscopy and photoelectron spectroscopy.

Calculations of the density of states of the valence band have been performed by the authors of the present monograph and their co-workers [5, 277–281] and compared with the data from X-ray emission spectroscopy, the purpose being to test the accuracy with which the form factors calculated by Cohen and Bergstrasser [138] provide a description of the structure of valence bands in crystals. The present authors [5, 148, 282] have shown that the X-ray emission spectra of light elements reflect the van Hove singularities in the electron density of states. Therefore, by using the theoretical and experimental data for the energy-band structure, it is possible to determine the regions in **k** space that lead to the observed features of the experimental spectra. It is at the same time possible to determine the energy position of states with different types of symmetry. It should be remembered that, for the simplest crystals, the symmetry properties of the energy bands determine—to some extent—the order of their distribution.

The results provided by X-ray emission spectroscopy should be used in conjunction with the optical data, to ensure that the derived empirical pseudopotential yields, first, a satisfactory trend in the distribution of electron density of states over the whole valence band of crystals, and, second, energy values of the optical transitions that are consistent with the experimental data. Nemoshkalenko *et al.* [277, 278] constructed a pseudopotential for β-SiC and for silicon that correctly reproduced the positions of van Hove singularities found from X-ray emission spectra. The consistency of this pseudopotential with the optical data was, however, rather poor. Use of the local approximation and of a potential independent of energy is possible only when applied to a limited energy region [120].

The necessity of using a nonlocal pseudopotential became apparent when it was found to be possible to measure energy levels difference between states with different quasimomenta **k**. Measurement of the parameters of X-ray photoemission allow the determination of these energies, and, therefore, in order to describe the experimental data, one should make corrections to the local pseudopotential. Pollak *et al.* [283] and Chelikowski *et al.* [284] have shown that, in using the local approximation, the nonlocal character of the pseudopotential may be taken into account approximately by replacing the electron mass m with an effective mass m^*. More accurate calculations have shown that, in spite of its simplicity, this replacement allows for the main effects of the nonlocal character of the pseudopotential and improves agreement with experiment for the upper part of the valence band. In order to obtain satisfactory agreement at the bottom of the valence band, it is necessary to make more accurate allowance for the nonlocal character of the pseudopotential.

Energy-Band Structure of Diamond, Graphite, Silicon, and Germanium

The crystals of elements in the fourth group of the periodic system have interesting physical properties. Silicon and germanium are semiconductors with small gaps. Diamond, although it has the same crystal structure, is a good insulator. However, at normal temperature and pressure, the thermodynamically stable phase is not diamond, but graphite—a semimetal that does not have any analogues in the fourth group of elements. Graphite, unlike diamond, is characterized by a trigonal coordination of carbon atoms.

Cora *et al.* [285] were the first to study the energy band structure of diamond by using the method of X-ray photoelectron spectroscopy. Unfortunately, their experiment was performed with poor resolution (1.5–1.7 eV) and a rather poor vacuum (2×10^{-7} torr). Though the specimen was cleaned with respect to adsorbed gases and carbon, the X-ray photoelectron spectra of the valence electrons still show a clear structure due to the presence of oxygen. Much better experimental conditions were used by Cavell *et al.* [286] and McFeely *et al.* [287], who measured the X-ray photoelectron spectra with an HP-5950A electron spectrometer, with a resolution of 0.55 eV. The vacuum level in the spectrometer was about 8×10^{-9} torr, and as the excitation source, the monochromatized $K\alpha_{1,2}$ radiation of aluminum was used. Without cleaning of the diamond surface, the intensity ratio of the 1*s* peaks of carbon and oxygen was 4. After cleaning in a nitrogen atmosphere, a better ratio, viz., 13, was attained. This permitted the determination of the valence band electron spectra of diamond without significant distortion. Since the 1*s* line of carbon is energetically closest to the valence electron band, it was used for the subtraction of the background of inelastically scattered electrons from the photoelectron spectrum of the valence electrons. For this purpose, the fine structure of the carbon 1*s* line was studied carefully in the region of high energies. The corrected spectrum, together with the density of valence band electronic states, is shown in Figure 105. The calibration of the spectrum was performed using the $4f_{7/2}$ line of gold, the binding energy of which was taken to be 84.0 eV.

The density of electronic states has been calculated by Painter *et al.* [288] using a modified tight-binding method (MLCAO). Here, in contrast to the ordinary tight-binding method, the many-center integrals were not calculated. Instead, the matrix elements $H_{ij} = \langle \varphi_i | H | \varphi_j \rangle$ and $S_{ij} = \langle \varphi_i | \varphi_j \rangle$ were calculated directly, where $|\varphi_i\rangle$ are the Bloch sums of atomlike functions. Account was also taken of the deviation of the potential from the MT potential. It turned out that, in order to obtain more accurate qualitative results, it was necessary to introduce corrections to the MT potential. For example, if the deviations from the MT potential are neglected, a width of the indirect gap is obtained for diamond that is five times smaller than that obtained when the parameter α in the exchange potential V_{ex}^{α} is changed from 1 to $\frac{2}{3}$.

Since in diamond the elementary cell contains two atoms, the valence band contains eight electrons distributed over four energy bands. Maximum (1) (Figure 105) in the density of states is determined by the states in the first energy band. This band has an *s*-like character, since it contains only a small number of *p* states. Along the *X–Z–W* directions in the Brillouin zone, the first and the second bands are degenerate. The second band has a higher content of *p* states compared to the first band, and the second maximum is determined by the details of its structure. The top of the valence band, where the third and the fourth bands are localized, appears as the third maximum (3) in the density-of-states curve. The states situated close to the top of the valence band have a high content of *p*-type states. This picture of the distribution of states in the valence band derived on the basis of their symmetry is confirmed by the investigation of the X-ray emission *K* line [289] resulting from electron transitions from the valence band to the 1*s* level. Since the shape of the *K* spectrum is determined by the density of states of electrons with a *p*-type symmetry in the valence band, it follows that the intensity will be highest in that energy region of the valence band in which *p*-type symmetry states are preponderant, i.e., at the top of the valence band. Since the width of the inner 1*s* level in the transitions leading to X-ray emission is 0.1–0.2 eV, the resolution of the X-ray emission spectra is determined mainly by the instrumental resolution. Wiech and Zöpf [289] maintained that the resolution in their work was approximately 0.6 eV, which is similar to the resolution of the HP-5950A spectrometer that was used for measuring the valence band spectra of diamond. However, the X-ray emission spectrum contains details of structure that were not present in the photoelectron spectrum.

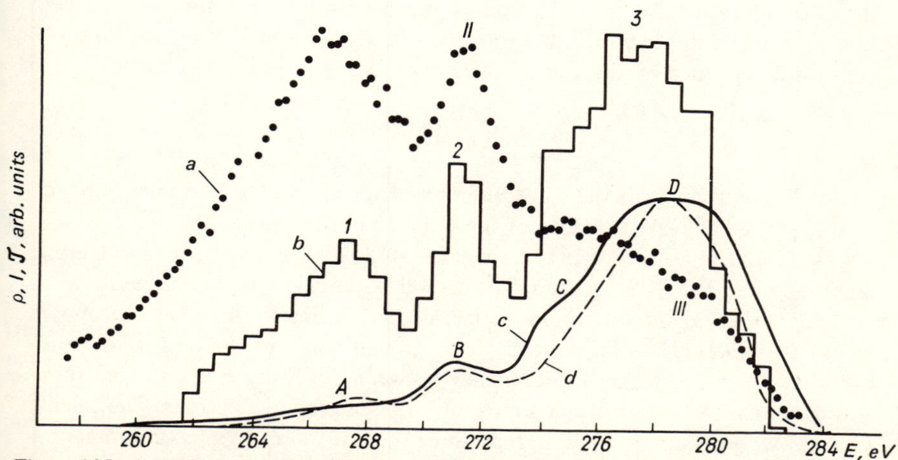

Figure 105. (a) Valence band photoelectron spectrum, (b) density of states, (c) X-ray emission K spectrum (experiment), and (d) X-ray emission K spectrum (theory) of diamond.

This indicates that the X-ray photoelectron spectrum was recorded with a greater spread in the intensity values than the emission spectrum.

The comparison of the X-ray emission and photoelectron spectra was performed as follows [287]. The maximum B is characterized by an energy of 271.1 eV of the emitted photon. In the photoelectron spectrum, it corresponds to the maximum II, since the difference between the binding energies of the carbon $1s$ line and of this maximum, measured with respect to the Fermi level of the spectrometer, is equal to $E_B^F(1s) - E_B^F(\text{II}) = 284.4 - 13.2 = 271.2$ eV. The two values are equal to within 0.1 eV, which lies inside the limits of experimental errors. From Figure 105 it can be seen that the width of the X-ray emission line is close to that calculated, while the bandwidth determined from the photoelectron spectrum is greater than the calculated width of the valence band. This is possibly related to the fact that the contribution of electron multiple scattering could not be correctly subtracted from the experimental photoelectron spectrum. Therefore, in order to determine the width of the valence band, the curve in the region of high binding energies was approximated to a parabola. In this way, the width of the valence band was estimated to be 24.2 eV. Although the width of the X-ray emission spectrum is close to the width of the valence band calculated by Painter *et al.* [288] (20.8 eV), it is evident from Figure 105 that its value is difficult to be determined with sufficient accuracy. This is caused above all by the fact that in the region of maximum 1, the X-ray emission spectrum does not have any structure, probably because of Auger-smoothing of the states situated at the bottom of the valence band [290]. The valence bandwidth determined from the X-ray emission spectra is approximately 21.0 ± 1.0 eV.

The shapes of the X-ray emission and X-ray photoelectron spectra are significantly different, although their structures are basically similar. This results from the different probabilities of electron excitation and photon emission. As we have mentioned in Chapter 2, the transition probability of X-ray emission processes is determined by the matrix element of the transition probability:

$$M(\mathbf{k}) = \langle \Psi_{n\mathbf{k}}(\mathbf{r}) \,|\, \nabla \,|\, \Psi_{c\mathbf{k}}(\mathbf{r}) \rangle,$$

where $\Psi_{n\mathbf{k}}(\mathbf{r})$ represents the wave function of the electron in the valence band and $\Psi_{c\mathbf{k}}(\mathbf{r})$ is the wave function of the electron in the conduction band.

Nemoshkalenko *et al.* [291] have calculated the X-ray emission spectrum of diamond using the APW method. The crystal potential for the configuration $1s^2 2s 2p^3$ of the diamond atom was obtained according to the scheme proposed by Mattheiss [211]. As Figure 105 shows, the influence of the transition probability results in a significant discrepancy between the X-ray emission spectrum and the curve of the density of states of the valence band electrons. The calculations, however, were not able to reproduce the flat top of the maximum D observed in the X-ray emission spectrum. This is possibly due to the fact that the interpolation of matrix elements was performed with only 89 **k** points in $\frac{1}{48}$ of the Brillouin zone. In spite of the great number of **k** points used in the Brillouin zone as a whole (approximately 10^6 points), the basis of principal points that was chosen is probably still insufficient for a detailed reproduction of the fine structure of the emission spectrum.

As was shown in Chapter 2, the intensity of X-ray photoemission may be written approximately:

$$I(E) \approx \rho^i(E)\,\rho^f(\hbar\omega - E)\,\sigma(\hbar\omega, E),$$

where $\rho^i(E)$ represents the density of valence electron states, $\rho^f(E)$ is the density of final states of photoelectrons, and $\sigma(\hbar\omega, E)$ is the photoionization cross section. When the $K\alpha_{1,2}$ line of aluminum is used as the source of exciting photons, the magnitude of $\rho^f(\hbar\omega - E)$ is practically the same as for free electrons, and therefore

$$I(E) \approx \rho^i(E)\,\sigma(\hbar\omega, E).$$

If the wave functions of electrons in the conduction band are chosen as plane waves, $\exp i(\mathbf{k}+\mathbf{Q})\cdot\mathbf{r}$, and the wave functions of electrons in the valence band as Bloch functions, $\Psi_k(\mathbf{r}) = \exp(i\mathbf{k}\cdot\mathbf{r})\,u_k(\mathbf{r})$, then the following is obtained:

$$\sigma(\hbar\omega, \mathbf{k}) \approx |\langle \Psi_{\mathbf{k}}(\mathbf{r})\,|\exp i(\mathbf{k}+\mathbf{Q})\cdot\mathbf{r}\rangle|^2.$$

The wave function $\Psi_k(\mathbf{r})$ may be represented as an expansion over spherical functions of s, p, and d type. The s functions are characterized by sharper variations than p functions, and, consequently, the magnitude of the overlap integral of the function $\Psi_k(\mathbf{r})$, with the plane wave corresponding to high kinetic energies, is greater for s functions than for p functions. The magnitude of $\sigma(\hbar\omega, \mathbf{k})$ is therefore greater for states in which the wave functions contain a significant contribution from s-type functions. In diamond, the photoelectron spectrum should have a greater intensity in the energy region in which the valence band mainly contains s-type states, than in the region where mostly p-type states exist. This conclusion is confirmed by the experimental curve shown in Figure 105.

The simultaneous use of X-ray emission and photoelectron spectra allows a determination of the degree of hybridization of the s and p states in diamond, since both spectra are determined by the behavior of the valence electron wave function in the immediate vicinity of atomic nuclei.

Let us consider the expression

$$\frac{[I'(E)/\rho(E)]_{\mathrm{I}}}{[I'(E)/\rho(E)]_{\mathrm{III}}} \approx 5.$$

If peak I was determined only by the $2s$ states and peak III only by the $2p$ states, then this ratio would have a value close to

$$\frac{\sigma(2s)}{\sigma(2p)} = 13.$$

In order to determine $R_{XPS} = I'(E)/\rho(E)$ and $R_X = \mathfrak{I}(E)/\rho(E)$ from the data in Figure 105, McFeely *et al.* [287] succeeded in extending somewhat the energy region in which the density of states $\rho(E)$ is finite. They were then able to smooth the rather rough curve obtained by direct calculation of these functions at points chosen on the energy scale. The smoothed functions R_{XPS} and R_X are shown in Figure 106. In order to use them further for the determination of the character of the hybridization of the s and p states in diamond, it is necessary to introduce the functions $f_s(E)$ and $f_p(E)$, which determine the contribution of these states to the states of the energy E in the valence band. These functions satisfy the relation:

$$f_s(E) + f_p(E) = 1.$$

Since the X-ray emission K spectra are determined by the p components of the wave function, it follows that

$$\frac{f_p(T)}{f_p(B)} = \frac{R_X(T)}{R_X(B)} = 5.6,$$

where T and B stand for the top and the bottom of the valence band, respectively. Since the ratio of the photoionization cross sections of the $2s$ and $2p$ electrons for the free ion is 13, the following is obtained:

$$\frac{f_p(B) + 13f_s(B)}{f_p(T) + 13f_s(T)} = \frac{R_{XPS}(B)}{R_{XPS}(T)} = 5.86.$$

Solving these two equations yields $f_p(B) = 0.16$ and $f_p(T) = 0.92$. By comparing $R_{XPS}(E)$ and $R_X(E)$ separately with these two values it is possible to determine the energy dependence of f_p on the basis of the data from X-ray emission and photoelectron spectroscopy. These two independent methods of determining f_p (Figure 107) lead to consistent results. For further calculations, it is convenient to make use of the mean value

$$\bar{f}_p = \frac{\int f_p(E)\,\rho(E)\,dE}{\int \rho(E)\,dE} = 0.695.$$

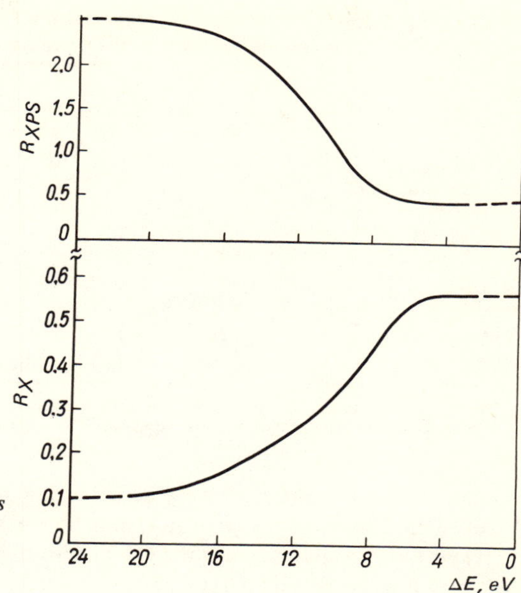

Figure 106. Transition probability as a function of energy in X-ray photoelectron spectra (R_{XPS}) and X-ray emission spectra (R_X).

Knowing $\bar{f}_p$, it is possible to calculate the occupation numbers n_p and n_s for the p and s electrons, since $\bar{f}_p = n_p/(n_s + n_p) = 0.695$, and $n_p + n_s = 4$.

Thus, the carbon atom in diamond is characterized by the configuration $s^{1.2}p^{2.8}$, which is much closer to the configuration sp^3 than to the configuration s^2p^2. In this way, X-ray photoelectron spectroscopy and X-ray emission spectroscopy give complementary information in an investigation of the s and p states in diamond. The position of critical points in the density of states of the

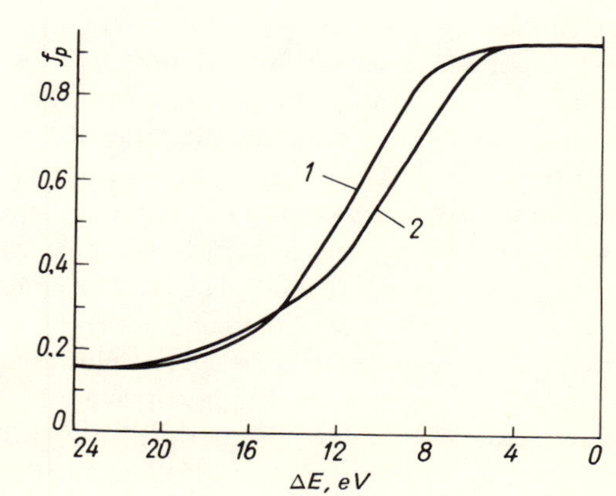

Figure 107. Energy distribution of p states in the diamond valence band as determined by the methods of (1) X-ray photoelectron spectroscopy and (2) X-ray emission spectroscopy.

TABLE 18. Comparison between Experimental and Theoretical Values of Energy for the High-Symmetry Points of the Brillouin Zone of Diamond[a]

Method of energy determination[b]	State					
	L'_2(I)	X_1(II)	$K'_1(C)$	X_4 (middle of the peak III)	L'_3	Width of valence band
OPW [292]	15.5	12.5	–	5.6	2.0	21.2
PPW [293]	19.6	11.5	7.3	5.2	2.4	19.6
MLCAO [288]	15.7	12.1	8.7	6.2	2.8	20.8
Pseudopotential [294]	22.6	18.1	9.4	6.6	2.8	27.5
XPS [287]	17.0	12.9	–	5.5	–	24.2
XES [289]	17.5	13.2	9.8	–	2.8	22.0

[a]The energy values are given in electron volts with respect to the top of the valence band.
[b]XPS = X-ray photoelectron spectroscopy; XES = X-ray emission spectroscopy.

valence band electrons can hardly be determined from the experimental data in Figure 105. However, the peak A(I) may be correlated to the point L'_2, the peak B(II) with the point X_1, and the peak C with the point K'_1. The experimental data and the calculated energy-band structure of diamond are given in Table 18. Van Haeringen and Junginger [294] found that the valence bandwidth of diamond, calculated by the pseudopotential method, was significantly greater than the experimental value, which indicates that the authors did not calculate the pseudopotential form factors correctly. In such cases, the experimental data obtained from X-ray emission and photoelectron spectroscopy may be used as a test of the validity of the theoretical models.

McFeely *et al.* [287] also studied the photoelectron spectra of microcrystalline graphite, and of amorphous carbon. The results of their experiment are shown in Figure 108. The position of the top of the valence band in diamond may be determined by extrapolating the decreasing intensity side of the photoelectron energy distribution toward lower energies, down to the background level. Graphite is a semimetal, and its metallic conductivity prevents it from becoming charged. Amorphous carbon is more difficult to investigate than diamond and graphite. Its band structure cannot be determined in the same terms as the band structure of diamond and graphite. In making the comparison of the photoelectron spectra of graphite and amorphous carbon (see Figure 108), it has been assumed that the binding energies of the most intensive peaks in the valence bands of graphite and of amorphous carbon coincide, and, consequently, the position of the Fermi level of graphite was chosen as the origin for the energy of amorphous carbon. Each of the spectra illustrated in Figure 108 may be characterized by the rather broad and intensive peak I, situated between 16 and 21 eV, the narrower and less intensive peak II between 10 and 15 eV, and the very broad, nearly structureless peak III between 10 eV and zero. The spectra shown

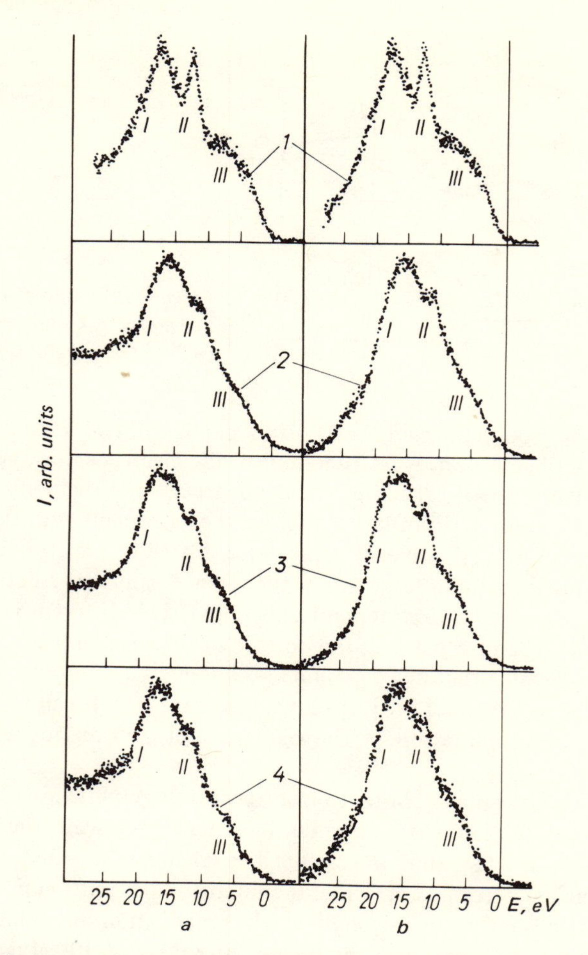

Figure 108. Valence band photoelectron spectra of (1) diamond; (2) fine-grained graphite; (3) crystalline graphite; (4) amorphous carbon; (a) without and (b) with subtraction of the background of inelastic electron scattering.

differ significantly from each other: the peaks I, II, and III are more clearly defined in the spectra of diamond than in the spectra of graphite and amorphous carbon. For graphite and microcrystalline graphite, the minimum between peaks I and II is still clearly seen, but in amorphous carbon it is less clear.

The elementary cell of graphite contains four atoms, and therefore its valence band contains eight filled subbands. The band structure of carbon has been calculated by Painter and Ellis [295] by the LCAO method. The carbon atoms in graphite are distributed in layers, and therefore the eight valence bands in

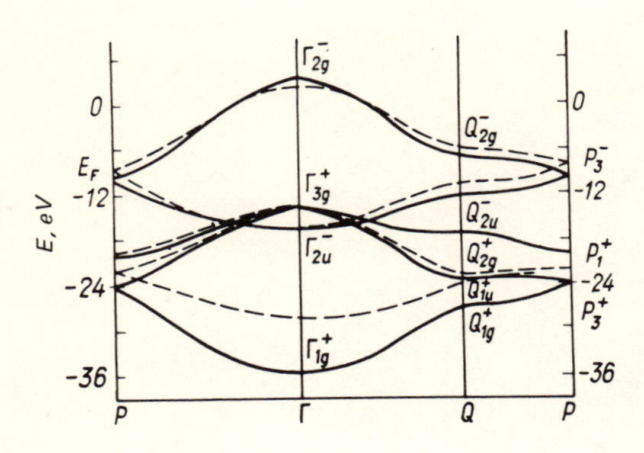

Figure 109: Energy-band structure of graphite, according to data from [295] (– – –) and [296] (——).

graphite may be grouped in two classes–six σ bands and two π bands. The latter are mainly described by $2p_z$ wave functions of the electrons in carbon atoms. Kortela and Manne [296] calculated the band structure of diamond by the semi-empirical LCAO method. In order to improve the agreement with the results of Painter and Ellis [295], the filled σ bands were shifted to 3.5 eV below the Fermi level. However, the total width of the valence band was still 6 eV higher than the value obtained by Painter and Ellis [295]. This difference is due to the structure of the σ band situated at the bottom of the valence band.

The energy-band structure of graphite, as obtained by Painter and Ellis [295] and by Kortela and Manne [296], is shown in Figure 109. It can be seen that the σ and π bands in graphite overlap. The magnitude of the overlapping is equal to 2.6 eV.

The X-ray emission and photoelectron spectra of graphite are shown in Figure 110. In order to place the spectra on the same energy scale, the binding energy of the 1s line of carbon in graphite was taken as $E_B^F = 284.7$ eV. Because of the extreme anisotropy of graphite, the intensity distribution in the X-ray emission spectra is influenced strongly by the polarization of X radiation. This effect was first observed experimentally by Borovski *et al.* [297] using an Mc-46 electron probe microanalyzer. Similar experiments have also been performed by Beyreuther and Wiech [298], Brümmer *et al.* [299], and Müller *et al.* [300]. Figure 111 shows the X-ray photoelectron spectra of graphite, measured by Beyreuther and Wiech [298] at various incidence angles of the exciting photons on the crystal monochromator of the spectrometer. The lower curve in this figure represents the emission spectrum corresponding to translations from the σ states. The radiation with energy between 280.7 and 284.5 eV is due to transitions from the π states. There exists an intermediate energy region in which radiation from both π and σ states is present. Figure 110 illustrates the π and σ emission spectra measured by Müller *et al.* [300]. The intensity of X-ray emission spectra calculated by Kortela and Manne [296] agree quite well with the structure of the experimental curves. However, the X-ray emission and photo-

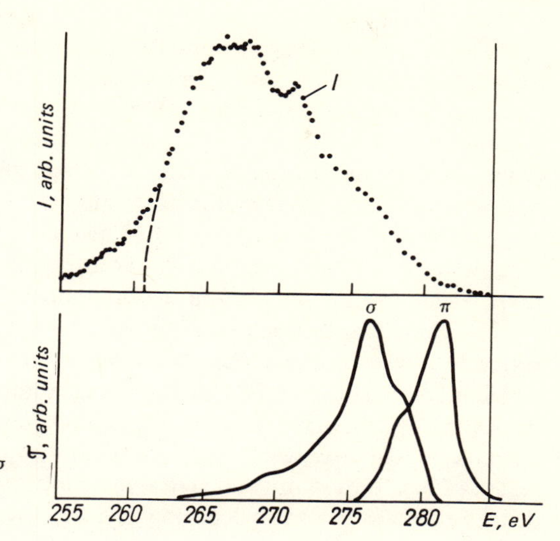

Figure 110. Valence band photoelectron spectrum of graphite. The σ and π components of the X-ray K emission spectrum are also shown.

electron spectra are significantly different from each other, due to the different influence of transition probabilities. Within an energy region extended over approximately 10–13 eV, at the bottom of the valence band, the energy distribution of photoelectrons reaches a maximum, whereas the intensity of X radiation

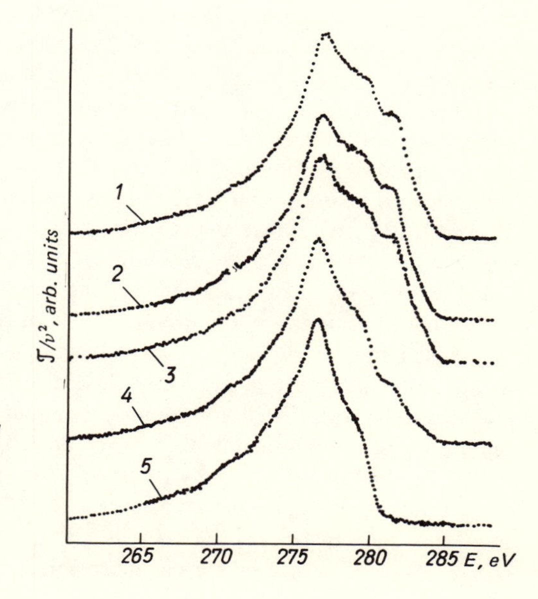

Figure 111. X-ray K emission spectra of graphite, corresponding to different incidence angles of the X rays on the crystal-monochromator: (1) 5°; (2) 10°; (3) 15°; (4) 45°; (5) 80°.

in the same region is insignificant. This indicates that at the bottom of the valence band, the σ states have s-type symmetry.

As in the case of diamond, the bottom of the valence band in graphite may be determined by extrapolating the photoelectron spectrum. This extrapolation is necessary because the background of inelastically scattered electrons could not be correctly subtracted from the experimental curve. The peak in the electron energy distribution, situated at 13.8 eV from the Fermi level and separated by a minimum from the flat maximum, is probably due to the high density of states in the vicinity of the point P_1^+ in the Brillouin zone.

The X-ray emission spectra have made it possible to establish that the energy region in which the σ and π bands overlap is situated 6 eV under the Fermi level, which is consistent with the results of calculations. The region of overlapping is determined by the states Γ_{2u}^- and Γ_{3g}^+. The rapid decrease of intensity as the Fermi level is approached, starting at about 4 eV, arises from the fact that in the region of localization of the π states, the contribution of s states is low, and the density of states $\rho(E)$ also decreases rapidly. The width of the valence band derived from the photoelectron spectroscopy data is about 24.0 eV, whereas the values calculated by Kortela and Manne [296] and by Painter and Ellis [295] were 25.6 eV and 19.3 eV, respectively. The width of the X-ray K emission band determined by Beyreuther and Wiech [298] is of the order of 21.7 eV.

The detailed interpretation of the structure of the energy bands of graphite is difficult because of the lack of thorough theoretical calculations. For graphite, the complementary use of X-ray emission and photoelectron spectra has proved to be useful. If account were not taken of the data from X-ray photoelectron spectroscopy, it might have been considered that the valence bandwidth calculated by Painter and Ellis [295] agreed rather well with experiment [298, 300]. However, the value obtained by Kortela and Manne [296] is more reliable.

Inspection of the X-ray photoelectron spectrum of the valence electrons in amorphous carbon (see Figure 108) shows that it is more like the spectrum of graphite than of diamond. In fact, the dip between the maxima I and II and the maximum III in amorphous carbon is less evident than in diamond. Weaire and Thorpe [301] have shown that the main features in the structure of the valence electron density of states are determined by the atomic properties and short-range ordering in the crystal, whereas long-range ordering is responsible for the fine structure. Therefore, the disappearance of the distinct minimum between the peaks I and II indicates that it is an example of long-range ordering. The details of the structure of the X-ray photoelectron spectra of amorphous carbon are consistent with the carbon atoms existing in trigonal and/or tetrahedral coordination. The most probable is, however, the trigonal coordination.

Figure 112 shows the X-ray photoelectron spectra of the 1s line of carbon for diamond, crystalline and microcrystalline graphite, and amorphous carbon. In all four spectra, besides the main line, a structure related to the characteristic energy losses of photoelectrons is observable. The energy position of the maxima

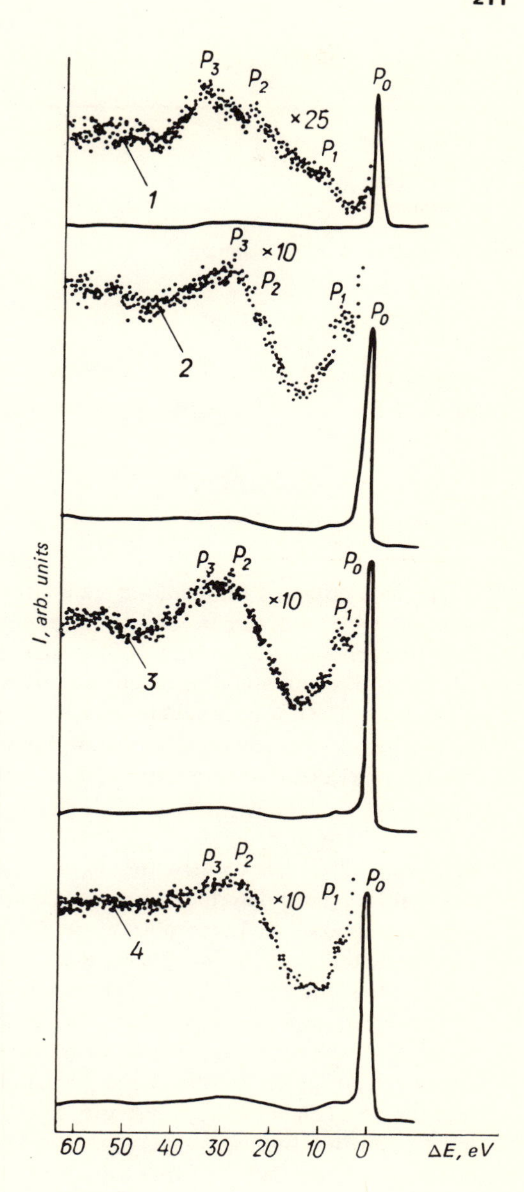

Figure 112. 1s photoelectron spectra of (1) diamond; (2) fine-grained graphite; (3) crystalline graphite; (4) amorphous carbon.

in this structure are given in Table 19. The characteristic loss spectra of amorphous carbon are more similar to those of graphite than to those of diamond. It is more probable that the maximum P_1 arises from interband transitions than from collective oscillations of πp electrons. This conclusion is based on the fact that the maximum P_1 is also observed in the spectrum of diamond.

TABLE 19. **Characteristic Losses in Diamond, Graphite, Microcrystalline Graphite, and Amorphous Carbon**

Method of energy determination	Sample	Energy position of the spectral structures, eV		
		P_1	P_2	P_3
XPS calculations	Diamond	11.3	25.4	34.1
		12.5	–	31.1
XPS calculations	Graphite	6.3	28.1	33.3
		7.5 [302]	25.1	–
		12.5 [303]		
XPS calculations	Microcrystalline graphite	5.6	22.0	30.3
		6.7–7.2	22.3–24.1	
XPS calculations	Amorphous carbon	5.6	26.5	31.6
		6.1	20.3	–

The data obtained for diamond are consistent with those provided by the measurement of reflectance [304]. For graphite, microcrystalline graphite, and amorphous carbon, the positions of the maxima P_1 and P_2 are in good agreement with other experimental data on characteristic losses [302, 305]. However, the maximum P_3 has not been observed earlier in graphite and amorphous carbon. The energy loss measurements for amorphous carbon provide added confirmation that its electron structure is closer to that of graphite than to that of diamond.

On the basis of X-ray diffraction studies, a series of models has been developed, which describes the structure of amorphous carbon [306, 307]. The results that have been described above are consistent with those models in which it was presupposed that amorphous carbon has both trigonal and tetrahedral bounding. Studies of the X-ray emission spectra of amorphous carbon have established that the carbon *K* spectrum has an intermediate energy position between the carbon *K* spectra of diamond and graphite.

Silicon and germanium are isoelectronic analogues of diamond. The X-ray photoelectron spectra of silicon and germanium, like that of diamond, have three maxima (Figure 113). Therefore, the first maximum (I) may be attributed to the first band, the second (II) to the second band, and the third (III) to the third and fourth bands. The first band is mainly of the *s* type, the third and the fourth bands are of the *p* type, and the second band contains a mixture of both *s* and *p* states. Therefore, any variation in the ratio $(S_{\mathrm{I}} + S_{\mathrm{II}})/S_{\mathrm{III}}$ of areas under the maxima should be related to variations in the ratio of photoionization cross sections $\sigma(s)/\sigma(p)$ of the atomic orbitals.

To study this problem in more detail, Cavell *et al.* [286] measured the X-ray photoelectron spectra of valence electrons in gaseous CH_4, SiH_4, and GeH_4 com-

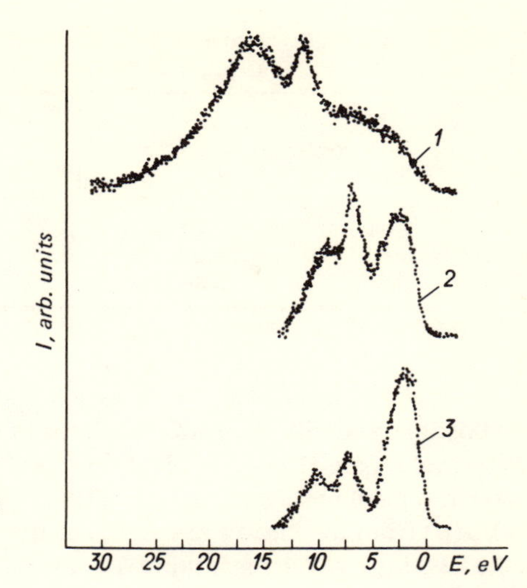

Figure 113. Valence band photoelectron spectra in (1) diamond; (2) silicon; (3) germanium.

pounds, using a spectrometer specially constructed for the study of gaseous samples and of free ions. The $K\alpha_{1,2}$ line of magnesium was used as excitation source. As Figure 114 shows, the X-ray photoelectron spectra in all three cases exhibited only two maxima, corresponding to the levels A_1 (s type) and T_2 (p type). The binding energies of these states were measured with respect to the binding energy

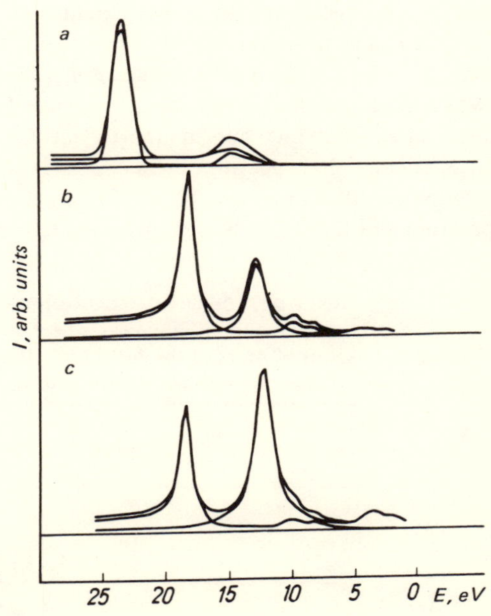

Figure 114. Valence band photoelectron spectra in gases: (a) CH_4; (b) SiH_4; (c) GeH_4.

TABLE 20. Values of Binding Energy and FWHM for the Molecular Orbitals of CH_4, SiH_4, and GeH_4 (in eV)

	A_1		T_2	
Sample	E_B	FWHM	E_B	FWHM
CH_4	23.08	1.71	14.5	2.8
SiH_4	18.01	1.16	12.67	1.69
GeH_4	18.46	1.17	12.28	1.75

of the 2s electrons of neon, taken as equal to 48.72 eV. With the LCAO method it is possible to derive the wave functions of these two states (the energy and HWHM values are given in Table 20) in terms of atomic s and p orbitals. This makes it possible to determine the ratio of photoionization cross sections $\sigma(s)/\sigma(p)$ from the known intensities of the maxima. According to the data in Table 21, this ratio changes more, in moving from diamond to germanium, than the ratio of peak areas $(S_{\rm I} + S_{\rm II})/S_{\rm III}$. This is because the energy bands in solids do not have pure s or pure p character.

A change of the excitation energy from that of Mg-$K\alpha$ to that of Al-$K\alpha$ does not have a significant effect on the photoionization cross section.

The X-ray photoelectron spectra of silicon and germanium will now be considered in more detail. The X-ray photoelectron spectrum of the valence electrons of silicon is illustrated in Figure 115. The intensity distribution in the $K\beta$ emission spectrum confirms the above conclusions regarding the localization of s- and p-type states in the valence band of silicon. The X-ray emission spectrum of silicon, measured by Nemoshkalenko *et al.* [277], is illustrated in Figure 115. It shows good agreement with the data of Länger [308], with the exception of the enhancement B', which appears more clearly.

The X-ray emission spectrum of silicon is characterized by a rather clear fine structure. Nemoshkalenko *et al.* [277, 278] used the information provided by this spectrum and that of β-SiC crystals, to determine the Fourier components

TABLE 21. Comparison between the Ratios of Atomic Photoionization Cross Sections and the Ratios of Areas of the Peaks I, II, III in the Photoelectron Spectra of Diamond, Si, and Ge

Sample	$\dfrac{S_{\rm I} + S_{\rm II}}{S_{\rm III}}$	$\dfrac{\sigma(s)}{\sigma(p)}$
Diamond	2.9 ± 0.3	12
Si	1.6 ± 0.2	3.4
Ge	0.7 ± 0.1	1.0

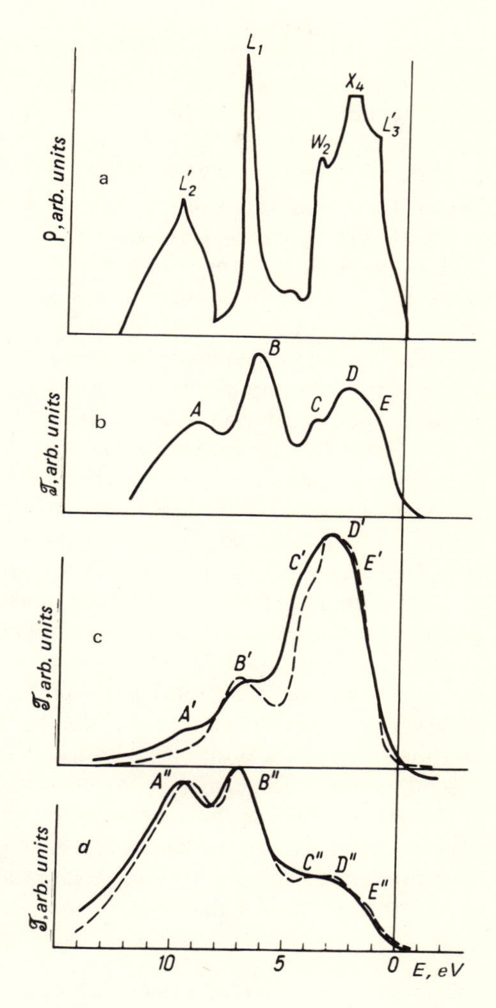

Figure 115. Electron density of states in the silicon valence band, and the corresponding photoelectron and X-ray emission spectra: (a) density of states; (b) photoelectron spectrum; (c) experimental (——) and calculated (– – –) X-ray emission K_β spectra of silicon; (d) experimental (——) and calculated (– – –) $L_{2,3}$ band of silicon.

of the pseudopotential. To find the magnitude of the potential form factors, it is only necessary to use the values of energy positions of the high symmetry points, obtained from the study of X-ray emission spectra. It turned out that such data were indeed sufficient for the determination of the structure of the valence bands of silicon and β-SiC crystals. However, the width of the direct and indirect gaps and the structure of the conduction band, determined from the X-ray data alone, did not agree with the corresponding optical data. Therefore, in determining the potential form factors, Nemoshkalenko *et al.* [277] also took into account the energies of several optical transitions.

From the position of the enhancement A, B, C, D, E and from the high-energy and low-energy limits of the X-ray emission K spectrum, as well as from the transition energies $E(\Gamma_{15c} - \Gamma_{25'v}) = 3.4, E(L_{1c} - L_{3'v}) = 3.2$, $E(X_{1c} - X_{4v}) = 4.1, E(L_{3c} - L_{3'v}) = 5.3$ eV, the values of pseudopotential form factors $V^s(3) = -0.233$, $V^s(8) = 0.045$, $V^s(11) = 0.080$ Ry were determined. These values are close to the results obtained by Cohen and Bergstrasser [138]. The density of states of silicon was calculated by using 1638 points in 1/48 of the Brillouin zone, which is equivalent to 61,776 points in the entire zone. The result of the calculation was consistent with that obtained by Kane [309], although Kane used the Animalu–Heine pseudopotential form factors [120], modified in such a way as to reproduce the experimental value of the width of the band gap in silicon (1.1 eV), and also the electron effective mass of the crystal, as determined by the method of cyclotron resonance.

Figure 115 demonstrates that the structure of the X-ray photoelectron spectrum is significantly different from that of the X-ray emission $K\beta$ spectrum. This can be explained, as in the case of diamond and graphite, by the significantly different influence of the transition probability on the shape of the photoelectron energy distribution and the intensity of emitted X-ray quanta. Comparison of the density of states with the photoelectron spectrum confirms that the photoionization cross section in silicon is higher for $2s$ electrons than for $2p$ electrons.

The position of the critical points in the density of states can be determined from the experimental curves. For this, it is first necessary to determine the relation between the positions of critical points in the smoothed and unsmoothed theoretical density-of-states curves, on one hand, and the width of the excitation line and the instrument aberrations, on the other hand. Table 22 gives the experimental and theoretical energy values of the high symmetry states in the Brillouin zone. These values agree rather well with each other. For the states at the bottom of the valence band of silicon, the best agreement with

TABLE 22. Theoretical and Experimental Values of Energy for the High-Symmetry Points

Method of energy determination	L'_3	X_4	W_2
XPS [310, 311]	–	2.5 ± 0.3	3.9 ± 0.2
UPS[b] [312]	1.2 ± 0.2	2.9 ± 0.3	–
UPS [313]	–	–	–
Pseudopotential [138]	1.1	2.8	4.0
Pseudopotential [314]	1.2	2.9	–
OPW [315]	2.7	2.7	–

[a]The energy values are given in electron volts with respect to the top of the valence band.
[b]UPS = ultraviolet photoelectron spectroscopy.

experiment is given by the results of Chelikowsky and Cohen [314] obtained by taking into account the nonlocal character of the potential. For the β-SiC crystal, the X-ray spectroscopy data also give good agreement with the structure of the energy bands.

For the germanium crystal, in contrast to silicon, the states near the point L_1 in the **k** space produce a maximum in the X-ray photoelectron spectrum that is much lower than the two other maxima. For germanium, good agreement exists between the photoelectron energy distribution and the calculated density of electronic states. In the calculation of the density of states by the pseudopotential model, an effective electron mass m^* was used to achieve optimum agreement with experimental data obtained using X-ray photoelectron spectroscopy and ultraviolet photoelectron spectroscopy. For germanium, $m/m^* = 1.089$. However, Pandey and Phillips [316] have expressed doubts regarding the validity of replacing the electron mass m by the effective mass m^* for germanium and GaAs crystals. First, there is no physical foundation for this procedure. Second, the procedure is unsatisfactory even from the empirical point of view, since the difference $m - m^*$ that has to be introduced to obtain the correct value of the valence bandwidth has the opposite sign to that which would be necessary to improve the agreement with the magnitudes of interband energies.

Figure 116 illustrates the photoelectron spectrum of the valence electrons in germanium, measured by Eastman *et al.* [317] using synchrotron radiation of 25 eV as the excitation source. The states situated in the two lowest subbands have in this case a somewhat higher excitation probability than the states in the upper subband of the valence band. The main peaks and the characteristic features in both of these curves agree well with each other. An exception is the structure at 0.4 eV observed in the spectrum measured by using photons of energy $h\nu = 25$ eV. This can be explained assuming the participation of surface states in the photoemission process [318]. The ultraviolet photoemission spectra

of the Valence Band of Si[a]

State				
Σ_1^{min}	L_1	W_1	L_2'	Γ_1
4.7 ± 0.3	6.8 ± 0.2	8.1 ± 0.3	9.3 ± 0.4	12.5 ± 0.6
4.4 ± 0.3	–	–	–	–
4.7 ± 0.2	6.4 ± 0.4	–	–	12.4 ± 0.6
4.5	7.2	8.1	10.2	12.6
4.5	7.0	–	9.5	12.4
–	6.7	–	9.4	11.7

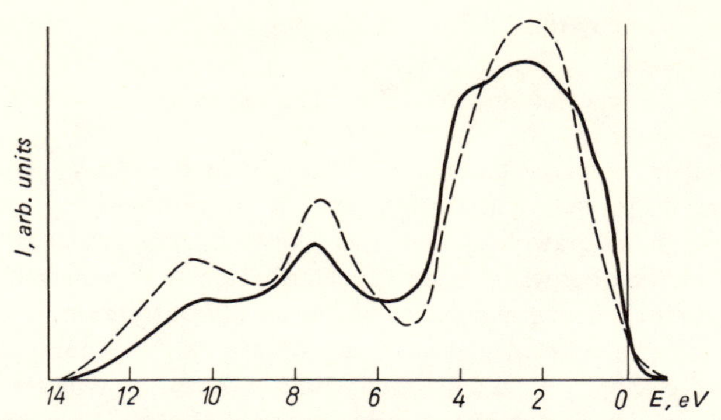

Figure 116. Valence band photoelectron spectra of germanium at 25 eV (——) and 1486.6 eV (– – –) excitation energy.

of germanium give information about the electronic states in the region 4–8 Å from the sample surface, whereas the X-ray photoelectron spectra characterize a layer of 15–30 Å thickness. The good agreement between the X-ray photoelectron spectra and the ultraviolet photoelectron spectra indicates that the bulk band structure is not changed significantly by the presence of surface states. However, the electron emission from the sample surface is superposed on that from the bulk, and this superposition seems to give rise to the structure at 0.4 eV.

In germanium, an interesting phenomenon is observed, namely, an increase in the yield of Auger-electron emission when the energy of ultraviolet quanta is changed. Figure 117 shows the curves of photoelectron distributions for different energies of exciting photons. At 27.5 eV excitation energy, ionization of d electrons of germanium does not occur (it starts at 29 eV). The spectrum changes significantly when the excitation energy is increased to 30 eV. This change is due to the $M_{4,5}VV$ Auger emission, which is superposed on the photoemission spectrum excited by the incident photons. The energy of electrons in

TABLE 23. Theoretical and Experimental Values of Energy for the High-Symmetry Points

Method of energy determination	L'_3	X_4	W_2
Pseudopotential [138]	1.1	2.4	3.3
Pseudopotential [284]	1.2	3.0	4.0
OPW [315]	1.2	2.8	–
UPS [312]	1.1 ± 0.2	–	–
XPS [284]	–	2.7 ± 0.3	3.9 ± 0.2

[a] The energy values are given in electron volts with respect to the top of the valence band.

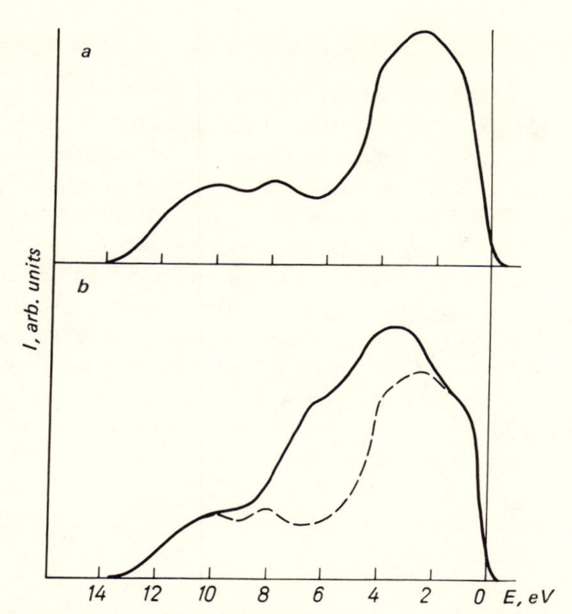

Figure 117. Valence band photoelectron spectra of germanium, at (a) 27.5 eV and (b) 30 eV excitation energy.

the $M_{4,5}VV$ Auger spectrum cannot exceed a maximum value determined by the energy position of the core *d* levels with respect to the top of the valence band. Therefore, the superposition of the Auger spectrum on the photoemission spectrum may be eliminated by further increasing the energy of exciting photons to about 40 eV.

As can be seen from Table 23, the theoretical and experimental energy values of the characteristic features in the band structure of germanium agree well with each other.

of the Valence Band of Ge[a]

State				
Σ_1^{min}	L_1	W_1	L_2^1	Γ_1
3.8	6.9	8.2	9.9	12.0
4.4	7.8	9.1	11.0	13.3
–	7.4	–	10.6	12.6
4.5 ± 0.2	7.7 ± 0.2	–	10.6 ± 0.3	12.6 ± 0.3
4.5 ± 0.3	7.4 ± 0.2	8.7 ± 0.3	10.5 ± 0.4	12.8 ± 0.4

Energy-Band Structure of A_3B_5- and A_2B_6-Type Compounds

The similarity of the structure of the energy spectra of binary semiconductors and insulators with a crystal lattice of the sphalerite type suggests that their X-ray photoelectron spectra reflect characteristics common to all compounds of this type. The photoelectron spectra of the compounds investigated in the energy region up to 50 eV are illustrated in Figures 118–121. The valence band photoelectron spectra are characterized by three maxima, which reflect the distribution of the outermost *s* and *p* electrons of the anion and cation. The cation *d* states are localized in the energy region 10–20 eV, and those of the anion in the energy region 35–40 eV. In indium and gallium compounds, the 4*d* states of indium and the 3*d* states of gallium are located below the nondegenerate subband of the valence band. In cadmium and zinc compounds the 4*d* states of cadmium and the 3*d* states of zinc prevent observation of the photoelectron lines corresponding to the lower subband of the valence band. The lines corre-

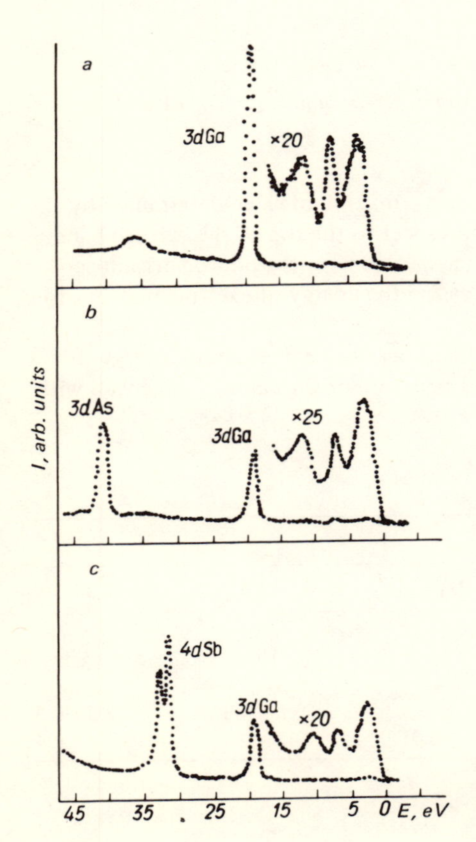

Figure 118. Valence band photoelectron spectra of: a—GaP; b—GaAs; c—GaSb.

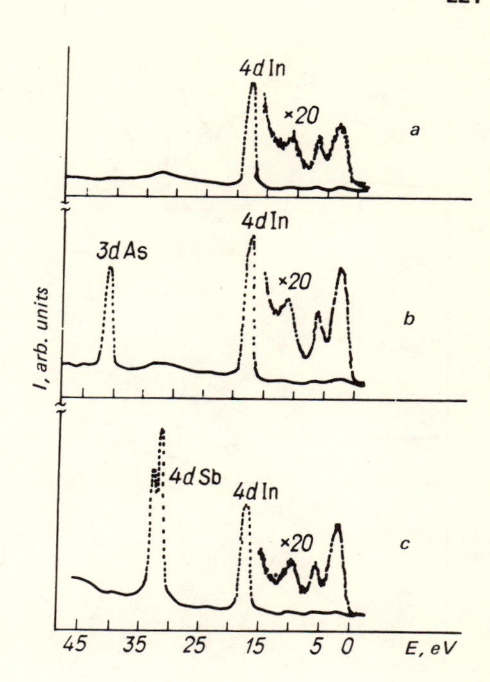

Figure 119. Valence band photoelectron spectra of a–InP; b–InAs; c–InSb.

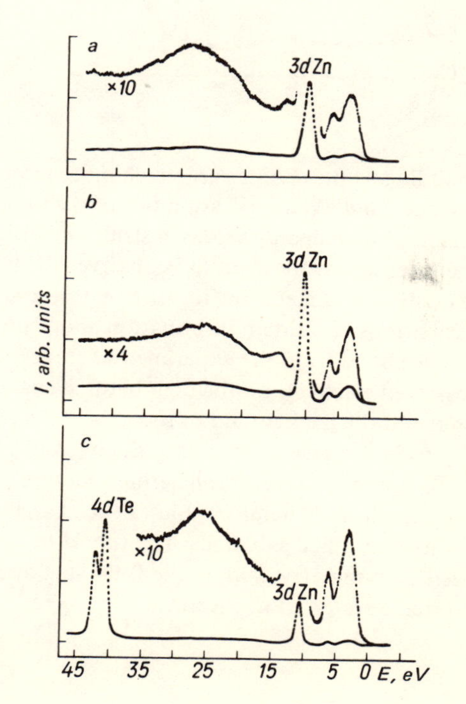

Figure 120. Valence band photoelectron spectra of a–ZnS; b–ZnSe; c–ZnTe.

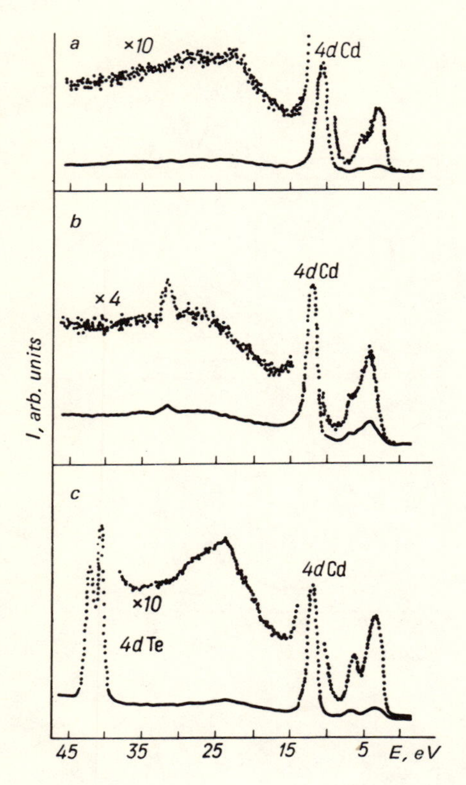

Figure 121. Valence band photoelectron spectra of a–CdS; b–CdSe; c–CdTe.

sponding to the d states are much more intense than the maxima related to the valence band. At a separation of up to about 15 eV from the d peaks, in the region of high energy values, a structure with an intensity of about 5–10% of the intensity of the d peaks is observed. This structure can be related to the excitation of d electrons together with the simultaneous excitation of plasma oscillations. To obtain information about the energy states of electrons in the valence band from the experimental curves, $I'(E)$, the distribution of elastically scattered electrons on the high-binding-energy side of the d states maxima was used and the background of secondary electrons was subtracted.

As in the case of diamond, silicon, and germanium, the unit cell in A_3D_5 and A_2B_6 compounds with sphalerite structure contains two atoms with eight valence electrons. Therefore, the valence band of these crystals consists of four completely filled subbands. The top of the valence band in all the compounds has Γ_{15} symmetry, and, in the following discussion, energy values will be expressed as referred to this state.

When establishing the correspondence of features observed in the experimental spectra with theoretically calculated critical points in the energy bands $E(\mathbf{k})$ and with the density of electronic states $\rho(E)$, it is necessary to take into account the magnitude of the spectrometer resolution. Consequently, calculated densities of states are smoothed by using the curve of instrumental aberrations of the spectrometer. Ley *et al.* [16] smoothed the curves $\rho(E)$ with a Gaussian having for the states at the bottom of the valence band a full-width at half-maximum (FWHM) of 0.8 eV and for the states at the top of the valence band, a HWHM of 0.7 eV. Aleshin [279] used for smoothing a dispersion curve with HWHM equal to 0.6 eV. As a result, the features of the curve $\rho(E)$ and of the smoothed curve $\rho'(E)$ will be somewhat shifted with respect to each other. This shift should be taken into account in the determination of the energy positions of features in the experimental curves $I'(E)$.

Let us consider the example of GaP (Figure 122), which is typical for this class of compounds, to see how relations can be established between the features observed in the experimental and theoretical curves. The first, rather flat maximum at the top of the valence band is situated between the energy states L_3 and X_5. This flat portion is even more smoothed in the curve $\rho'(E)$, but the peaks related to L_3 and X_5 are distinguishable in the photoelectron spectrum. In order to determine the energy of the state Σ_1^{min} (the high-energy limit of the third peak), it can be assumed to a good approximation that Σ_1^{min} is situated at an energy where the intensity is the average of the intensity at W_2 and at the minimum.

Peak II appears because at the surface of the Brillouin zone, the states in the second band have rather close energy values. The top of this peak coincides with the position of point W_1 within up to 0.2 eV. It is worth noting that the energy position of the top of this peak is usually one of the most accurately determined structural features in the photoelectron spectrum. Point X_3 corresponds to the high-energy edge of peak II in the photoelectron spectrum. The first energy band is correlated with peak I.

The energy values at points X_1 and W_4 of this band are nearly coincident. To a good approximation, it can be considered that, at these points, the intensity is $\frac{3}{4}$ of the intensity of peak I. The top of peak I agrees rather accurately with the position of point L_1.

In the remainder of this chapter, the energy band structure of A_3B_5 and A_2B_6 compounds will often be discussed in terms of the data from ultraviolet photoelectron spectroscopy. The results obtained from X-ray photoelectron spectroscopy and ultraviolet photoelectron spectroscopy may be different. Besides the difference in resolution, there are three other basic reasons for this discrepancy, namely, the variation of the density of states in the conductivity band, the variation of photoionization cross section, and the relaxation effects. In ultravio-

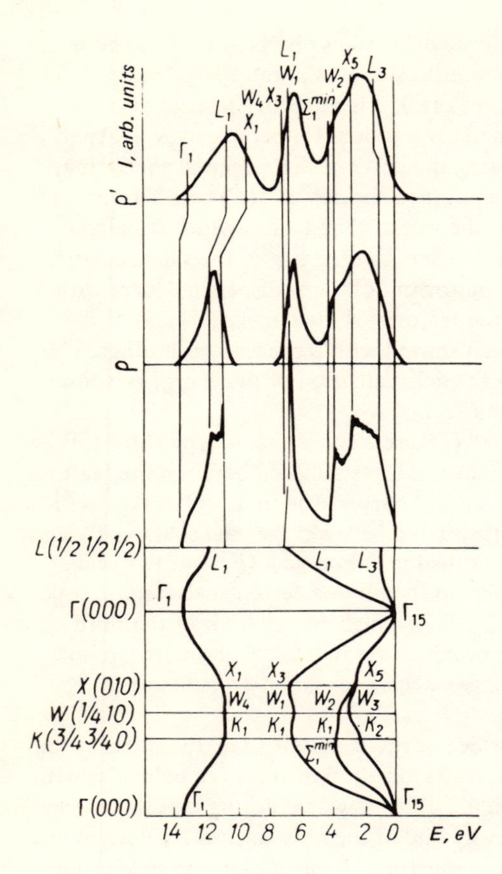

Figure 122. Correlation between the singular points in the photoelectron spectrum I(E), the density of states ρ(E), the smoothed density of states ρ'(E), and the electron energy spectrum of crystalline GaP.

TABLE 24. Theoretical and Experimental Values of Energy for the High-Symmetry Points

Method of energy determination	L_3	X_5	W_2
XPS [287]	1.2 ± 0.3	2.7 ± 0.2	3.6 ± 0.2
UPS [320]	0.8	–	–
Pseudopotential [284]	1.0	2.5	3.7
OPW [315] (with adjusted parameters)	0.9	2.3	–
OPW [315] ($V_{ex}^{\alpha=2/3}$)	0.9	2.3	–
ROPW [321] ($V_{ex}^{\alpha,\beta}$)[b]	0.9	2.2	–

[a]Values of energy (eV) are given with respect to the top of the valence band.
[b]ROPW–relativistic OPW method.

let photoelectron spectroscopy, the distribution of the density of states in the conduction band is superimposed on the distribution of the density of states in the valence band of the crystal. By increasing the photon energy up to values characteristic for X-ray photoelectron spectroscopy, the influence of the structure of the conduction band can be efficiently eliminated. The photoionization cross sections for s, p, d, and f electrons are different, and they may change differently when the excitation energy is increased. The ultraviolet photoelectron spectra are determined by the behavior of the wave function of valence electrons in the outer atomic region, and the X-ray photoelectron spectra are determined by the behavior of the wave function of valence electrons in the vicinity of the nucleus. The problem of final-state relaxation effects is completely neglected in photoemission spectroscopy, although Spicer [319] has drawn attention to these effects. Since, in the calculation of energy bands, the exchange potential is not taken to be of the Hartree–Fock type, but of the Slater type, the Koopmans' corrections should be added to the energy values obtained. For deeply lying valence bands, the relaxation in the final state results in a shift of the structural particularities upwards, towards the top of the valence band. Particularly large will be the shift of the states situated at the bottom of the valence band. This shift upward is revealed by $\rho(E)$ curves derived by any self-consistent energy-band calculation, without introduction of corrections related to the relaxation of electron states in the valence band of crystals.

From Table 24, it is evident that differences exist between the data from X-ray photoelectron spectroscopy and from ultraviolet photoelectron spectroscopy. The difference between energy values increases toward the bottom of the valence band. For GaP, rather good agreement is observed between the theoretical and experimental energy values. This is also the best agreement of all the calculations performed by the pseudopotential method.

of the Valence Band of Crystalline GaP[a]

State					
Σ_1^{min}	W_1	$X_3(L_1)$	X_1	L_1	Γ_1
4.0 ± 0.2	6.5 ± 0.2	6.9 ± 0.2	9.6 ± 0.3	10.6 ± 0.3	13.2 ± 0.4
4.1	–	6.9	9.7	–	11.8
4.1	6.6	6.9	10.9	11.7	13.6
–	–	6.1	9.2	10.0	11.8
–	–	6.1	9.4	10.1	11.9
–	–	6.1	9.5	10.3	12.0

Figure 123 illustrates the photoelectron spectra of valence electrons in GaAs, and also the X-ray emission $K\beta$ spectrum of gallium in GaAs. It is seen that the transition probability affects the shape of the X-ray photoelectron spectrum and the X-ray emission spectrum differently. Since transitions from *s* states to the 1*s* level are forbidden, the presence of transitions from the nondegenerate subband of the valence band confirms the existence of hybridization between *s* and *p* states in this nondegenerate subband. However, the detailed structure of the valence band of the GaAs crystal is less clearly represented in the X-ray emission spectra than in the X-ray photoelectron spectra, because of the significant width of the inner *K* level (half-width of the order of 0.9 eV). For the GaAs crystal, rather good agreement is observed between the structure of the electron spectra and the density-of-states curve, as well as between the experimental and theoretical values of the energy position of the structural particularities of the valence band (see Table 25). Characteristic for the photoelectron spectrum of GaAs is the absence of a deep minimum between the degenerate and nondegenerate parts of the valence band, which indicates that the background of inelastically scattered electrons cannot be adequately subtracted from the X-ray photoelectron spectrum.

In the case of GaSb crystal (Figure 124), a significant discrepancy (1.2 eV) in the position of peak II in the degenerate part of the valence band is observed. The data in Table 26 show that the calculated energy values of the critical points in the GaSb crystal are consistent with the experimental values.

For the InP crystal, as can be seen from Figure 125, rather good agreement is observed between the X-ray photoelectron spectra and the calculated structure of the valence band. The density of electron states was calculated using the form

TABLE 25. Theoretical and Experimental Values of Energy for the High-Symmetry Points

Method of energy determination	L_3	X_5	W_2
XPS [16]	1.4 ± 0.3	2.5 ± 0.3	4.0 ± 0.2
UPS [320]	0.8	–	–
Pseudopotential [284]	0.9	2.5	3.5
OPW [315] (with adjusted parameters)	0.9	2.3	–
OPW [315] ($V_{ex}^{\alpha=2/3}$)	1.0	2.3	–
ROPW [321] ($V_{ex}^{\alpha,\beta}$)	1.1	2.4	–
SOPW[b] [322] ($V_{ex}^{\alpha=2/3}$)	1.0	2.5	3.4
SOPW [322] ($V_{ex}^{\alpha=1}$)	1.0	2.3	3.0

[a]Values of energy (in electron volts) are given with respect to the top of the valence band.
[b]SOPW–self-consistent OPW method.

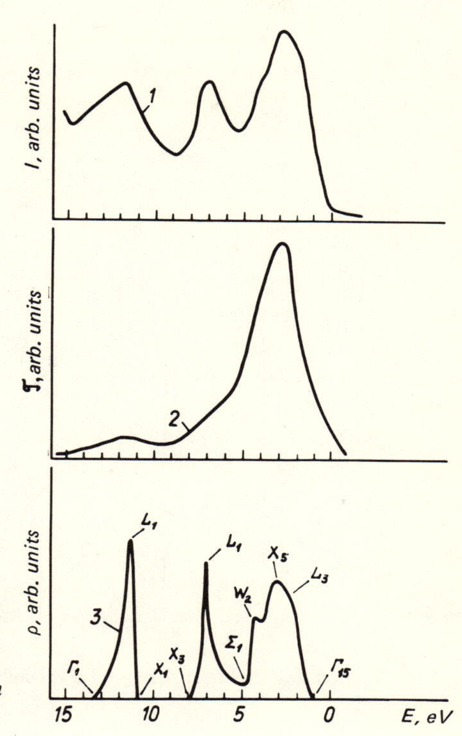

Figure 123. Density of states, X-ray emission spectrum, and photoelectron spectrum of crystalline GaAs: 1–valence band photoelectron spectrum of crystalline GaAs; 2–X-ray emission K spectrum of gallium in crystalline GaAs; 3–electron density of states.

of the Valence Band of Crystalline GaAs[a]

State					
Σ_1^{min}	W_1	$X_3(L_1)$	X_1	L_1	Γ_1
4.4 ± 0.2	6.1 ± 0.1	7.1 ± 0.2	10.7 ± 0.3	12.0 ± 0.5	13.8 ± 0.4
4.1	–	6.9	10.0	–	12.9
3.9	6.6	6.8	11.4	–	13.8
–	–	5.5	10.7	11.1	12.4
–	–	6.3	9.7	10.4	12.0
–	–	6.4	10.2	10.9	12.4
4.0	6.2	6.6	9.2	10.1	11.9
3.3	6.0	6.3	9.5	10.2	11.8

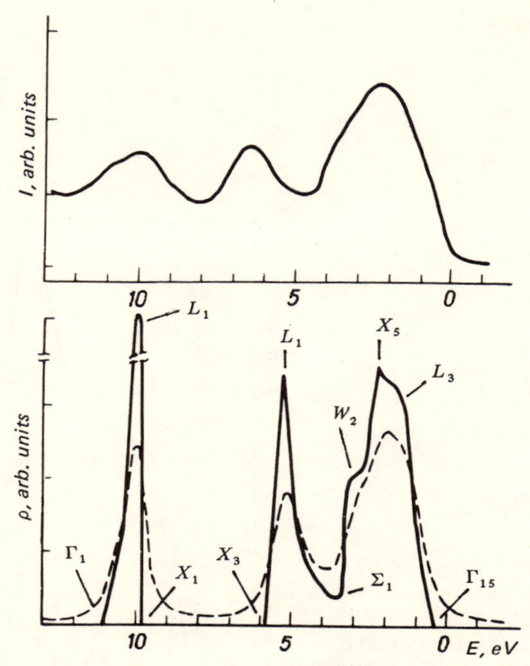

Figure 124. Electron density of states and valence band photoelectron spectrum of crystalline GaSb.

factors of Cohen and Bergstrasser [138]. Investigation of the X-ray photoelectron spectra leads to an interpretation of some of the structural particularities of the X-ray K and L emission spectra of phosphorus in crystalline InP [323, 324]. In Figure 125, the main features in the K and L spectra are shifted on the energy scale by about 0.7 eV, which may, for example, be due to incorrect scaling of the K and L spectra by Wiech [323], and also to the different influence of tran-

TABLE 26. Theoretical and Experimental Values of Energy for the High-Symmetry Points

Method of energy determination	L_3	X_5	W_2
XPS [16]	1.3 ± 0.2	2.7 ± 0.2	2.6 ± 0.2
OPW [315] (with adjusted parameters)	0.9	2.2	–
OPW [315] ($V_{ex}^{\alpha=2/3}$)	1.1	2.4	–
ROPW [321] ($V_{ex}^{\alpha,\beta}$)	1.2	2.5	–

[a] Values of energy (in electron volts) are given with respect to the top of the valence band.

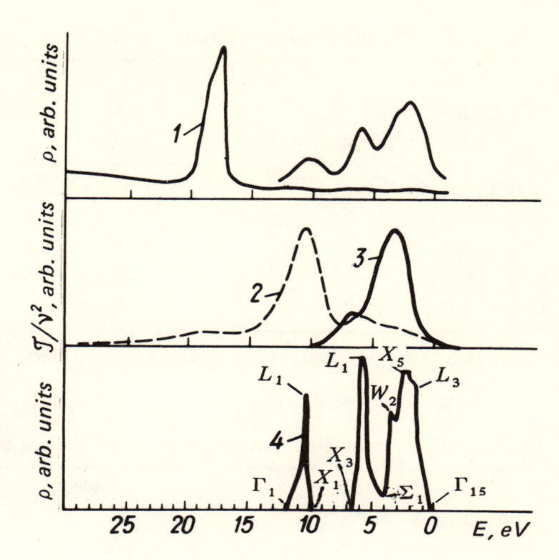

Figure 125. Electron density of states, valence band X-ray emission spectrum, and valence band photoelectron spectrum of crystalline InP: 1—valence band photoelectron spectrum of crystalline InP; 2—X-ray emission $L_{2,3}$ band of phosphorus in crystalline InP; 3—electron density of states.

sition probabilities on the shape of these spectra. In the X-ray emission L spectrum of phosphorus, an enhancement on the high-binding-energy side of the main peak is observed. The nature of this enhancement remained unclear as long as data from X-ray emission spectroscopy alone were used. Investigation of the photoelectron spectra of valence electrons in InP suggested that this enhancement might arise from a contribution from the $4d$ electrons of indium to the X-ray emission spectra of phosphorus. This would confirm that the overlapping integral of $2p$ electron wave functions of phosphorus and $4d$ electron wave functions of indium in InP compound is finite. As the data in Table 27 show, good agreement exists between theoretical and experimental results.

The X-ray photoelectron spectra and the valence band density of states of the compound InAs are shown in Figure 126. The density-of-states calculation was performed using the form factors taken from the work of Cohen and Bergstrasser [138]. Again, good agreement is observed between the photoelectron

of the Valence Band of Crystalline GaSb[a]

State					
Σ_1^{min}	W_1	$X_3(L_1)$	X_1	L_1	Γ_1
3.8 ± 0.2	6.4 ± 0.1	6.9 ± 0.3	9.4 ± 0.2	10.3 ± 0.3	11.6 ± 0.3
—	—	5.2	9.8	9.9	11.1
—	—	6.3	7.9	9.0	10.7
—	—	6.9	8.9	9.7	11.3

TABLE 27. Theoretical and Experimental Values of Energy for the High-Symmetry Points

Method of energy determination	L_3	X_5	W_2
XPS [16]	1.0 ± 0.3	2.0 ± 0.2	2.5 ± 0.2
OPW [325] (with adjusted parameters)	0.6	1.7	–
OPW [325] ($V_{ex}^{\alpha=2/3}$)	0.6	1.7	–
ROPW [321] ($V_{ex}^{\alpha,\beta}$)	0.7	1.6	–

[a] Values of energy (in electron volts) are given with respect to the top of the valence band.

TABLE 28. Theoretical and Experimental Values of Energy for the High-Symmetry Points

Method of energy determination	L_3	X_5	W_2
XPS [16]	0.9 ± 0.3	2.4 ± 0.3	2.7 ± 0.3
OPW [325] (with adjusted parameters)	0.6	1.7	–
OPW [325] ($V_{ex}^{\alpha=2/3}$)	0.7	1.8	–
ROPW [321] ($V_{ex}^{\alpha,\beta}$)	0.8	1.9	–

[a] Values of energy (in electron volts) are given with respect to the top of the valence band.

spectra and the density of states of valence electrons. Table 28 shows the calculated and experimental values of the energy of singular points of crystalline InAs.

The last indium compound to be discussed in the present context is InSb. Figure 127 illustrates the valence band photoelectron spectra as well as the structure of the valence band, calculated by the pseudopotential method taking into account spin-orbit interactions. The theoretical and experimental energy values are given in Table 29.

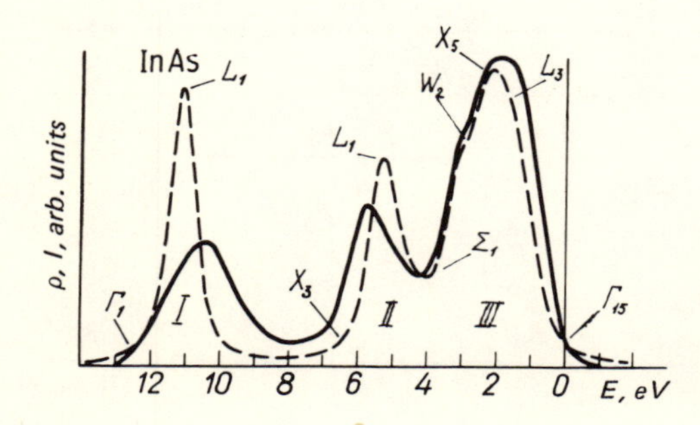

Figure 126. Valence band density of states (– – –) and photoelectron spectrum (——) of InAs. The theoretical curve is smoothed by a dispersion curve with 0.3 eV half-width.

of the Valence Band of Crystalline InP[a]

State					
Σ_1^{min}	W_1	$X_3(L_1)$	X_1	L_1	Γ_1
3.2 ± 0.2	5.4 ± 0.2	5.9 ± 0.2	8.9 ± 0.3	10.0 ± 0.3	11.0 ± 0.4
–	–	4.6	9.7	10.1	11.1
–	–	4.5	9.0	9.4	10.6
–	–	4.6	9.2	9.7	10.8

of the Valence Band of Crystalline InAs[a]

State					
Σ_1^{min}	W_1	$X_3(L_1)$	X_1	L_1	Γ_1
3.3 ± 0.2	5.8 ± 0.2	6.3 ± 0.2	9.8 ± 0.3	10.6 ± 0.3	12.3 ± 0.4
–	–	4.7	10.3	10.6	11.5
–	–	4.7	9.4	9.8	10.8
–	–	5.1	10.0	10.4	11.4

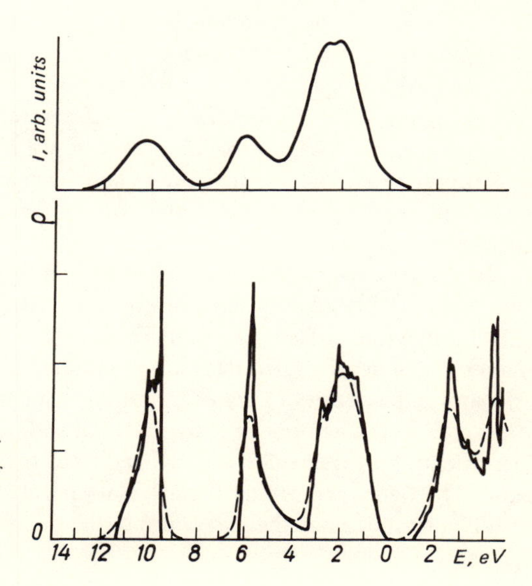

Figure 127. Valence band density of states and photoelectron spectrum of crystalline InSb. The theoretical curve is smoothed by a dispersion curve of 0.3 eV half-width.

TABLE 29. Theoretical and Experimental Values of Energy for the High-Symmetry Points

Method of energy determination	L_3	X_5	W_2
XPS [16]	1.4 ± 0.3	2.4 ± 0.4	3.1 ± 0.2
UPS [320]	1.05	–	–
Pseudopotential [284]	1.2	2.1	2.8
OPW [325] (with adjusted parameters)	0.7	1.8	–
OPW [325] ($V_{ex}^{\alpha=2/3}$)	0.8	1.9	–
ROPW [321] ($V_{ex}^{\alpha,\beta}$)	1.1	2.1	–

[a] Values of energy (in electron volts) are given with respect to the top of the valence band.

Let us now consider the compounds of zinc (ZnS, ZnSe, ZnTe). Their X-ray photoelectron spectra are shown in Figure 128, with the calculated and the experimental energy values of the singular points of the valence zone given in Tables 30–32. Comparison of Figure 128 with Figure 120 shows that in Figure 128 the contribution of *d* states to the photoelectron spectrum has already been subtracted. In Figure 120, however, owing to the overlapping of the intense *d*-electron spectrum with the nondegenerate subband of the valence band, this contribution is difficult to estimate correctly. In such a case, use of the method of X-ray emission spectroscopy is particularly useful. As Figure 129 shows, the $K\beta$ line of zinc in the compound ZnSe allows the determination of the energy position of the bottom of the valence band. In compounds with a crystal lattice of the sphalerite type, the *s* and *p* states at the bottom of the band are hybridized, and the presence of the mixture of *p* states allows identification of the *K* X-ray emission spectrum from this subband. It is to be noted that, in the energy region of the *d* state of zinc, the X-ray emission spectrum of zinc in the compound ZnSe does not exhibit any structure. The $d \rightarrow s$ transitions are of quadrupole type. The absence of any enhancement in the emission spectrum that would correspond to this transition confirms that no hybridization of *d* and *p* states occurs.

In the series of cadmium compounds CdS, CdSe, CdTe, the density of states have been calculated only for CdTe by using the pseudopotential method, with [284] and without [148] account being taken of spin–orbit effects. The shapes of the density-of-states curves obtained in these works are similar. Figure 130 shows the density of states in the valence band of CdTe, as calculated by Aleshin [279]. The peak of the *d* states of cadmium does not allow a better identification of the region of localization of the lower subband of the valence band. The figure shows that the calculation does not give the correct position of this subband, since the maximum in the density of states which corresponds

of the Valence Band of Crystalline InSb[a]

State					
Σ_1^{min}	W_1	$X_3(L_1)$	X_1	L_1	Γ_1
3.4 ± 0.2	5.9 ± 0.2	6.4 ± 0.2	9.5 ± 0.2	10.5 ± 0.3	11.7 ± 0.3
3.65	–	6.5	9.0	–	11.2
3.2	5.7	6.2	9.5	10.1	11.3
–	–	4.7	9.0	9.3	10.2
–	–	5.0	7.7	8.3	9.6
–	–	5.7	8.8	9.3	10.5

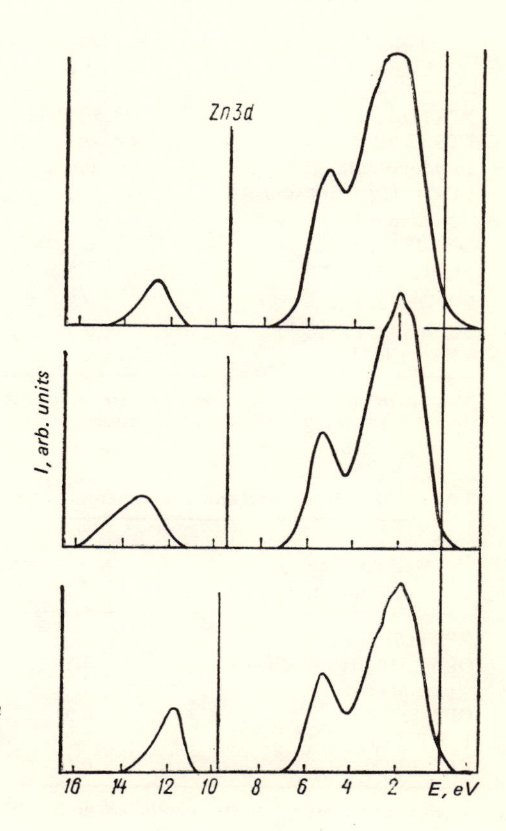

Figure 128. Valence band photoelectron spectra of (a) ZnS; (b) ZnSe; (c) ZnTe after subtraction of the background of inelastic electron scattering.

TABLE 30. Theoretical and Experimental Values of Energy for the High-Symmetry Points

Method of energy determination	L_3	X_5	W_2
XPS [16]	1.4 ± 0.4	2.5 ± 0.3	3.0 ± 0.2
OPW [326] (with adjusted parameters)	0.3	1.1	–
OPW [326] ($V_{ex}^{\alpha=2/3}$)	0.4	1.2	–
ROPW[16] ($V_{ex}^{\alpha,\beta}$)	0.5	1.3	–
SOPW [326]	0.6	1.6	2.0
KKR [327]	0.6	1.4	–
APW [328]	0.9	1.7	–

[a]Values of energy (in electron volts) are given with respect to the top of the valence band.

TABLE 31. Theoretical and Experimental Values of Energy for the High-Symmetry Points

Method of energy determination	L_3	X_5	W_2
XPS [16]	1.3 ± 0.3	2.1 ± 0.3	2.6 ± 0.2
UPS [320]	0.7	–	–
Pseudopotential [284]	0.9	2.1	3.3
OPW [315] (with adjusted parameters)	0.4	1.4	–
OPW [315] ($V_{ex}^{\alpha=1}$)	0.4	1.3	–
ROPW [16] ($V_{ex}^{\alpha,\beta}$)	0.7	1.6	–
SOPW [322] ($V_{ex}^{\alpha=2/3}$)	0.7	2.0	2.7
SOPW [322] ($V_{ex}^{\alpha=1}$)	0.7	1.6	2.3
KKR[b] [327]	0.6	1.3	–

[a]Values of energy (in electron volts) are given with respect to the top of the valence band.
[b]KKR = the method of Green's functions.

TABLE 32. Theoretical and Experimental Values of Energy for the High-Symmetry Points

Method of energy determination	L_3	X_5	W_2
XPS [16]	1.1 ± 0.3	2.4 ± 0.2	2.7 ± 0.2
OPW [315] (with adjusted parameters)	0.5	1.4	–
OPW [315] ($V_{ex}^{\alpha=1}$)	0.6	1.5	–
ROPW [16] ($V_{ex}^{\alpha,\beta}$)	1.0	2.0	–
KKR [327]	0.6	1.6	–

[a]Values of energy (in electron volts) are given with respect to the top of the valence band.

of the Valence Band of Crystalline ZnS[a]

State					
Σ_1^{min}	W_1	$X_3(L_1)$	X_1	L_1	Γ_1
3.4 ± 0.3	4.9 ± 0.2	5.5 ± 0.2	12.0 ± 0.3	12.4 ± 0.3	13.5 ± 0.4
–	–	3.5	–	–	–
–	–	3.5	–	–	–
–	–	3.8	11.2	11.5	12.2
2.1	3.7	4.2	10.0	10.6	11.7
2.1	–	3.3	11.9	12.1	12.6
–	–	3.4	13.4	13.4	14.0

of the Valence Band of Crystalline ZnSe[a]

State					
Σ_1^{min}	W_1	$X_3(L_1)$	X_1	L_1	Γ_1
3.4 ± 0.2	5.2 ± 0.2	5.6 ± 0.3	12.5 ± 0.4	13.1 ± 0.3	15.2 ± 0.6
3.4	–	5.3	–	–	–
3.8	5.3	5.9	14.2	14.5	15.8
–	–	3.7	–	–	–
–	–	3.8	–	–	–
–	–	4.2	11.6	11.9	12.6
3.1	4.5	4.7	10.4	10.8	11.8
4.5	4.2	4.4	10.5	10.9	11.8
2.2	–	3.6	12.0	12.2	12.6

of the Valence Band of Crystalline ZnTe[a]

State					
Σ_1^{min}	W_1	$X_3(L_1)$	X_1	L_1	Γ_1
3.2 ± 0.3	5.1 ± 0.2	5.5 ± 0.2	11.6 ± 0.3	12.0 ± 0.3	13.0 ± 0.4
–	–	3.7	–	–	–
–	–	4.3	–	–	–
–	–	5.0	10.2	10.6	11.5
2.8	–	4.3	9.6	9.7	10.5

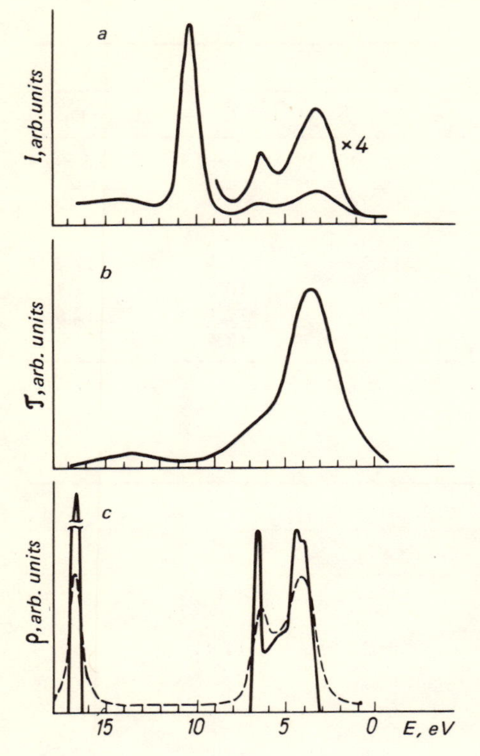

Figure 129. Valence band density of states, X-ray emission spectrum, and photoelectron spectrum of crystalline ZnSe: (a) photoelectron spectrum; (b) X-ray emission K spectrum of zinc in crystalline ZnSe; (c) electron density of states. The theoretical curve is smoothed by a dispersion curve of 0.3 eV half-width.

TABLE 33. Theoretical and Experimental Values of Energy for the High-Symmetry Points

Method of energy determination	L_3	X_5	W_2
XPS [16]	0.9 ± 0.3	1.8 ± 0.2	2.2 ± 0.3
UPS [320]	0.7	–	–
Pseudopotential [284]	1.0	1.5	2.0
OPW [315] (with adjusted parameters)	0.4	1.1	–
OPW [315] ($V_{ex}^{\alpha=1}$)	0.4	1.1	–
ROPW [16] ($V_{ex}^{\alpha,\beta}$)	0.8	1.6	–
KKR [327]	0.6	1.4	–

[a] Values of energy (in electron volts) are given with respect to the top of the valence band.

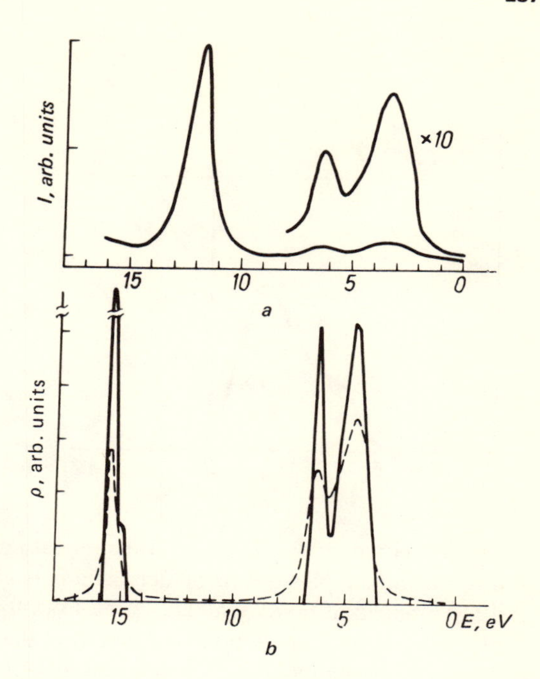

Figure 130. Valence band photoelectron spectrum (a) and electron density of states (b) of crystalline CdTe. The theoretical curve is smoothed by a dispersion curve of 0.3 eV half-width.

to this subband is not observed at energies greater than that of the maximum in the d states in the photoelectron spectrum. Table 33 indicates that, for the degenerate part of the valence band, the calculated energy values agree with the experimental values.

Figure 131 shows the photoelectron spectra of CdS and CdSe after subtraction of the background of electron inelastic scattering and the contribution of d states.

of the Valence Band of Crystalline CdTe[a]

State					
Σ_1^{min}	W_1	$X_3(L_1)$	X_1	L_1	Γ_1
2.7 ± 0.3	4.5 ± 0.2	5.1 ± 0.2	–	–	–
2.8	–	4.7	8.8	–	–
2.7	4.3	4.6	10.6	–	11.8
–	–	3.0	–	–	–
–	–	3.1	–	–	–
–	–	3.9	10.1	10.3	10.8
2.1	–	3.5	8.7	9.2	10.3

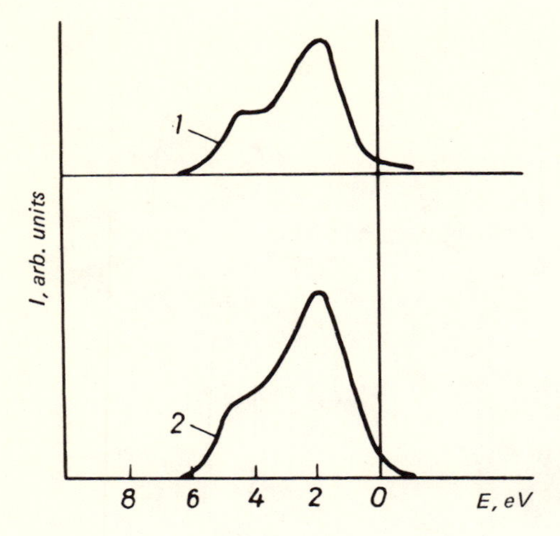

Figure 131. Valence band photoelectron spectra of (1) CdS and (2) CdSe after subtraction of the background of inelastic electron scattering.

Binary compounds differ from the monoatomic crystals of diamond, silicon, and germanium in having a forbidden band between the top of the valence band and the subband of nondegenerate states. For monatomic crystals the symmetry group includes inversion, and therefore the states at the point X are degenerated. In compounds, however, the magnitude of band splitting at this point depends—as has already been pointed out by Cohen and Bergstrasser [138]—on the antisymmetric form factors of the pseudopotential and, consequently, on the ionic character of the compounds. A general tendency has been observed, namely, an increase of the splitting when passing from A_3B_5 compounds to A_2B_6 compounds. Grobman *et al.* [329] used the two-band model to show that the magnitude of the splitting is close to the parameter C introduced by Phillips [330], which is determined by the magnitude of the antisymmetric Fourier component of the potential.

A more detailed study of this problem has been performed by Harrison [331] with a model in which hybridized s and p states were taken into account. The author constructed for each atom the sp^3 hybrid orbitals directed toward the four nearest neighbors. Thus, for the anion, such an orbital may be expressed as follows:

$$|\varphi_a\rangle = \frac{1}{2}(|s^a\rangle + \sqrt{3}\,|p^a\rangle),$$

where $|p^a\rangle$ is a linear combination of p-type states. The orbitals of each atom in the crystal are orthogonal to each other. The mean value of the Hamiltonian for the φ_a orbital is given by the expression

$$\varepsilon^a = \langle\varphi_a|H|\varphi_a\rangle = \frac{1}{4}(\varepsilon_s^a + 3\varepsilon_p^a),$$

where $\epsilon_s^a = \langle s^a | H | s^a \rangle$ and $\epsilon_p^a = \langle p^a | H | p^a \rangle$. Similar expressions may be written for the wave function $|\varphi_c\rangle$ and the energy ϵ^c of the cation. $V_3 = \frac{1}{2}(\epsilon^c - \epsilon^a)$ is one of the basic parameters of the model, and it describes the change in the polarity of the binding. It is to be expected that the electrons would be transferred toward the anion if V_3 has positive values. Though the hybrid orbitals of any atom are orthogonal, two hybrid orbitals are correlated by a nonzero matrix elements of the Hamiltonian. The second important parameter is

$$V_1^a = -\langle \varphi_a | H | \varphi_{a'} \rangle =$$

$$= \frac{1}{4}(\varepsilon_p^a - \varepsilon_s^a),$$

where the sign is chosen in such a way that V_1^a will be positive. The parameter V_1^c can be determined in a similar way.

Let us now find the matrix element of the Hamiltonian between two hybrid orbitals of a single binding:

$$V_2 = -\langle \varphi_c | H | \varphi_a \rangle.$$

Here, the sign is chosen so that the parameter V_2 may be considered as positive. Let us consider for each binding the linear combination of hybrid orbitals that yields the minimum energy

$$|\varphi\rangle = u_a | \varphi_a \rangle + u_c | \varphi_c \rangle,$$

where $u_a^2 + u_c^2 = 1$. Here, u_a and u_c are given by the following expressions:

$$u_a = \left[\frac{1}{2}(1 + \alpha_p)\right]^{1/2},$$

$$u_c = \left[\frac{1}{2}(1 - \alpha_p)\right]^{1/2},$$

where

$$\alpha_p = \frac{V_3}{(V_2^2 + V_3^2)^{1/2}}.$$

The matrix element $\langle \varphi | H | \varphi \rangle$ is given by

$$\langle \varphi | H | \varphi \rangle = \frac{1}{2}(\varepsilon^a + \varepsilon^c) - (V_2^2 + V_3^2)^{1/2}.$$

By using the matrix elements between two binding orbitals of the anion and cation, the following parameters can be determined:

$$A = -\langle \varphi | H | \varphi' \rangle^a = \frac{1}{2}(1 + \alpha_p) V_1^a,$$

$$C = -\langle \varphi | H | \varphi' \rangle^c = \frac{1}{2}(1 - \alpha_p) V_1^c.$$

Knowing the matrix elements, the energy bands can also be calculated. For this purpose, Harrison [331] used, as basis functions, combinations of binding orbitals having the orientation α:

$$\Psi^a(\mathbf{k}, \mathbf{r}) = \frac{1}{\sqrt{N}} \sum_j \exp i\mathbf{k} \cdot \mathbf{R}_j | \varphi^\alpha (\mathbf{r} - \mathbf{R}_j)\rangle,$$

where the sum is performed over N positions of the vector $\mathbf{R}_j$ defining the middle point of the binding α. For each state $\mathbf{k}$, the wave function is expressed as the sum of four Bloch functions with different α values. By analyzing the secular equation obtained, the magnitude of the splitting of energy bands can be determined. In the above model, this splitting is given by $\Delta E(x_1 - x_3) = 4|A - C|$. Figure 132 shows that the experimental values agree satisfactorily with the theoretical ones.

The spin-orbit term in the Hamiltonian is given by the expression

$$H_{\text{s-o}} = \frac{\alpha^2}{2}\left(\frac{1}{r}\frac{\partial V}{\partial r}\right)\mathbf{L} \cdot \mathbf{S},$$

where α is the fine-structure constant, V is the electrostatic potential, and $\mathbf{L}$ and $\mathbf{S}$ are the orbital and spin momenta, respectively.

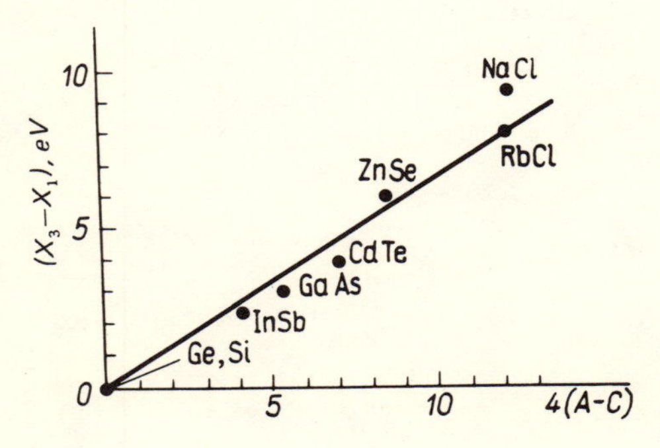

Figure 132. Correlation between experimental and theoretical values of valence band energy splitting at the point X.

Comparison of the magnitudes of the splitting ΔE_{s-0} for different charge states of the free ions [332] has shown that for the given configuration of d electrons, the spin–orbit splitting increases with increasing ionic charge, but not more than 2% per unit change in charge. Transition from the configuration d^9 to the configuration d^8 results in a significantly greater change in the magnitude of spin–orbit splitting (typically 7%). The magnitudes of spin–orbit splitting in the d states of cations and anions in A_3B_5 and A_2B_6 compounds given in Tables 34 and 35 show that, for core d electrons of atoms in a tetrahedral structure, ΔE_{s-0} does not increase significantly (it increases less than 1%) as compared with the corresponding values in free ions. The compounds of tellurium are an exception. For these the spin–orbit splitting increases by 2.8%. In metals (indium, cadmium, and lead) ΔE_{s-0} is 7%, 4%, and 1% greater than the values obtained for free ions, respectively. An even greater difference is observed for zinc. In zinc $\Delta E_{s-0} = 0.54$ eV [333], although in free ions of zinc (Zn II) its value is only 0.34 eV [334]. For lead and antimony no increase of the spin–orbit interaction was observed when passing from free ions to compounds. The observed variations in the magnitude of ΔE_{s-0} arise from multiplet splitting in the crystal field. Thus, because of the spin–orbit interaction and the octahedral crystal field, the energy levels of d electrons are split into the doubly degenerated level Γ_7 with energy $E(\Gamma_7) = -4Dq + \xi$, and the two Γ_8, four-times-

TABLE 34. Spin-Orbit Splitting of the Cation d States in the Compounds A_3B_5 and A_2B_6

Sample	Level	Lattice type	Magnitude of the splitting (eV)	Ref.
GaP	3*d*Ga	Sphalerite	0.4	[336]
GaAs	3*d*Ga	Sphalerite	0.4	[336]
GaSb	3*d*Ga	Sphalerite	0.4	[336]
InP	4*d*In	Sphalerite	0.78	[336]
InAs	4*d*In	Sphalerite	0.82	[336]
InSb	4*d*In	Sphalerite	0.84	[336]
In (metallic)	4*d*In	Tetragonal	0.91	[336]
In III (free ion)	4*d*In	–	0.85	[336]
CdTe	4*d*Cd	Sphalerite	0.70	[320]
CdS	4*d*Cd	Sphalerite	0.76	[337]
CdSe	4*d*Cd	Sphalerite	0.87	[337]
Cd (metallic)	4*d*Cb	Hexagonal	0.95	[333]
Cd II (free ion)	4*d*Cd	–	0.67	[334]
PbS	5*d*Pb	NaCl type	2.58	[338]
PbSe	5*d*Pb	NaCl type	2.61	[338]
PbTe	5*d*Pb	NaCl type	2.61	[338]
Pb (metallic)	5*d*Pb	Face-centered cubic	2.66	[333]
Pb IV (free ion)	5*d*Pb	–	2.64	[334]

TABLE 35. Spin-Orbit Splitting of the Anion *d* States in the Compounds A_3B_5 and A_2B_6

Sample	Level	Lattice type	Magnitude of the splitting (eV)	Ref.
GaSb	4*d*Sb	Sphalerite	1.21	[16]
InSb	4*d*Sb	Sphalerite	1.22	[16]
Sb (metallic)	4*d*Sb	Rhomboid	1.25	[339]
Sb V (free ion)	4*d*Sb	–	1.24	[334]
ZnTe	4*d*Te	Sphalerite	1.47	[16]
CdTe	4*d*Te	Sphalerite	1.44	[16]
HgTe	4*d*Te	Sphalerite	1.44	[16]
Te (metallic)	4*d*Te	Hexagonal	1.51	[339]
Te VII (free ion)	4*d*Te	–	1.41	[334]
HgTe	5*d*Hg	Sphalerite	1.77	[16]
HgSe	5*d*Hg	Sphalerite	1.81	[337]
HgS	5*d*Hg	Sphalerite	1.79	[337]
Hg (liquid)	5*d*Hg	–	1.83	[16]
Hg II (free ion)	5*d*Hg		1.80	[334]

degenerate levels

$$E_1(\Gamma_8) = 6Dq + \sqrt{\frac{3}{2}}\,\xi\cos\theta,$$

$$E_2(\Gamma_8) = -4Dq - \frac{1}{2}\xi - \sqrt{\frac{3}{2}}\,\xi\cos\theta,$$

where ξ is the magnitude of spin-orbit interaction and θ is given by the equation

$$\tan 2\theta = -\frac{\sqrt{6}\xi}{10Dq + \frac{1}{2}\xi}.$$

These results may also be used for the tetrahedral field since it also splits the *d* levels into the t_{2g} and e_g components, which in this case are in the reverse order. In the above formula, $-10Dq$ can be replaced by the empirical splitting parameter B. Moreover, all the constant factors can be included in B. The spin-orbit interaction constant is negative for hole states, while B may change sign in A_3B_5 and A_2B_6 compounds. Figure 133 shows ΔE_d [in energy units $(5/2)|\xi|$] versus B for hole states in a tetrahedral field. Here, ξ is fixed and B is variable. If the lines are narrow, then the photoelectron spectrum should exhibit three peaks. However, if the lines are rather broad, the effects caused by the crystal field result in a redistribution of the intensity of two photoelectron lines.

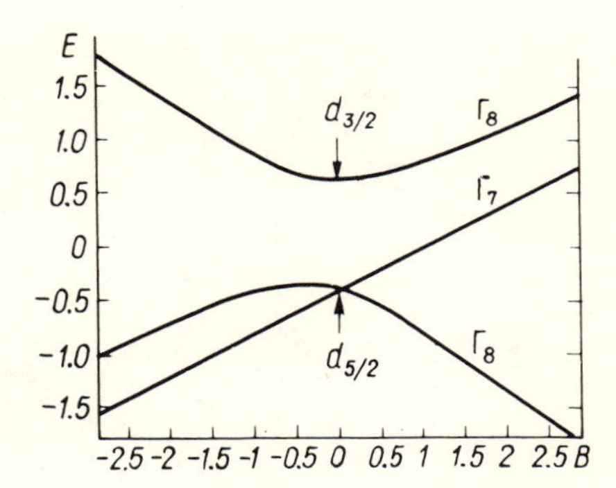

Figure 133. Spin-orbit splitting and crystal field splitting of d states in crystals with a sphalerite-type lattice. [B is given in units of (5/2) |ξ|.]

Evidently, for tellurium compounds the ratio $|B/\xi|$ is approximately 0.5, and B is negative. Ley *et al.* [332] have analyzed the possible splittings for hexagonal crystals. They have also used the experimentally determined values of ΔE_{s-0} to evaluate the constants that characterize the crystal field splitting of energy levels.

To illustrate the influence on ΔE_{s-0} of the symmetry of surrounding atoms in crystals, Figure 134 shows the $4d$ doublet of cadmium for both the alloy $Ag_{90}Cd_{10}$ and pure cadmium. The latter was evaporated onto a gold substrate in a sufficient amount to form a monolayer of cadmium. Cadmium atoms in a cubic lattice are characterized by a ΔE_{s-0} value close to that corresponding to free atoms. For cadmium evaporated onto gold, the magnitude of spin-orbit splitting is also small, and therefore, because of the low dissociation energy of cadmium atom clusters, it follows that in this case ΔE_{s-0} reflects the properties of separate cadmium atoms. When atomic clusters are formed on the surface, the magnitude of ΔE_{s-0} increases.

Thus, the variation of the magnitude of spin-orbit interaction in compounds, as compared to free atoms and ions, is determined by the influence of the crystal field on ΔE_{s-0}. The acquisition of quantitative data regarding this problem, however, still requires a large amount of work.

Chemical Shifts of Core Levels

As we have shown in Chapter 2, the binding energy values of core-level electrons are sensitive to the charge distribution of valence electrons and can thus be used for the determination of the ionic character of compounds. In the determination of binding energies of core electrons in compounds, however, account must be taken of the possible oxidation of the surface layer of the sample under investigation, as well as effects arising from electrostatic charging. As a convenient reference point for estimating the values of binding energies, the top of the valence band of the crystal is often used. In this case, the absolute value of binding energies may be found by adding the value of the work function of the sam-

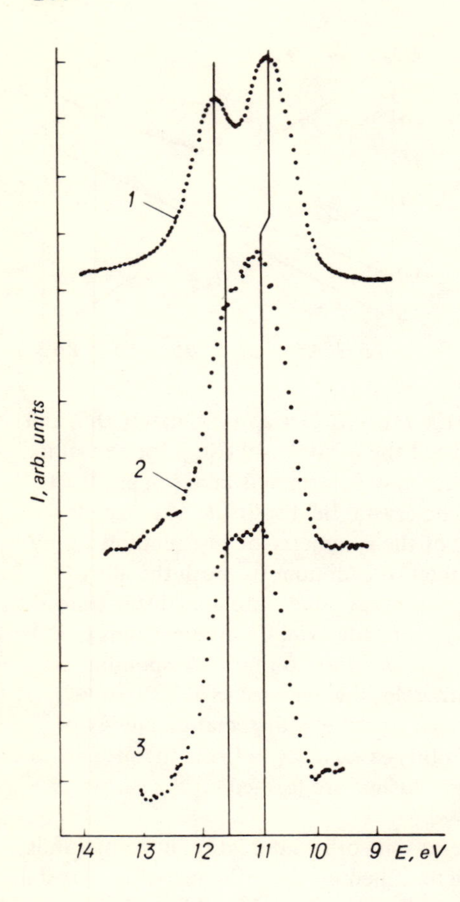

Figure 134. Valence band photoelectron spectra of (1) cadmium; (2) $Ag_{90}Cd_{10}$; (3) cadmium on gold substrate.

ple. Shevchik *et al.* [336] proposed a formula for the chemical shift of cation core levels, which is much simpler than the formulas (30) and (31):

$$\Delta E_c = \Delta q e \left(\frac{A(\Gamma)_c}{r_c} - \frac{\alpha}{R} \right),$$

where r_c has a value close to the ionic radius of the cation; R is the distance between the cation and the nearest anion; $A(\Gamma)$ is a geometrical factor that accounts for the particularities of the distribution of the density of electronic charge; and Δq is the magnitude of the charge transferred from the cation to the anion. The magnitude of cation charge in compounds with a sphalerite-type lattice can, according to the work of Phillips [330], be expressed as follows:

$$q_c = 4e(1 - f_i),$$

where f_i is the ionicity. Then

$$\Delta q = Z - 4e(1 - f_i).$$

The magnitude of the shift ΔE_c is given by

$$\Delta E_c = [2 - 4(1 - f_i)]\, e^2 \left[\frac{A[\Gamma]_c}{r_c} - \frac{\alpha}{R} \right].$$

It should be noticed that this formula describes the shift of the given cation core level with respect to the vacuum level of the sample. The similar expression for the shift of anion core levels is

$$\Delta E_a = -[2 - 4(1 - f_i)]\, e^2 \left[\frac{A(\Gamma)_a}{r_a} - \frac{\alpha}{R} \right].$$

The model described above for calculating chemical shifts gives less accurate results for anions, because in this case it is more difficult to determine the extension and the shape of the charge distribution at the anion. The expression $A(\Gamma)$ for a charge distribution shaped like a spherical shell with radii $\Gamma + r$ and r may be written as follows:

$$A(\Gamma) = \frac{1 - \Gamma^2}{1 - \Gamma^3}.$$

Estimation of Γ on the basis of the electron charge distribution in atoms yields the value $\Gamma \approx 0.5$. Table 36 gives the data obtained by Shevchik *et al.* [336] on the shift of $3d$, $4d$, and $5d$ core levels in pure metals as compared to those in a

TABLE 36. Energy Shifts of Cation $3d$, $4d$, $5d$ Levels with Respect to Their Position in Metals

Compound	Δf	ΔE_{VB}	ΔE_B^V	$A\ (\Gamma = 1)$	$B\ (\Gamma = 0)$	$C\ (\Gamma = 1/2)$
GaAs	0.24	−0.3	−1.7	−0.22	−1.41	−0.90
GaSb	0.045	−0.4	−1.1	−0.06	−0.29	−0.19
InP	0.69	0.05	−1.6	0.39	−2.49	−1.12
InAs	0.47	0.25	−1.0	0.15	−1.77	−0.93
InSb	0.27	0.05	−0.7	−0.02	−1.17	−0.69
ZnSe	0.7	0.81	−1.65	−0.43	−3.85	−2.48
ZnTe	0.18	0.37	−1.19	−0.24	−0.86	−0.67
CdS	0.72	1.54	−1.56	−0.11	−2.98	−1.65
CdSe	0.79	1.16	−1.34	−0.12	−3.50	−2.1
CdTe	0.69	0.66	−1.44	−0.52	−3.50	−2.2
HgS	–	−0.6	−3.0	–	–	–
HgSe	0.66	−1.9	−2.9	0.06	−2.46	−1.35
HgTe	0.72	−2.3	−3.7	−0.32	−3.37	−2.08

number of compounds of A_3B_5 and A_2B_6 type. In the third column the energies of the core levels were measured with respect to the top of the valence band (E_{VB}), whereas in the fourth column they were measured with respect to the vacuum level (E_B^V). The table also shows the values of the shifts predicted by the potential model for three particular cases A, B, C corresponding to values of Γ of 1, 0, and $\frac{1}{2}$.

Shifts determined with respect to the top of the valence band have small negative values for gallium compounds, small positive values for indium compounds, large positive values for zinc and cadmium compounds, and large negative values for mercury compounds.

The change in sign indicates that in the determination of core-level shifts it is wrong to use binding energy values measured with respect to the top of the valence band. In fact, the cation transfers part of its charge to the anion, and, consequently, the binding energy of the cation core electrons is increased. Using the first model (model A) for the study of shifts, the values obtained for the shifts in InP, InAs, and HgSe are positive, because of geometrical factors. If the binding energies are measured with respect to the vacuum level it follows that all the experimental shifts are negative. Model C describes the compounds of indium, zinc, and cadmium well, and model B describes the compounds of mercury successfully. However, for mercury the chemical shifts were determined with large error (± 1.0 eV), although for all the other compounds studied the error amounted to ± 0.1 eV. The large values of the chemical shifts in HgTe and HgSe compounds are explained in both model B and model C as being due to the geometrical factor, because the magnitudes of the charges Δq for these compounds practically coincide. The anomalously small value of the predicted shift for GaSb arises from the low value of the charge Δq determined from the ionicity f_i.

Influence of the Disordered Distribution of Atoms on the Density of States of Valence Electrons

Disordered solids may be grouped into two classes. The first class consists of solids for which the disorder has a more qualitative character, the atoms being distributed randomly at the sites of the crystal lattice, so that each site can take an atom of any type. If the system contains only two types of atoms, A and B, then the electron structure can be studied by using the theoretical models developed within the theory of disordered binary alloys.

For solids in the second class, the disorder has a spatial character. For example, the system may contain atoms of only one type, which, however, do not exist as a periodic crystal lattice; that is, a disorder exists in the spatial distribution of the atoms. A particular case of spatial disorder, in which the disordered system exhibits the same coordination of nearest neighbors but different binding

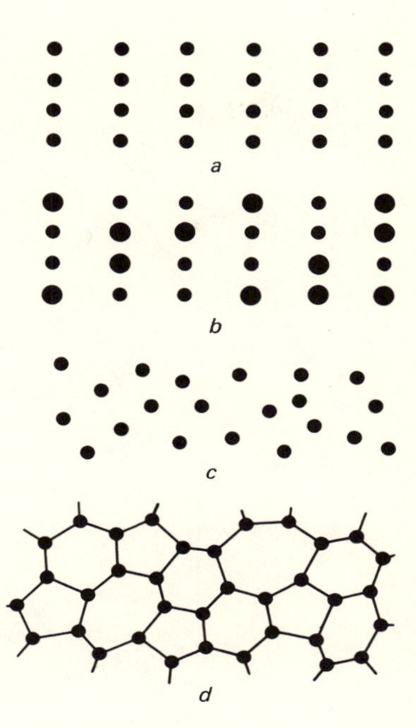

Figure 135. Various disorder types in solid state materials: (a) ideal order; (b) disorder of a binary alloy; (c) space disorder; (d) topological disorder.

lengths, is the topological disorder. The types of disorder mentioned above are illustrated in Figure 135.

The amorphous semiconductors–silicon and germanium–are characterized by topological disorder. In describing the amorphous semiconductors, it can be assumed that the deviation in binding lengths and angles from their values in crystals amounts to, say, 10%.

Weaire and Thorpe [301] studied the electron structure of amorphous silicon and germanium by using the tight-binding approximation. Joanopoulos and Cohen [340] used the pseudopotential method to calculate the simple tetragonal structure of silicon and germanium with 12 atoms in each elementary cell ($ST = 12$). As Figure 136 shows, the X-ray photoelectron spectra of crystalline and amorphous silicon and germanium obtained by Pollak [341] and Ley *et al.* [342] using an HP-5950A electron spectrometer are significantly different. The *s*-type bands in amorphous germanium and silicon do not exhibit two separated peaks in the density-of-states curve. The density of states calculated by Joannopoulos and Cohen [340] confirms the experimentally determined structure of the X-ray photoelectron spectra of materials in the amorphous state.

Ley *et al.* [342] measured the X-ray photoelectron spectra of arsenic, antimony, and bismuth in the crystalline as well as the amorphous states, by using an HP-5950 A electron spectrometer. The monocrystalline samples were cleaned

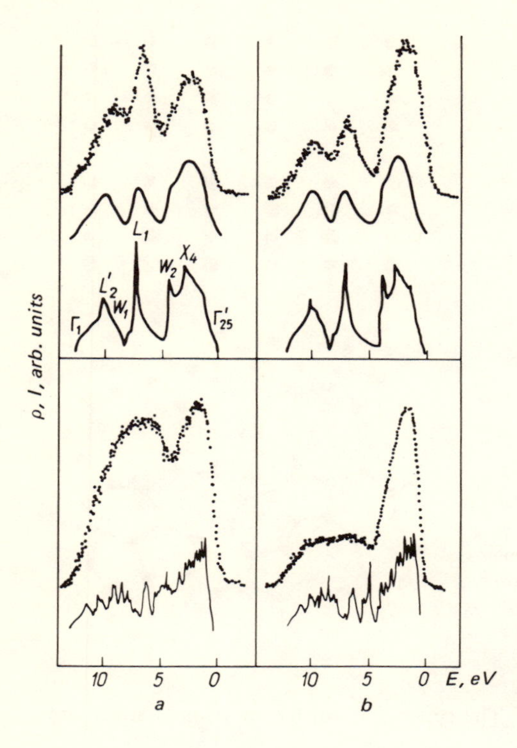

Figure 136. Valence band photoelectron spectra and density of states in crystalline and amorphous silicon (a) and germanium

in a nitrogen atmosphere followed by argon ion bombardment. Throughout the measurement the pressure in the spectrometer was no lower than 5×10^{-8} torr. The amorphous arsenic sample was obtained by *in situ* evaporation onto a gold substrate. Amorphous antimony and bismuth were prepared by argon ion bombardment of monocrystal surfaces. The valence band protoelectron spectra obtained are shown in Figure 137. The similarity of the photoelectron spectra for crystalline samples demonstrates the similarity of their elementary structure (electron configuration s^2p^3). Since the elementary cell contains two atoms, ten valence electrons will fill up five valence subbands. In free atoms, the binding energies of the *s*-type valence electrons are 8-10 eV greater than those of electrons with *p*-type symmetry. The peaks 1 and 2 in the X-ray photoelectron spectra are situated at about 7-8 eV from the peaks 3 and 4, and therefore they can be attributed to the *s* electrons. The existing theoretical calculations of the energy band structure [343-348] show that in these crystals there exist two energetically separated *s* bands and three *p* subbands situated at the top of the valence band. However, since the calculations [343-348] have not been extended as far as to calculate the electron density of states in the valence band,

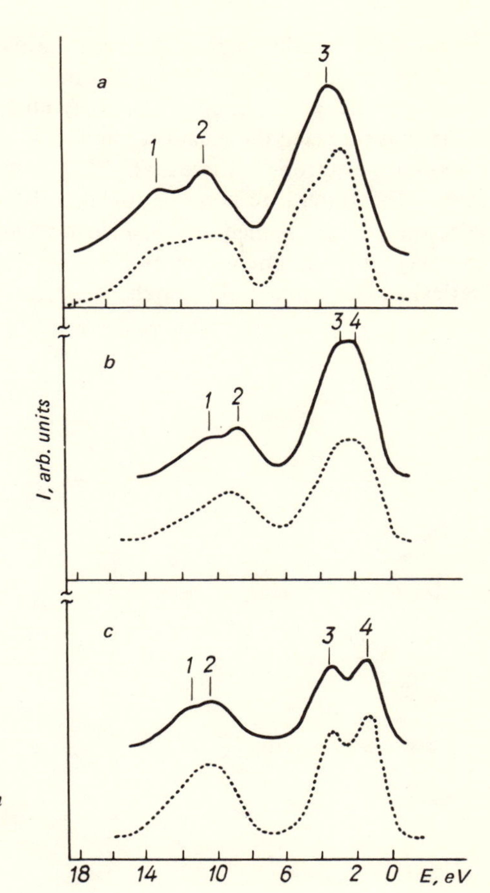

Figure 137. Valence band photoelectron spectra in crystalline (——) and amorphous (· · ·) (a) arsenic; (b) antimony; (c) bismuth.

it is difficult to make a detailed comparison of the experimental and theoretical results. The magnitude of the splitting of *s* bands is consistent with the theoretical values. In bismuth, peak 3 is split into two peaks, 3 and 4, separated by 2.2 eV. This is consistent with the value 2.16 eV for the spin–orbit splitting of the 6*p* states in free atoms [349].

The density of states for amorphous samples is significantly different from that of crystals. As in the case of silicon and germanium, the density of *s* states does not exhibit two distinct peaks, and the structure of the *p* states is also more smeared out. The replacement of the two peaks, 1 and 2, in the photoelectron spectra of crystalline samples by a single peak in the spectra of amorphous samples does not result from an increase of the width of each of the peaks, since the total widths of the corresponding peaks are practically the same in both amorphous and crystalline states. Not even the energy separation between the sites of localization of the *s* and *p* states is significantly changed.

Thus, the investigation of the X-ray photoelectron spectra of amorphous solids has provided the means to show that the structure of the energy spectra of valence electrons in these materials is determined mainly by the short-range order. This order is responsible for the existence of structure in the density of electronic states, namely, the van Hove singularities and the forbidden bands between different energy bands. However, these problems require further theoretical and experimental research.

5

Halides of Alkali and Alkaline-Earth Metals

In recent years alkali halide crystals have been the object of intensive studies. They form a group of the simplest binary compounds, consisting of positive and negative ions bound into the crystal lattice by attractive Coulomb forces. For many alkali halide crystals, the experimental studies have included X-ray emission and absorption spectra, as well as spectra of emission and reflection in the far ultraviolet. However, in spite of the extensive investigations, some of the detailed features of the X-ray and optical absorption spectra remain unclarified. It can be hoped that data regarding the energy distribution of the electronic states, determined by X-ray photoelectron spectroscopy, will provide some new information, since such data make it possible to place the spectra obtained in various energy regions on a common energy scale. Crystals of the fluorite group, unlike the alkali halide crystals, have been studied to a much smaller extent. A simultaneous use of X-ray emission spectra and X-ray absorption and reflection spectra in the far ultraviolet has not only allowed clarification of the main features of the reflectivity spectra, but also revealed the structure of the lower part of the conduction band and the existence of excitonic states in the optical and X-ray spectra.

X-Ray Photoelectron Spectra and the Band Structure of Alkali Halide Crystals

Siegbahn *et al.* [1] reported the first X-ray photoelectron spectra of a number of alkali halide crystals. More complete studies have been published subsequently [350-354]. The X-ray photoelectron spectra of the valence electrons of alkali halide crystals, as obtained by Pollak [354], are shown in Figures 138-142, and the corresponding values of binding energies are given in Tables 37-41. Pollak [354] performed the measurements using an IEE-15 electron spectrometer with a resolution of 0.9 eV. As photon source, the Mg $K_{\alpha 1,2}$

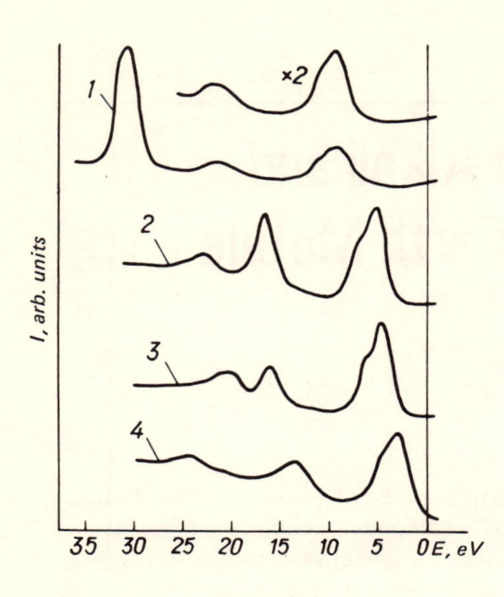

Figure 138. Valence band photoelectron spectra of alkali halide crystals: (1) LiF; (2) LiCl; (3) LiBr; (4) LiI.

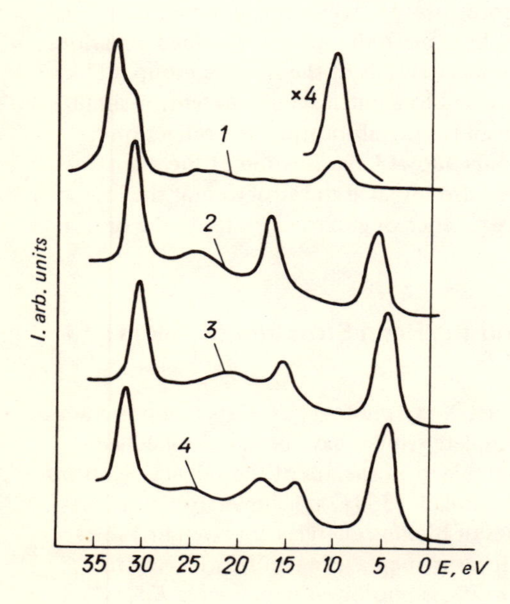

Figure 139. Valence band photoelectron spectra of alkali halide crystals: (1) NaF; (2) NaCl; (3) NaBr; (4) NaI.

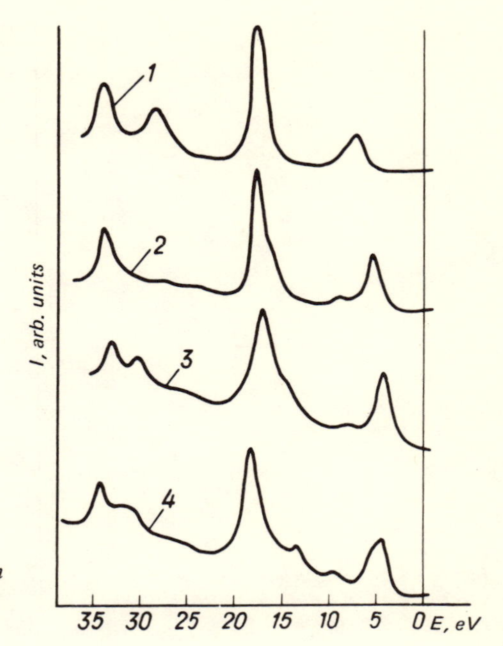

Figure 140. Valence band photoelectron spectra of alkali halide crystals: (1) KF; (2) KCl; (3) KBr; (4) KI.

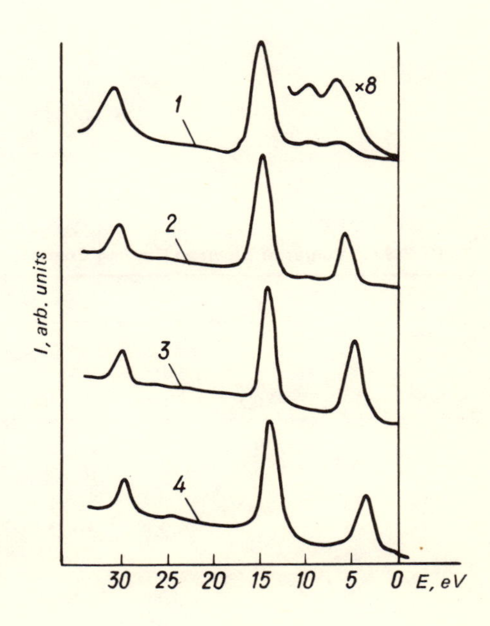

Figure 141. Valence band photoelectron spectra of alkali halide crystals: (1) RbF; (2) RbCl; (3) RbBr; (4) RbI.

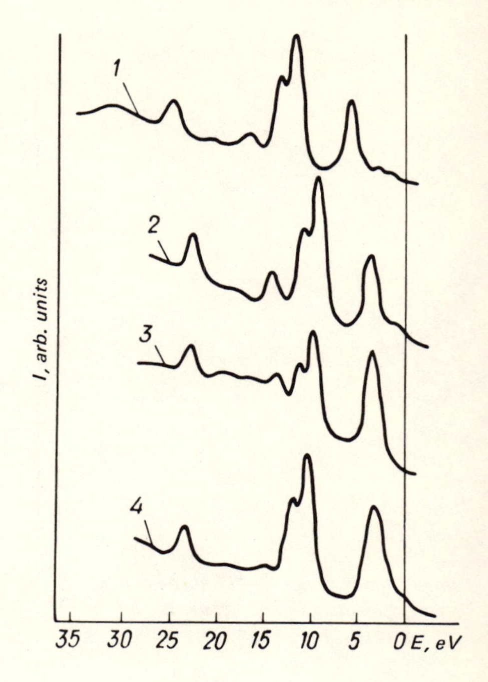

Figure 142. Valence band photoelectron spectra of alkali halide crystals: (1) CsF; (2) CsCl; (3) CsBr; (4) CsI.

TABLE 37. Values of Electron Binding Energies of LiF, LiCl, LiBr, and LiI Cyrstals

Sample	Ion	$1s_{1/2}$	$2s_{1/2}$	$2p_{1/2}$	$2p_{3/2}$	$3s_{1/2}$	$3p_{1/2}$	$3p_{3/2}$
LiF	Li^+	56.5						
	F^-	685.6	30.7					
LiCl	Li^+	56.5						
	Cl^-		269.8	200.5	198.9	16.6		
LiBr	Li^+	56.6						
	Br^-					256.0	189.2	182.4
LiI	Li^+	55.8						
	I^-					1070.9	930.8	874.8

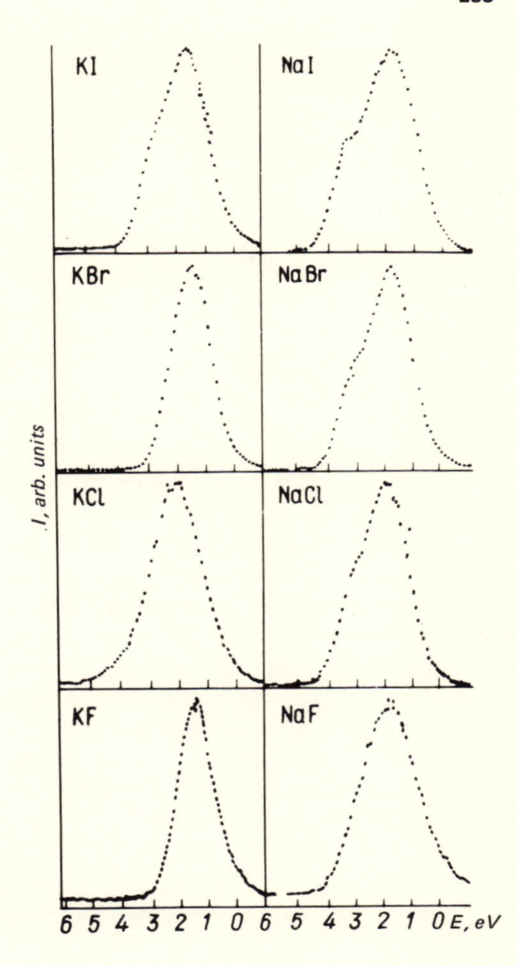

Figure 143. Valence band photoelectron spectra of p electrons in some alkali halide crystals.

State								
$3d_{3/2}$	$3d_{5/2}$	$4s_{1/2}$	$4p_{1/2}$	$4p_{3/2}$	$4d_{3/2}$	$4d_{5/2}$	$5s_{1/2}$	Valence band (eV)
								9.4
								6.9; 5.2
69.2	16.0							4.6; 6.3
								3.1
631.6	619.2	186.3	122.1		50.8	49.2	13.5	

TABLE 38. Values of Electron Binding Energies of NaF, NaCl, NaBr, and NaI Crystals

Sample	Ion	$1s_{1/2}$	$2s_{1/2}$	$2p_{1/2}$	$2p_{3/2}$	$3s_{1/2}$	$3p_{1/2}$	$3p_{3/2}$
NaF	Na^+	1074.4	65.7	33.0				
	F^-	687.1	31.3					
NaCl	Na^+	1072.0	63.9	31.2				
	Cl^-		269.4	200.7	199.1	16.8		
NaBr	Na^+	1071.6	63.7	30.8				
	Br^-					255.9	188.9	182.2
NaI	Na^+	1072.9	64.7	31.9				
	I^-					1072.7	931.5	875.5

line was used. Kowalczyk *et al.* [353] obtained the valence band photoelectron spectra (Figure 143) using the monochromatized Al K_α line with an HP-5950A electron spectrometer, having a resolution of the order of 0.6 eV. Since the alkali halide compounds are strongly bound ionic crystals, the energy width of their valence bands is not large. For many purposes, it is sufficient to consider that the upper valence band consists mainly of the *p* states of the anion. However, as is demonstrated in Figure 143, the structure of the photoelectron spectra of compounds in which the cation is potassium is significantly different from that of compounds with sodium as cation. Consequently, in discussing the structure of the energy bands of the alkali halide crystals, it

TABLE 39. Values of Electron Binding Energies of KF, KCl, KBr, and KI Crystals

Sample	Ion	$1s_{1/2}$	$2s_{1/2}$	$2p_{1/2}$	$2p_{3/2}$	$3s_{1/2}$	$3p_{1/2}$	$3p_{3/2}$
KF	K^+		377.6	295.8	293.0	33.6	17.4	
	F^-	683.7	28.0					
KCl	K^+		377.7	295.8	293.0	33.6	17.5	
	Cl^-		269.4	200.2	198.6	16.3		
KBr	K^+		377.7	296.0	293.2	33.0	17.0	
	Br^-					256.1	189.3	182.6
KI	K^+		378.0	296.1	293.3	34.0	18.0	
	I^-					1071.5	931.2	874.0

State								
$3d_{3/2}$	$3d_{5/2}$	$4s_{1/2}$	$4p_{1/2}$	$4p_{3/2}$	$4d_{3/2}$	$4d_{5/2}$	$5s_{1/2}$	Valence band (eV)
								10.2
								5.6
69.2		15.5						4.5
631.5	620.0	187.4	123.3		51.9	50.3	14.0	4.5

should be remembered that even in the highest subband of the valence band the states are hybridized. A number of calculations indicate [355-360] that in the majority of crystals the width of the upper valence band is of the order of 1-2 eV. Therefore, it is not possible to determine the fine structure of the energy bands of alkali halide crystals using presently available electron spectrometers, which have a resolution of the order of 0.6-0.9 eV.

Citrin and Thomas [352] calculated the electron binding energy for a number of alkali halide crystals by using the electrostatic model of point ions. The theoretical and experimental values of the binding energies are in good agreement, which argues in favor of the application of this simple model to the alkali

State								
$3d_{3/2}$	$3d_{5/2}$	$4s_{1/2}$	$4p_{1/2}$	$4p_{3/2}$	$4d_{3/2}$	$4d_{5/2}$	$5s_{1/2}$	Valence band (eV)
								7.2
								5.4
68.8		15.0						4.2
								5.2
630.5	618.9	186.4	122.1		52.2	49.6	13.5	4.2

TABLE 40. Values of Electron Binding Energies of RbF, RbCl, RbBr, and RbI Crystals

Sample	Ion	$1s_{1/2}$	$2s_{1/2}$	$2p_{1/2}$	$2p_{3/2}$	$3s_{1/2}$	$3p_{1/2}$	$3p_{3/2}$
RbF	Rb^+					322.6	247.8	238.9
	F^-	686.9	28.8					
RbCl	Rb^+					322.2	247.8	239.0
	Cl^-		269.8	200.7	199.1	16.3		
RbBr	Rb^+					321.7	247.4	238.4
	Br^-					255.9	188.8	182.1
RbI	Rb^+					321.1	246.6	237.8
	I^-					1070.7	930.3	873.5

TABLE 41. Values of Electron Binding Energies of CsF, CsCl, CsBr, and CsI Crystals

Sample	Ion	$1s_{1/2}$	$2s_{1/2}$	$2p_{1/2}$	$2p_{3/2}$	$3s_{1/2}$	$3p_{1/2}$	$3p_{3/2}$	$3d_{3/2}$
CsF	Cs^+						1065.3	997.1	738.6
	F^-	685.1	32.6						
CsCl	Cs^+						1064.2	996.6	737.4
	Cl^-		269.0	199.1	197.5	14.8			
CsBr	Cs^+						1065.0	997.4	738.0
	Br^-					254.5	187.6	180.9	68.0
CsI	Cs^+						1065.0	996.5	737.8
	I^-					1070.4	929.6	873.6	630.0

halide crystals. Kowalczyk *et al.* [353] also determined the energies of satellite lines in the energy region up to about 40 eV apart from the main lines. Satellite lines in the spectra of alkali halide crystals may be due either to the presence of some additional final states appearing in the process of photoemission or to electron energy losses in the process of plasmon excitation. Best [361] and Creuzberg [362] studied the spectra of electron energy losses in alkali halide crystals. With these independent data on the energy of plasmon excitations, it is, in principle, possible to identify the satellite lines observed in X-ray photoelectron spectra. The interpretation is difficult, however, because of the differences in the structure and energy characteristics of satellite lines originating from cations and anions.

State							
$3d_{3/2}$	$3d_{5/2}$	$4s_{1/2}$	$4p_{1/2}$ $4p_{2/3}$	$4d_{3/2}$	$4d_{5/2}$	$5s_{1/2}$	Valence band (eV)
112.3	111.1	30.8	14.7				6.6
							9.7
112.1	110.9	30.3	14.6				
							5.7
111.5	110.2	29.8	14.1				
69.0	69.0						4.6
111.3	110.0	29.7	13.9				3.4
630.1	618.6	186.1	121.7	50.6	49.0	13.9	

State									
$3d_{5/2}$	$4s_{1/2}$	$4p_{1/2}$	$4p_{3/2}$	$4d_{3/2}$	$4d_{5/2}$	$5s_{1/2}$	$5d_{1/2}$	$5p_{3/2}$	Valence band (eV)
724.6	230.3	170.7	159.6	78.3	76.1	25.4	13.8	12.2	6.3
723.4	229.3	169.4	158.3	77.0	74.7	23.1	11.3	9.7	4.0
724.0	229.4	169.7	158.6	77.2	75.0	23.2	11.8	10.2	3.7
	14.1								
723.6	229.5	170.1	158.8	74.7	75.4	23.6	12.0	10.4	3.3
618.5	185.0	123.3		50.6	49.0	14.6			

Characteristics of X-Ray Absorption Spectra of Alkali Halide Crystals

The problems related to the structure of optical and X-ray absorption spectra have been discussed by a number of authors [363–369]. Recently, interest in these problems has revived [370–372]. The X-ray photoelectron spectroscopy offers the possibility to place the X-ray and optical absorption spectra on a common energy scale. Gudat *et al.* [370] combined the spectra of LiF in this way, while Menzel *et al.* [372] and Kunz *et al.* [373] have calculated the *K* absorption spectra of the same compound. Gudat *et al.* [370], Menzel *et al.* [372], Kunz *et al.* [373], and Pantelides [374] used three different approximations for the interpretation of the K absorption spectra, namely: the pure band approximation, the band approximation including the exciton structure, and the band approximation including the energy shift of the valence band states as a whole. The results of calculations performed using the first and the third approximations for K absorption spectra are shown in Figure 144. The position of the theoretical main peak in the imaginary part of the dielectric function differs by approximately 2 eV from that of the experimental one.

Figure 144 also shows the quantum yield curve. This reproduces the shape of the absorption curve quite well. Gudat and Kunz [375] have shown that the agreement between the quantum yield and absorption curves has a general character and exists for a large number of solid materials. They also measured [370] the energy distribution of photoelectrons, and this allowed determination of the position of the valence band with respect to the 1s level of lithium. In Figure 145 are shown both the optical [376] and the X-ray absorption spectra on the same scale. This figure shows that the difference between the X-ray and optical absorption edges is of the order of 8 eV. The band model cannot explain this effect satisfactorily because of electron–hole interactions. According to the results obtained by Kunz *et al.* [373], an electron situated in the conduction band interacts more strongly with a hole in the 1s level of lithium than with a less localized hole in the valence band of the crystal. The different degree of hole localization is the cause of the difference of about 8 eV in the absorption edges. It can be considered that the main maximum in the optical absorption spectrum has an exciton origin [373, 376], and consequently the bottom of the conduction band is situated at an energy of 13.6 eV. Therefore, in the K X-ray absorption spectrum, the low absorption is related to the fact that the transition in the s–s exciton band is forbidden due to selection rules while the main maximum at 61.9 eV is related to transitions into the allowed p exciton band. Transitions

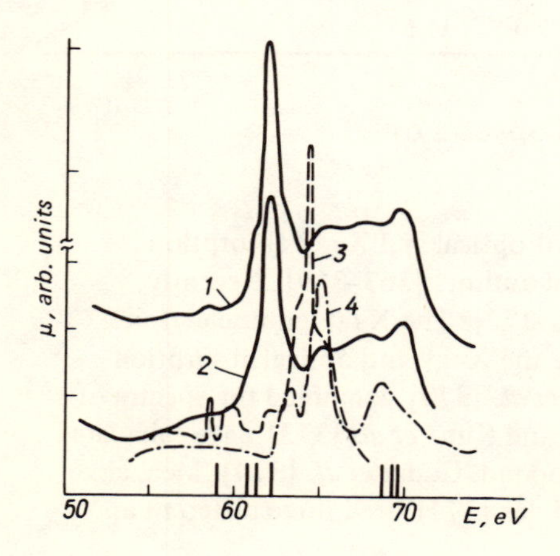

Figure 144. Theoretical and experimental X-ray absorption K spectra of lithium in crystalline LiF: (1) quantum yield K spectrum of lithium; (2) experimental X-ray absorption K spectrum of lithium; (3, 4) calculated X-ray absorption K spectrum of lithium [372, 373].

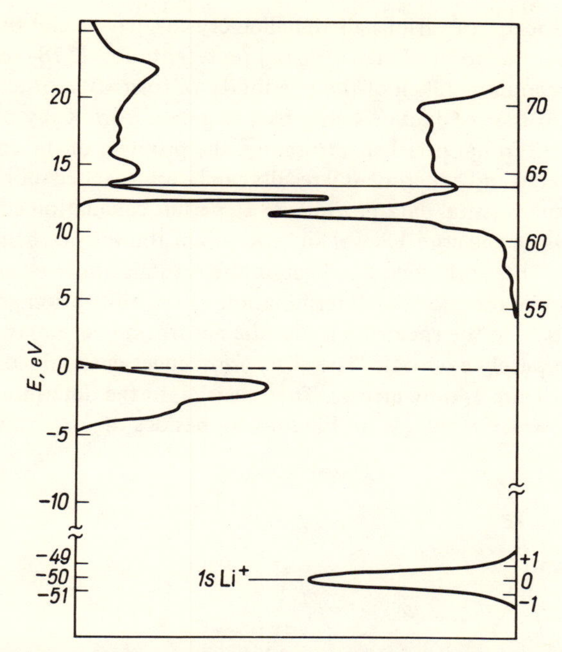

Figure 145. Comparison, on the same energy scale, of the data from optical, X-ray absorption, and photoelectron spectroscopy of crystalline LiF.

into the low conduction band start at 63.8 eV where a pronounced minimum occurs in the *K* absorption spectrum. Therefore the exciton binding energy of LiF is approximately 10 eV.

The third approximation used for the interpretation of the absorption spectra does not take sufficient account of the electron-hole interaction, since it merely reduces the interaction to the same energy shift for all of the states of the conduction band. This shift is related to the Coulomb interaction in an elementary crystal cell, whereas the energy bands are calculated for the states of the whole crystal. As can be seen from Figure 144, the energy position of the lower excitation levels of the free (Li^+) ion [377] and the position of the main maximum in the absorption spectrum of LiF are in good agreement. On this basis, the agreement between the shapes of the experimental and calculated spectra of Kunz *et al.* [373] should be considered accidental. Pantelides [371] compared the experimental [368] and calculated [372, 373, 378–381] *K* absorption spectra of lithium in LiF, LiCl, LiBr, and LiI. It is difficult to calculate the absolute values of transition energies, and therefore it is usual to compare the theoretical and experimental spectra with respect to various maxima and minima that are common to both spectra. This method is, however, rather arbitrary.

The X-ray spectra of various alkali halide crystals have so far only been interpreted from the point of view of band representation [378–381]. In this approach, no account is taken of the possibility of formation of exciton states at the bottom of the conduction band. By using data from X-ray photoelectron spectroscopy and from optical spectroscopy, the problem of the correspondence between calculated and experimental results can be approached more completely.

Existing experimental data [1, 350, 353] permit calculation of the energy difference between the core 1*s* level and the top of the valence band. By knowing the extent of the forbidden band gap in the crystals under investigation, the optical data can be used for the determination of the absorption edges. Any structure observed in the spectrum under the absorption edge may be interpreted as being purely excitonic. The absorption edges determined in this way are marked in Figure 146 by arrows. The precision in the determination of these values is of the order of 0.5 eV. In Figure 146, besides the experimental *K*

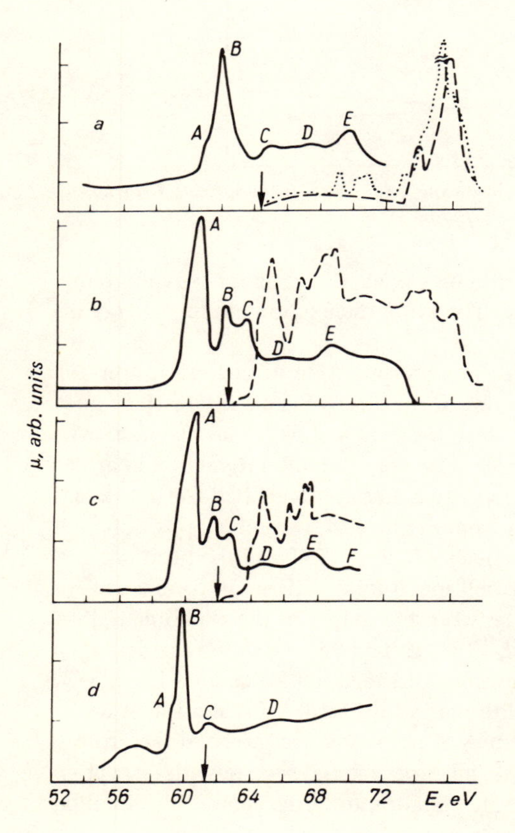

Figure 146. Comparison, on the same energy scale, of the experimental (——) and theoretical data for the K absorption spectra of lithium in (a) LiF, (b) LiCl, (c) LiBr, and (d) LiI crystals. The calculated K spectra are taken from [372] (· · ·) and [373] (– – –) for LiF, from [378] for LiCl, and from [381] for LiBr.

absorption spectra of lithium in LiF, LiCl, LiBr, and LiI, the results of theoretical calculations for LiF, obtained by Menzel *et al.* [372] and Kunz *et al.* [373], and for LiCl and LiBr, obtained by Brown *et al.* [363] and Sagawa *et al.* [364-366] are also shown. As can be seen from Figure 146, the main line in the absorption spectrum is situated 1.5-2.5 eV lower than the calculated absorption edge, which is an indication of the exciton character of this peak. Since the bottom of the conduction band is characterized by a symmetry of *s* type (Γ_1), and as a result of the conditions imposed by the selection rules, the possibility of formation of *s* excitons should be excluded. Therefore, it can be assumed that the observed exciton originates from higher *p*-type states of the conduction band. There is practically no agreement between the theoretical and the experimental data shown in Figure 146 for the energy positions of the observed structural features. A correlation exists, however, between the shapes of the experimental and theoretical spectra. Therefore, the nature of *K* absorption spectra is to a great extent excitonic, reflecting partially the structure of the band states.

Pantelides [382] used the X-ray photoelectron spectra [353] and the optical spectra [383] of NaBr in order to determine the position of the bottom of the conduction band with respect to the core levels. Figure 147 shows the ultrasoft X-ray absorption spectra and the results of a theoretical calculation of the interband density of electron states. In the calculation of the interband density of states it was considered that, in the given energy region, transitions to the conduction band are possible for both the 2*s* electrons of sodium and the 3*d* electrons of bromine. The energy differences between the corresponding core

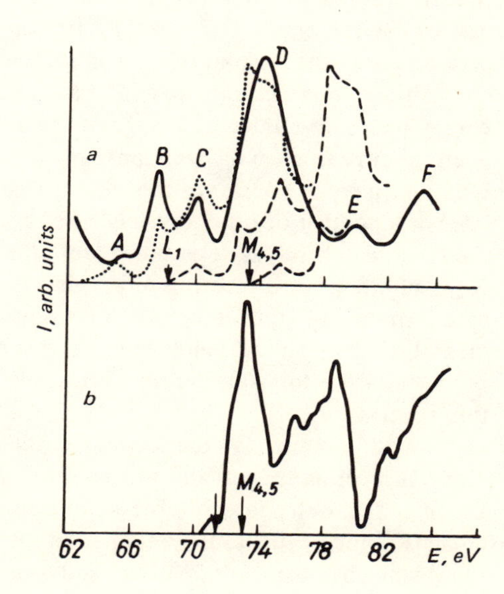

Figure 147. X-ray absorption spectra in the energy region 62-84 eV for NaBr (a) and KBr (b) crystals. The calculated spectra (· · · and – – –) are according to [382].

levels were taken from the work of Kowalczyk *et al.* [353]. However, since the absolute values of the transition energies were not determined, the curve of the interband density of states was shifted on the energy scale until the best agreement with the experimental data was obtained (see Figure 147). On this basis, the bottom of the conduction band is found to be located at 63 eV. However, the X-ray photoelectron spectra and the optical absorption spectra indicate that the bottom of the conduction band is situated at 68.5 eV. In Figure 147, the absorption edge for transitions from the 2*s* shell of the sodium ion is marked by a line labeled L_1. Consequently, the starting point of the curve of the interband density of electron states should be shifted up to this point. In spite of the fact that by doing so some agreement is found with respect to the position of the maxima (but not their intensities), the maxima A and B should be interpreted as being of excitonic origin.

The $M_{4,5}$ absorption edge of bromine in crystalline NaBr has been determined from photoelectron spectroscopy data (see Figure 147). Any structure appearing at lower energy values should be related to excitons arising from $M_{4,5}$ transitions and to the structure of the L_1 band. In order to give an idea of the contribution of $M_{4,5}$ transitions to the X-ray absorption spectrum, the $M_{4,5}$ absorption spectrum of bromine in KBr is also shown in Figure 147. It can be seen that the maxima A and B are actually related to the L_1 spectrum of sodium in NaBr. Maximum C, corresponding to a small decrease at 71–72 eV in the spectrum of NaBr, probably appears as a result of transitions of valence electrons to the 2*s* states of sodium and the 3*d* states of bromine. Thus, the structure observed in the vicinity of the optical absorption edge is also in this case of exciton origin. Excitons of *p* type appear as a result of the existence of *p* states above the bottom of the conduction band, and their occurrence demonstrates the strong electron–hole interaction that results in an overlapping of the energy bands over an energy range of the order of 1 Ry. It should be noted that excitons appear when halogen ions are excited, and that they have a lower binding energy as a result of the weaker electron–hole interaction.

For a number of other crystals it has been observed [354] that the width of the forbidden band, as determined from the X-ray absorption spectra and by X-ray photoelectron spectroscopy, is less than the width of the forbidden band, as determined by optical methods. Therefore, in the interpretation of experimental results on alkali halide crystals, it is necessary to take into account the presence of exciton states in the optical spectra, as well as in the X-ray absorption spectra.

As will be shown later, electron excitonic states are also apparent in the X-ray absorption spectra and the quantum yield spectra of crystals belonging to the fluorite group. However, before discussing the photoelectron spectra of these crystals, some of the results obtained in the study of the energy structure of the valence band and conduction band using optical methods will be considered.

Optical Constants of Crystals in the Fluorite Group

The optical constants of solids may be determined from reflection data by using the Kramers-Kronig transformation. This method has been largely used in the optical spectroscopy of solids [384-388] because it is simple and rather accurate.

In order to determine the phase $\theta(\omega_0)$ with formula (100), it is necessary to know the reflectivity spectrum over an infinitely large frequency range. Rather accurate values of the optical constants can be obtained if the measurements cover the region below the edge of direct optical transitions in nonmetals and a substantial part of the structure of the reflectivity at high energies.

It is easy to make an extrapolation to zero frequency, since for a large class of crystals the dispersion in the region situated under the direct transitions is well known.

No simple rules exist for the extrapolation of the reflection coefficient R into the region of higher energies. Often used in the literature is the method of extrapolation based on the asymptotic behavior of the reflection coefficient [386, 387]:

$$R(\omega) = C\omega^{-K},$$

where the constants C and K are determined by the condition that the function $\theta(\omega)$ is continuous and agrees with the refractive index in the region of transparency. This method is completely satisfactory for crystals for which the reflection coefficient undergoes a pronounced decrease in the high-energy region. The method is not suitable for ionic crystals, since for energies greater that 20 eV their reflection spectra are characterized by a pronounced structure.

The optical properties of fluorites in the region of self-absorption have been the subject of extensive investigation [389-392]. Frandon *et al.* [393] measured the characteristic loss spectra, as well as the spectrum of plasmon oscillations of valence electrons in thin crystalline films of alkaline-earth fluorites and, by using the Kramers-Kronig method, determined independently the optical constants and the dielectric susceptibility. However, in spite of the fact that there exists a large amount of experimental data, the interpretation of optical spectra has encountered great difficulties. The authors of the works mentioned have either based their interpretation on the erroneous assumption that the valence bands of fluorites and of alkali halide crystals are identical, or attempted to correlate the main features of the spectra with the transition energies of the free ion.

The reflection spectra of the crystals of the fluorite group, measured by Nemoshkalenko *et al.* [394] at the temperature of liquid helium, are shown in Figure 148. The positions and intensities of the main features in the structure of the spectra agree well with those obtained by other authors [390-392]. The

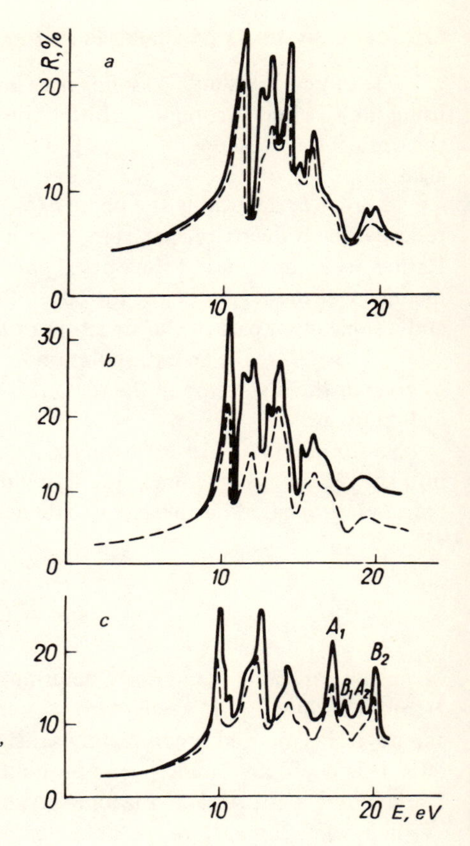

Figure 148. Reflectivity spectra of (a) CaF_2, (b) SrF_2, and (c) BaF_2 crystals at liquid nitrogen (– – –) and liquid helium (——) temperature.

reflection spectra of crystals exhibit a strong temperature dependence, which, as the temperature decreases, results in an increase of the intensity or a shift of some of the peaks. By comparing these spectra with the spectra measured by Rubloff at the temperature of liquid nitrogen [390], it can be seen that lowering of the temperature of the crystal to 4.2 K, reveals an additional structure. Among the characteristics of the above-mentioned spectra is one that is common for the whole group of crystals, namely, the temperature dependence of the first peak. When the temperature decreases, the intensity of the first peak increases, and its position is shifted toward higher energies: by 0.17 eV for CaF_2, by 0.18 eV for SrF_2, and by 0.2 eV for BaF_2.

For CaF_2 and SrF_2 at the temperature of liquid helium, the intensity of the group of bands situated in the energy region 12–13 eV increases strongly, and a fine structure appears. Large maxima are observed in the reflection spectra of

CaF_2 and SrF_2 at 14eV, as well as in the spectrum of BaF_2 at 12.5 eV. The reflection spectra of CaF_2 and SrF_2 at 15–17 eV and of BaF_2 at 14–16 eV are characterized by a broad band that has a similar structure for all of the crystals. At higher photon energies, the intensity of the reflection spectra of CaF_2 and CrF_2 decreases appreciably, and at energies around 19 eV, only a weak band appears. In the spectrum of BaF_2, in the energy region 17–20 eV, two doublets A_1-A_2 and B_1-B_2 should be observed, separated by 1 eV, and with 2 eV separation between their components. By lowering the temperature, a shift by 0.1 eV is observed toward lower energies for the doublet A_1-A_2 and in opposite direction for the doublet B_1-B_2. The spectra of dielectric permeability are shown in Figure 149.

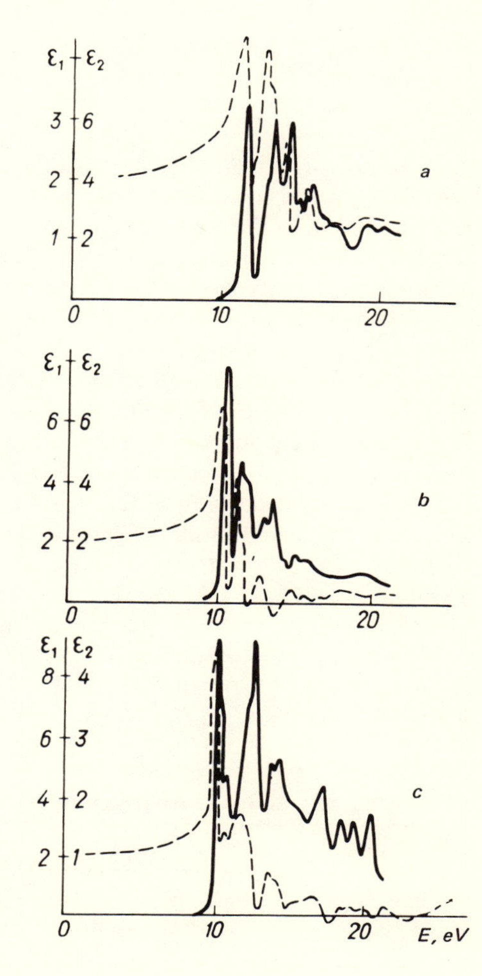

Figure 149. Dielectric permeability of (a) CaF_2, (b) SrF_2, and (c) BaF_2 crystals: ϵ_1(– – –), ϵ_2(——).

In order to interpret the optical spectra, a number of assumptions should be made about the structure of the conduction band of fluorites. Phillips [163] has shown that a good approximation for the description of the conduction band structure of alkali halide crystals is provided by the model of nearly free electrons. This has been confirmed by the calculations of Kunz [358]. In this approximation, the center of weight of the states X_1, X_3 and L_1, L_2' should be situated above the bottom of the conduction band Γ_1, with an energy difference equal to the electron kinetic energy: $E_f(X) = h^2/2ma^2$, and $E_f(L) = 3h^2/8ma^2$. The calculated energy parameters for the conduction band are given in Table 42.

The Γ_{12} and Γ_{25}' states should contribute to the spectra of longitudinal waves in fluorites, since the *ns* and *nd* levels of the free ion of the metal have approximately the same energy.

CaF_2. According to the results of theoretical calculations performed by Starostin *et al.* [395-397], the top of the valence band mainly contains the 2*p* levels of fluorine ions, and is split into odd and even subbands, the odd one, Γ_{15}, being situated 2.87 eV higher than the even one, Γ_{25}. The maximum splitting of the valence band occurs at the point X of the Brillouin zone. The energy difference between the even states X_5 and X_1 is 8.97 eV, and the difference between the odd states X_2' and X_5' is 6.89 eV. At the point L, the width of the valence band, equal to 4.18 eV, is determined by the states L_3' and L_1, L_2'. The top of the valence band corresponds to the point X_1, and is situated 1.28 and 2.04 eV above the Γ_{15} and L_3', respectively. The first maximum in the spectrum of the imaginary part of the dielectric susceptibility ϵ_2 is situated at an energy of 11.15 eV. The shape of the line and its temperature dependence have confirmed that it is due to exciton generation in the center of the band. The structure at 11.70 eV is related to the transition $\Gamma_{15} \rightarrow \Gamma_1$. In this case, the exciton binding energy is equal to 0.5-0.6 eV, which is in good agreement with the calculated exciton spectra of fluorites [398].

In the region of the absorption edge, it can be assumed that there is a strong transition Γ_{15}-$(\Gamma_{12}, \Gamma_{25}')$. This transition gives rise to the structures at 12.3 and 13 eV in the ϵ_2 spectrum. In this case, the magnitude of the splitting of the conduction band, corresponding to the states Γ_1 and Γ_{12}, is equal to 0.6 eV. The calculations of Starostin and Shepilov [396] have shown that the state Γ_{12} is situated under the state Γ_{25}'. According to the theoretical calculations, the

TABLE 42. Energy Parameters of the Conduction Band

Crystal	$a \times 10^{-8}$ (cm)	$E(X)$ (eV)	$E(L)$ (eV)
CaF_2	5.45	5.06	3.80
SrF_2	5.78	4.50	3.38
BaF_2	6.18	3.94	2.96

TABLE 43. Theoretical and Experimental Values of the Optical Transitions between the 2p Level of the F Ion and the Conduction Band at the Points of High Symmetry[a]

Sample	Method of energy determination	Γ_{15}-Γ_1	Γ_{15}-Γ_{12}	Γ_{15}-Γ'_{25}	X'_2-X_3	X'_2-X_1	X'_5-X_1	L'_3-L_1	L_2-L_1
CaF_2	Theory	11.06	–	–	14.0		21.8	15.60	19.80
	Experiment	11.70	12.3	13.0	14.70	15.30	25.0	15.81	19.79
SrF_2	Theory	10.82	–	–	14.5		20.4	14.40	17.00
	Experiment	11.15	11.4	12.0	14.30	15.00	19.3	15.70	16.80
BaF_2	Theory	10.10	–	–	13.9		17.4	12.80	14.00
	Experiment	11.00	12.0		14.2		–	13.61	15.40

[a]According to our data, the vacuum level for the CaF_2, SrF_2, BaF_2 crystals is situated very close to the bottom of the conduction band.

optical transitions at the point X-(X'_2-(X_3, X_1)) of the Brillouin zone should be situated in the region of 15 eV (Table 43). By using the spectra of the imaginary part of the dielectric susceptibility, it is possible to attribute to this transition the maxima at 14.7 and 15.3 eV. The order of the states X_3 and X_1 inside the conduction band cannot be determined uniquely. Calculations of the conduction bands for the alkali halide crystals [358] show that the state X_3 is situated below the state X_1. Therefore, the transition X'_2-X_3 probably corresponds to the energy 14.7 eV, and the transition X'_2-X_1 to the energy 15.3 eV. The strong maximum at 14 eV may be interpreted as corresponding to the X_3 exciton, this being in good agreement with the shape of the absorption line and its temperature dependence.

The structure at 15.81 eV in the ϵ_2 spectrum may be attributed to the transition L'_3-L_1, for which the theoretical energy value is 15.6 eV. In the dipole approximation, the transition L'_3-L'_2 is forbidden, and it appears in the spectrum as a singularity at 16.5 eV. Thus, in the conduction band, the lowest state is L_1, with the state L'_2 situated 0.7 eV above it. At the point L, one more transition can be expected to occur, namely, L'_2-L_1. Since the valence band is split at the point L by 4.18 eV, it follows that the corresponding singularity in the spectrum should appear at an energy of approximately 20 eV. In the spectrum of the dielectric susceptibility ϵ_2, a band is observed having a maximum at 19.79 eV, which may be identified as resulting from the transition L'_2-L_1. The transition X'_5-(X_1, X_3) should be rather weakened, since the width of the odd valence band at the point X is 6.89 eV. However, owing to the limitations of the optical measurements, investigations could not be performed in this region of the spectrum.

According to literature data [390, 391], a band having some structure at about 25 eV is observed in the reflection spectrum of crystalline CaF_2 (at room temperature). We attribute this band to the transitions X_5'-(X_3, X_1). Results of calculation show that the width of the $3p$ band is 3.3 eV, with the top of the band situated at the point X_4'. In the reflection spectra of crystalline CaF_2 [390, 391], a band having two broad maxima is observed in the energy region 30–36 eV. The first maximum at 30 eV is related to the transition Γ_{15}-Γ_1, for which the theoretically calculated energy value is 32.8 eV. According to the calculations, the transitions X_4'-(X_3, X_1) should have an energy of 35.1 eV. Therefore, in the reflection spectrum, these transitions appear as a bump on the line at 32.85 eV. The transition L_2'-L_1 should involve an excitation energy of 34.9 eV. In the experimental spectrum, the corresponding maximum is observed at 34.5 eV. The transitions L_3'-L_1 and X_5'-(X_3, X_1) (for which the experimental energy values have been measured as 35.85 eV, 36.31 eV, and 36.37 eV, respectively) appear as weak structures on the decreasing side of a broad band. In Table 44 are shown the possible transitions from the $3p$ band of the Ca^{2+} ion to the conduction band, and their corresponding excitation energies. In the dipole approximation, transitions from the $2s$ band of the F^- ion are forbidden, and therefore they should appear in the reflection spectra as very weak structures in the energy region 35–41 eV.

SrF_2. Calculations of the valence band of crystalline SrF_2 show that its band structure is similar to that of CaF_2. The width of the odd part of the $2p$ band of the F^- ion is determined by the magnitude of the splitting of the X_2' and X_5' levels, and is equal to 5.8 eV. At the point L, the width of the valence band is 2.7 eV, which corresponds to the energy separation between the states L_3' and L_2'. The point (X_2', X_1) is closest to the vacuum level (10.1 eV) among all the points of the valence band of the crystal. The state Γ_{15} is situated 10.8 eV apart from the vacuum level, and the state L_3' is situated 11.0 eV apart from it. Therefore, on the basis of the above-mentioned assumptions regarding the

TABLE 44. Theoretical and Experimental Energy Values of the Optical Transitions between the *np* Band of Me^{2+} and the Conduction Band[a]

Sample	Method of energy determination	$\Gamma_{15}\to\Gamma_1$	$X_4'\to X$		$X_5'\to X$	$L_2'\to L_1$	$L_3'\to L_1$
			$X_4'\to X_3$	$X_4'\to X_1$			
CaF_2	Theory	32.80	35.1		37.9	34.9	37.2
	Experiment	30.00	34.50		–	–	–
SrF_2	Theory	25.69	28.7		30.1	27.4	26.8
	Experiment	24.70	27.24	28.13	29.70	–	–
BaF_2	Theory	20.00	22.0		23.5	25.6	24.3
	Experiment	20.6–21.3	22.13	22.72	23.69–24.90	26.00	–

[a]According to our data, the vacuum level for the CaF_2, SrF_2, BaF_2 crystals is situated very close to the bottom of the conduction band.

conduction band of fluorites, it can be concluded that the transition Γ_{15}-Γ_1 determines the absorption edge. The centers of gravity of the transitions X_2'-(X_3, X_1) are situated at about 14.5 eV. At the point L of the Brillouin zone, it can be expected that the transitions L_3'-L_1, with an energy of 14.4 eV, and L_2'-L_1, with an energy of 17.0 eV, will occur.

Tables 43 and 44 show the main features of the spectra and their possible interpretation. It should be noted that in the center of the Brillouin zone, the conduction band is split into three states, namely, Γ_1, Γ_{12}, and Γ_{25}. From the data of Table 43 it can be seen that, at the point X, the splitting of the conduction band is equal to 0.7 eV. The structures of the short-wavelength part of the reflection spectra of SrF_2 [390] and CaF_2 are similar. In this region, the SrF_2 spectrum (with the exception of two exciton peaks at 22.47 eV and 22.56 eV) is determined by the interband transitions from the 4p states of the Sr^{2+} ion to the conduction band (the energy region 25–30 eV). In analogy with the spectrum of crystalline CaF_2, it can be assumed that the three maxima of the broad band appearing in this energy region correspond to the transitions from the upper 4p band of the Sr^{2+} ion to the corresponding points of the conduction band. The reflection coefficient in the region of energies greater than 32 eV is determined by the transitions from the 2s band of the F^- ion.

BaF_2. The structure of the 2p band of the F^- ion in crystalline BaF_2 exhibits a series of detailed features that probably determine the slightly different optical spectrum of this crystal, as compared to the corresponding spectra of CaF_2 and SrF_2. The closest point of the vacuum level in the valence band is the point L_3 (9.8 eV). The width of the odd subband of the valence band is determined, as in the case of CaF_2 and BaF_2, by the magnitude of the splitting at the point X, which amounts to 4.6 eV. According to calculations, the width of the forbidden band at the point Γ is equal to 10.1 eV. This value determines the absorption edge of BaF_2 and corresponds to the transition $\Gamma_{15} \rightarrow \Gamma_1$. The spectrum of BaF_2 in the long-wavelength region is different from the corresponding spectra of CaF_2 and SrF_2. In the spectrum of ϵ_2, at the temperature of liquid helium, there appears a structure that has been interpreted by Tomiki and Miyata [391] as being an exciton series with $n = 1, 2$. In this case, the absorption edge can be determined by using the formula $E_n = E_g - R/n^2$. Using the energy values of each of the terms of the series, it results in $E_g = 10.75$ eV. By analyzing the ϵ_2 spectrum of BaF_2 from the value of the energy at which absorption begins, beyond the exciton series, a value $E_g = 10.12$ eV is obtained.

The structure at 11.8 eV is determined by the transitions from the state Γ_{15} to the d states of the conduction band, Γ_{12}, Γ_{25}'. All the other features in the long-wavelength part of the spectrum are interpreted in the same way as for the CaF_2 and SrF_2 crystals. In the region of energies of the order of 20 eV, the structure of the spectrum is mainly determined by the excitation of Γ and X excitons from the 5p band of Ba^{2+}. The spin–orbit splitting for the 5p states of Ba^{2+} is approximately equal to 2 eV [399], which is in good agreement with the doublet splitting of excitons. The temperature dependence of the line shapes show that the first exciton doublet is a Γ exciton, which is split as a result of the spin–orbit interaction. The transitions from the 5*p* band of the Ba^{2+} ion to the conduction band determine the character of the reflection spectrum in the energy region between 21 and 30 eV [390].

X-Ray Emission Spectra, Quantum Yield Spectra, and Photoelectron Spectra of CaF_2, SrF_2, and BaF_2

The *K* emission spectra and the quantum yield spectra of F^- in CaF_2, SrF_2, and BaF_2 were measured using the RSM-500 spectrometer-monochromator equipped with a diffraction grating with radius of curvature $R = 6$ m and a line density of $N = 600$ lines/mm [402, 403]. The emission spectra were recorded with an anode potential of 6 kV and a current of 2 mA. The quantum yield spectra were obtained by using the bremsstrahlung radiation, with a voltage of 6 kV on the tungsten anode, and a current of 160 mA. Both the emission and the quantum yield spectra were recorded with a resolution of 1.7 eV, and the energy was measured with an accuracy of the order of 0.5 eV.

The *K* X-ray emission spectra and the quantum yield spectra of the F^- ion in CaF_2, SrF_2, and BaF_2 crystals are shown in Figure 150. The emission bands exhibit a maximum (*a*), a fine structure (*b* and *c*) on the higher–energy side of this maximum, and a bump on its lower-energy side. The quantum yield spectrum reproduces the structure of the absorption spectrum previously obtained

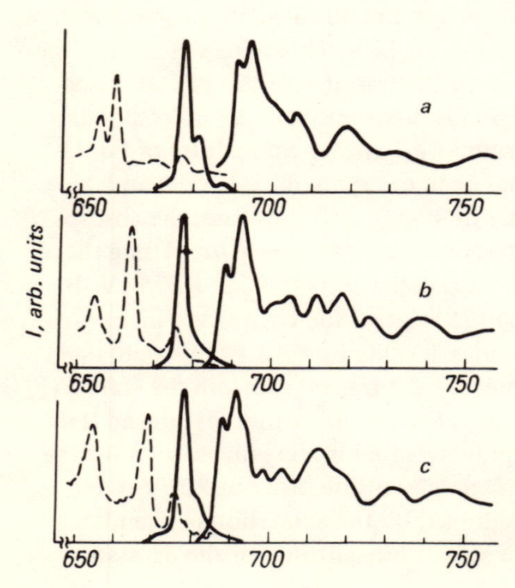

Figure 150. X-ray emission and photoelectron spectra of (a) CaF_2, (b) SrF_2, and (c) BaF_2 crystals: (——) X-ray emission K spectra and quantum yield spectra of fluorine; (– – –) photoelectron spectra.

by Zimkina and Vinogradov [400]. Since transitions from the even states are forbidden by the restrictions imposed by the selection rules, it follows that the maximum in the X-ray intensity will correspond to the transition from that region of the valence band in which mainly odd states are concentrated. The study of the conditions of excitation of the peaks *b* and *c* has shown that they are satellites from multiple ionization [401].

For crystalline CaF_2, an attempt was made to reveal the X-ray emission transitions from a band containing mainly the 2*s* states of the F^- ion. The absence of such a band demonstrates the strong localization of the 2*s* states of the F^- ion in crystalline CaF_2 and the insignificant hybridization of the 2*s* wave functions of F^-. As we have already mentioned, the bottom of the conduction band in CaF_2, SrF_2, and BaF_2 crystals contains mainly the *s* and *d* electronic states of the metal ion. Comparison of the quantum yield spectra of CaF_2, SrF_2, and BaF_2 with the absorption spectra of the F^- ion in NaF and MgF_2 [400], in which at the bottom of the conduction band there is no significant contribution of *d* states, permits the conclusion that the first maximum in the quantum yield spectra actually reflects the contribution of *d* states with Γ'_{25} and Γ_{12} symmetry. As was found from the previous analysis of the optical spectra, the bottom of the conduction band Γ_1 is situated below the center of gravity of the *d* states at a distance of 0.9 eV in CaF_2, 0.6 eV in SrF_2, and 1.0 eV in BaF_2. Consequently, the position of the states *L* and *X* may be determined by using the model of nearly free electrons. It can be seen from Figure 150 that the calculated position of the state *X* agrees well with the energy position of the maximum *B* in all of the crystals that have been investigated. The existence of transitions in the energy region in which states of the *s* and *d* type are localized demonstrates that in this region there exists a significant contribution of *p*-type states. This can be understood by taking into account the fact that for the *s* and *d* states the wave function of the metal ion is quite extended and has an appreciable amplitude at the points in which F^- ions are located.

The X-ray photoelectron spectra of the valence and core electrons in CaF_2, SrF_2, and BaF_2 were obtained using an IEE-15 spectrometer [402]. Binding energies were determined with respect to the Fermi level of the spectrometer material. In the case of sharp lines, the accuracy in the determination of binding energies was about 0.1–0.2 eV. The resolving power, with a voltage of 80 V on the analyzer, amounted to 1.5 eV. Calibration of the spectra was performed with respect to the 1*s* line of carbon (284.0 eV). The X-ray photoelectron spectra of valence electrons of CaF_2, SrF_2, and BaF_2 are shown in Figure 150. Knowing the values of the binding energies of the 1*s* electrons of the F^- ion (686.5 eV for CaF_2, 686.0 eV for SrF_2, and 686.7 eV for BaF_2), it is possible to place the X-ray emission spectra, the quantum yield spectra, and the photoelectron spectra on the same energy scale. As has been shown in Chapter 4, the X-ray photoelectron spectra reproduce quite well the distribution of the density of states in

TABLE 45. Calculated and Experimental Values of the Energy of the Center of Gravity of the Bands for the Ions F^- and Ca^{2+}, Sr^{2+}, and Ba^{2+} in the Compounds CaF_2, SrF_2, and BaF_2

Sample	Method of energy determination	State		
		$2p$ F^-	np Me^{2+}	$2s$ F^-
CaF_2^2	Theory	13.6	32.8	38.0
	Experiment	13.4	30.3	34.3
SrF_2	Theory	13.1	23.6	32.2
	Experiment	11.9	23.5	32.5
BaF_2	Theory	12.4	19.7	36.5
	Experiment	11.2	19.2	32.6

the valence band of dielectrics. Owing to the absence of strong selection rules, they reflect the even as well as the odd states. However, because of the rather low resolving power of the method of electron spectroscopy, no fine structure of the valence band of the investigated crystals could be observed.

The calculated data given in Table 45 reproduce well the energy position of the center of gravity of states with different types of symmetry. The widths of the valence subbands, which contain mainly the $2p$ and $2s$ states of the F^- ions, determined from X-ray photoelectron spectra are in quite good agreement with the theoretical calculations. Much worse agreement is observed between the widths of subbands containing mainly the *np* states of the metal ions. This is because the calculations have not taken into account the spin–orbit splitting of these states. The experimental data also confirm the fact that, in BaF_2, the energy bands of the $2s$ electrons of F^- and those of the $5s$ electrons of Ba^{2+} are close to each other. The information provided by electron spectroscopy is also important because it permits an interpretation of the optical spectra in the far-ultraviolet region, by allowing the determination of the energy position of the states of the Γ_1 bands containing mainly the *np* states of the metal ions and the $2s$ states of the F^- ion.

The results obtained by Nemoshkalenko *et al.* [394, 402] concerning the important role of metal *d* states in the formation of the bottom of the conduction band have subsequently been confirmed [403] in studies of the $L_{2,3}$ and $M_{2,3}$ X-ray emission bands, of the quantum yield of calcium in CaF_2, as well as of the $N_{4,5}$ emission spectra and quantum yield of barium in BaF_2. Figure 151 shows the $L_{2,3}$, $M_{2,3}$ emission spectra and the quantum yield spectra of calcium in CaF_2, and also the reflection spectrum of CaF_2. All of these spectra are presented on a common energy scale using the core-level energy values of the transition metals, as determined by the method of electron spectroscopy.

A characteristic feature of the spectra is the coincidence of the peak *a* in the $L_{2,3}$ and M_{23} spectra of calcium in CaF_2 and in the optical spectrum, displayed

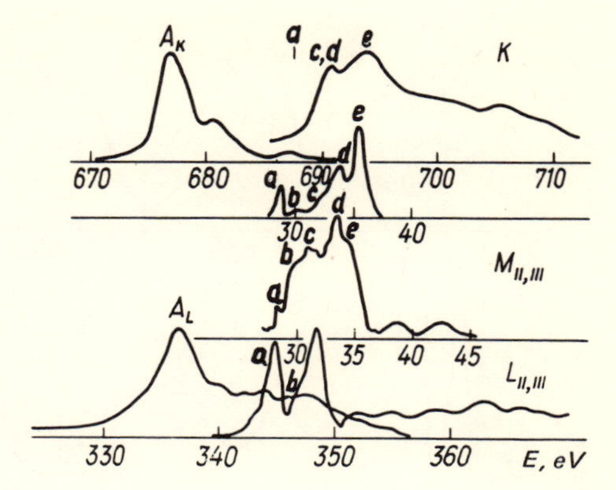

Figure 151. X-ray emission K spectra and quantum yield K spectra of fluorine; $M_{2,3}$ and $L_{2,3}$ emission spectra and quantum yield spectra of calcium as well as the optical spectra of crystalline CaF_2 are also shown.

on a common energy scale, and also the absence of a similar peak in the *K* spectrum of fluorine. The nature of the peak *a* in the optical spectrum was clearly determined in the study of the reflection coefficient at low temperatures [394]. It was shown that in the region of high energies, the peak *a* shifts, its intensity increases, and its width decreases, which confirms its exciton nature.

The energy coincidence of the above-mentioned peaks for the whole series confirms that, here, the peaks *a* in the *M* and *N* spectra of calcium in CaF_2 are of exciton origin and are characterized by an *s* symmetry. This is also confirmed by the absence of the peak in the quantum yield *K* spectrum of fluorine. It is interesting to compare the energy position of the valence band (its position is marked by a line in the *K* spectrum) with the position of excitons of the internal levels. As can be seen from Figure 151, the optical exciton is shifted 1 eV toward higher energies as compared with the excitons of the internal levels. In view of the experimental accuracy (0.6 eV in the experimental X-ray spectra), this agreement may be considered satisfactory.

The bottom of the conduction band is situated at *b* and has an *s*-type symmetry (it contains the *ns* functions of the metal). It is natural that in the *K* spectra, such a peak is absent. In this case, the exciton binding energy is 1 eV, which is in good agreement with the optical and theoretical data [394, 397]. The *d* states of the conduction band appear in the *L* and *M* spectra as the structures *c* and *d*, separated by an energy difference of 1.5 eV. They may be correlated to the states Γ_{12} and Γ'_{25}. In the *K* spectrum of fluorine in CaF_2, these states (*c* and *d*) do not appear so clearly. The peak *e* in the *M* spectra may be correlated to the *d* states at the point *X* of the conduction band.

The *K* X-ray emission spectrum and the quantum yield spectrum of fluorine in BaF_2 are shown in Figure 152. The $N_{4,5}$ spectrum of barium in BaF_2 exhibits a sharp structure at about 75 eV, which can be correlated to the structure of the barium 5*p* band, split as a result of spin–orbit interaction. According to the data from X-ray spectroscopy, the magnitude of this splitting is equal to 1.8 eV. Be-

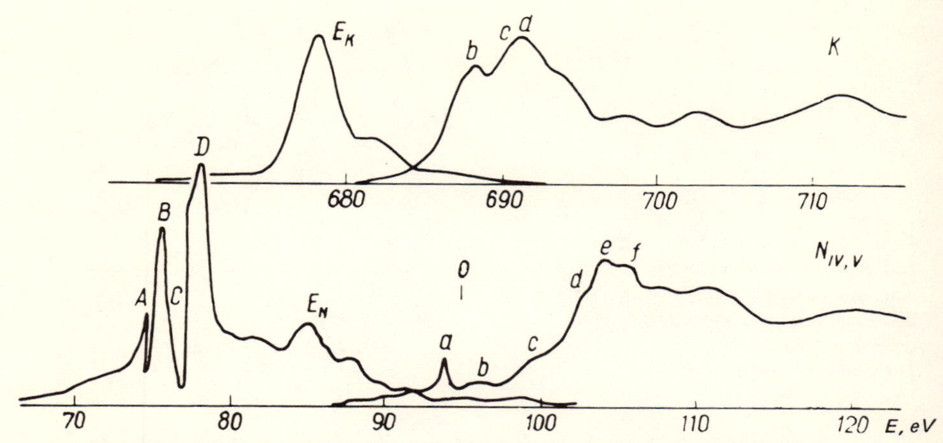

Figure 152. X-ray emission K spectra and quantum yield K spectra of fluorine. $N_{4,5}$ emission spectra and quantum yield spectra of barium in crystalline BaF_2 are also shown.

tween 85 and 90 eV, the $N_{4,5}$ spectrum reflects the states of the valence band. The quantum yield spectra of the $N_{4,5}$ level of barium in BaF_2 are essentially different from the K spectra of fluorine. Taking into account that the $N_{4,5}$ spectrum reflects mainly the states with p symmetry, it can be seen that the $N_{4,5}$ spectrum of the quantum yield of barium in BaF_2, shown in Figure 152, confirms the conclusions of Nemoshkalenko *et al.* [394, 402] regarding the structure of the bottom of the conduction band in this compound, which contains mainly the s and d states of the metal.

Plasmon Excitations in the Optical and X-ray Spectra of Crystalline CaF_2, SrF_2, and BaF_2

Additional information about the electron spectra of dielectrics can be obtained by calculating the functions:

$$N_{\mathrm{eff}}(\omega) = \frac{m}{2\pi^2 N e^2} \int_0^{\omega} \omega' \varepsilon_2(\omega')\, d\omega',$$

$$\epsilon_{\mathrm{eff}}(\omega) = 1 + \frac{2}{\pi} \int_0^{\omega} \frac{\varepsilon_2(\omega')}{\omega'}\, d\omega',$$

where N_{eff} represents the effective number of electrons that contribute to the optical transitions characterized by an energy lower than $\hbar\omega$. The function $N_{\mathrm{eff}}(\omega)$ allows the determination of the distribution of oscillator strengths for

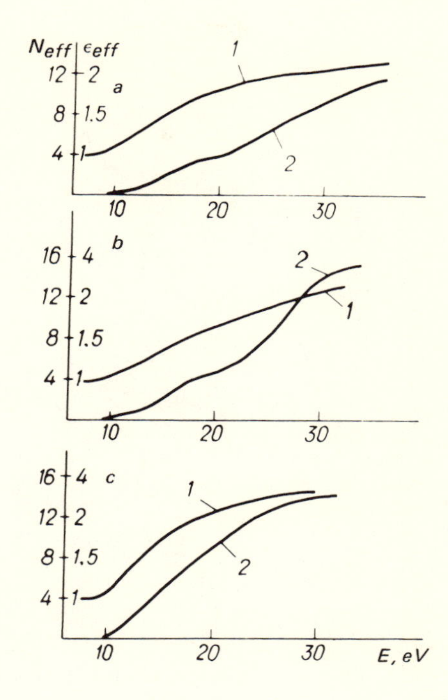

Figure 153. Effective values of (1) the dielectric permeability and (2) the number of electrons involved in optical transitions for (a) CaF_2, (b) SrF_2, and (c) BaF_2 crystals.

interband transitions, while $\epsilon_{eff}(\omega)$ characterizes the contribution of electrons from different bands to the static dielectric permeability. Figure 153 shows the functions $N_{eff}(\omega)$ and $\epsilon_{eff}(\omega)$ calculated over the energy region up to 36 eV [394]. The values of the reflection coefficient in the energy region over 21 eV were taken from the work of Rubloff [390]. Analysis of the behavior of the function $N_{eff}(\omega)$ for all three crystals indicates that only few valence electrons take part in the transitions from the valence $2p$ band of the F^- ion. The main part of the oscillator strengths of the $2p$ band of the F^- ion is preserved up to high energies, until the electrons from the up-bands of the metal ion are excited, when N_{eff} increases. The function ϵ_{eff} tends toward its limiting value, equal to the optical static dielectric function ϵ_0. The effective dielectric susceptibility ϵ_{eff} (see Figure 153) reaches the value ϵ_0 in the region of the spectrum in which the contribution of interband transitions from the *np* bands of the metal ion are still significant. This demonstrates that the polarizability of the investigated crystals is not only determined by the valence electrons of the $2p$ band of fluorine, but also to a significant extent by those of the *np* band of the metal ion.

The conditions necessary for the excitation of plasmons in CaF_2, SrF_2, and BaF_2 have been discussed by Frandon *et al.* [393]. The plasmons are not excited by light under normal incidence. However, the basic information about this

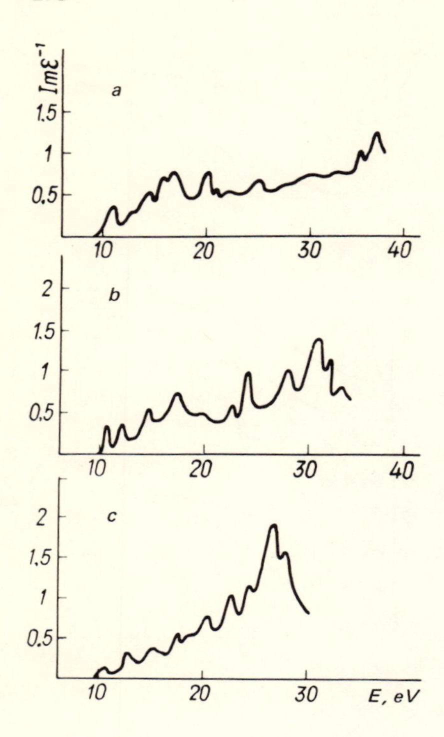

Figure 154. Loss function, Im [I/ε(k)], for (a) CaF_2, (b) SrF_2, and (c) BaF_2 crystals.

type of oscillation may be obtained by calculating the loss function Im $[1/\epsilon(\omega)]$ (Figure 154). However, not all of the maxima of the loss function correspond to plasmon excitation. It is necessary that the condition for the existence of plasmons is also fulfilled: $\epsilon_1(\omega) = 0$, $\epsilon_2(\omega) \ll 1$. The maxima of the loss function that satisfy this condition are given in Table 46. In the second column of the table are given the values of the plasmon frequency for 12 valence electrons of the $2p$ band of the F^- ion, derived on the assumption that these electrons behave as if they were free. The values in the third column correspond to

TABLE 46. Plasma Frequencies for Valence Electrons

	$2p$ band F^-				np band Me^{2+}		
Sample	ω_p	Ω_{opt}	Ω_{x-n}	$\Omega_{Rönt}$	Ω_{opt}	Ω_{x-n}	$\Omega_{Rönt}$
CaF_2	20.2	17.3	17.0	17.0	36.3	35.8	30–35
SrF_2	16.7	17.5	17.2	17.6	31.1	30.6	30
BaF_2	18.5	–	–	–	26.8	26.4	25

the optical plasmon frequency obtained from the spectra Im $[1/\epsilon(\omega)]$. The fourth column includes the values of plasmon frequencies, calculated from the experimental data on electron characteristic losses [393], and the fifth column, the plasmon frequencies determined from quantum yield spectra. These data show that the plasmon frequency of the oscillations of $2p$ electrons of the F^- ion is lower than that of free electrons. This can be explained by the influence of the interband transitions from the *np* band of the metal ion on the frequency of oscillation. In BaF_2, no maximum could be revealed corresponding to oscillations of electrons from the $2p$ band of the F^- ion. This demonstrates that the oscillator strengths of the $2p$ band of the F^- ion and the *np* band of the metal are strongly mixed. Therefore, it is possible to speak in terms of a common oscillation of electrons of the $2p$ band of the F^- ion and the *np* band of the metal.

In the region of the spectrum in which the oscillator strengths of the transitions from the *np* band of the metal ions are negligible, the function Im $[1/\epsilon(\omega)]$ exhibits a maximum. The position of the maxima agree satisfactorily with the values of plasmon frequencies determined from the characteristic loss spectra. The study of the spectra of plasmon oscillations in fluorites has shown that here the plasma includes the interacting *p* electrons of the fluorine and metal ions. In the quantum yield spectra, the energy position of some of the maxima, as can be seen from Table 46, coincides with the frequency of plasma oscillations.

X-Ray Photoelectron Spectra and $L_{2,3}$ Quantum Yield Spectra of Chlorine in Crystalline $CaCl_2$, $SrCl_2$, and $BaCl_2$

Figure 155 shows the $L_{2,3}$ quantum yield spectra and the X-ray photoelectron spectra of the upper valence bands of chlorine in the compounds, $CaCl_2$, $SrCl_2$, and $BaCl_2$, displayed on the same energy scale [403]. The first subband of the valence band contains mainly the $3p$ wave functions of chlorine. The second contains the $3s$ wave functions of chlorine, while the subband containing the *np* levels of the metal shifts toward higher energies in passing from $CaCl_2$ to $BaCl_2$. In $BaCl_2$, a common broad *np*–3*s* band is already formed. The $L_{2,3}$ quantum yield spectra of chlorides are characterized by a doublet peak, which has an exciton character. The absorption limit is situated between the L_2 and L_3 peaks. The magnitude of the forbidden band amounts to 10.5 eV in $CaCl_2$, and 8.0 eV in $BaCl_2$. Thus, the simultaneous investigation of the optical spectra, X-ray emission spectra, and X-ray quantum yield spectra offers the possibility of obtaining more complete information on the electron structure of crystals. Use of various spectroscopic methods in the study of alkali halide crystals shows that the widths of the forbidden bands determined by optical and X-ray methods are different. This indicates that in the excitation process in crystals the interaction between the hole and the electrons in the conduction band depends on whether the hole

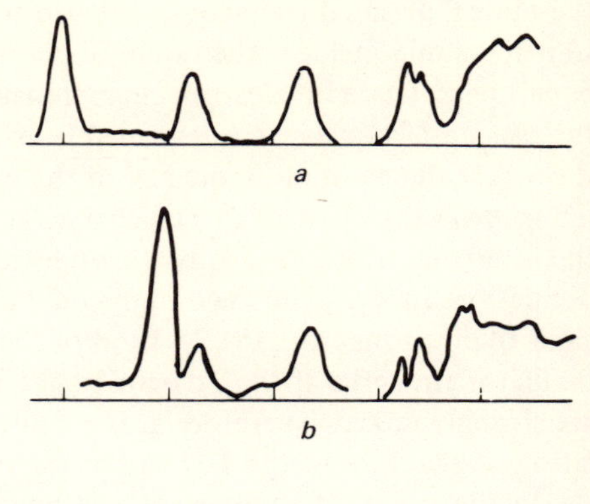

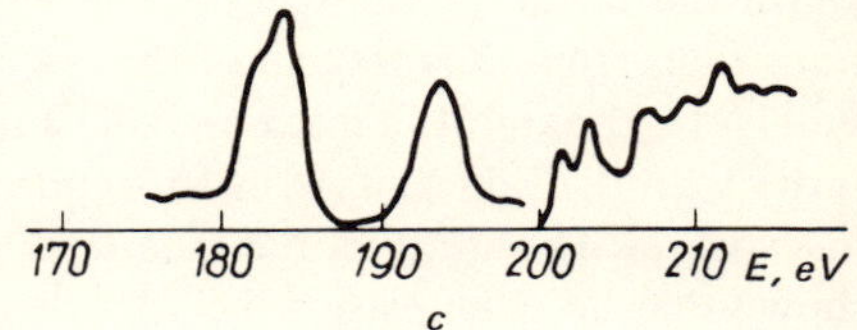

Figure 155. Valence band photoelectron spectra and $L_{2,3}$ quantum yield spectra of chlorine in (a) $CaCl_2$, (b) $SrCl_2$, (c) $BaCl_2$ crystals.

exists in a core level or in a valence band level. This interaction is stronger in the case of core holes.

Experimental data also allow the determination of the symmetry of the states situated at the bottom of the conduction band and the symmetry of exciton states. Analysis of the data obtained from X-ray photoelectron spectroscopy has shown that the calculations of the X-ray absorption spectra of alkali halide crystals are doubtful. For the crystalline halides of the alkaline-earth metals, the optical and X-ray methods have been used simultaneously to interpret the optical spectra in a large energy range [394, 402, 403]. This permitted a determination of the structure of the conduction band. It was found that the conduction band of these crystals can be described satisfactorily within the pseudopotential approach. The symmetry of the states situated at the bottom of the conduction band, the space localization, and the hybridization of the wave functions of electrons in the valence band and the conduction band have also been determined. For the crystals in the fluorite group, the energy position of the exciton states observed in the optical absorption spectra coincides rather well with the position of the excitons appearing in the X-ray quantum yield spectra.

6

Energy Location of the Valence *s, p,* and *d* States in Transition Metal Compounds

In spite of the fact that intensive studies are currently being performed, the existing information on the electron structure of the compounds of transition metals is insufficient to permit a clear interpretation of their basic physical properties. Attempts to treat the valence electrons of these compounds within the framework of the usual band model [114] lead to the conclusion that the energy bands are not completely filled, and therefore all of these compounds should be metallic. Of course, this conclusion contradicting the experimental facts, further development of the theory is necessary. Mott [404] has pointed out the importance of taking into account the correlation between *d* electrons in the interpretation of the band properties of transition metal compounds. The problem of electron–electron correlations and their influence on the physical properties of many-particle systems is one of the most important problems of solid state physics. If the correlation corrections are small, they can be neglected or treated within the framework of the perturbation theory. In many cases (for example, in the case of nontransition metals and semiconductors) this approach is satisfactory [114]. However, in some cases, inclusion of electron interactions plays a principal role since it significantly modifies the properties of the system. In nontransition metals these interactions are strong if the average distance between the electrons is greater than the radius of their orbitals. In this case, the electron correlation will determine whether the given compound is a metal or an insulator.

Let us consider the compound NiO. This is an ionic crystal with a crystal structure of the NaCl type. Two *s* electrons of nickel are transferred to the oxygen atom and the Ni^{2+} ion remains with eight 3*d* electrons. The distance between the nickel ions in the crystal is much greater than the radius of the

3*d* orbitals. From the point of view of the usual band representations, the compound NiO should be metallic since, for the *d* band, in the calculation for each elementary cell, out of ten possible states, only eight *d* electrons occur.

Therefore, the *d* states in the valence band are only partially filled. Consequently, the analogous compounds MnO and CaO should also be metallic, since for them the number of *d* electrons per ion is odd. However, it has been shown experimentally that all these compounds are good insulators. The oxides TiO, ReO_3, and Cr_2O_3 exhibit metallic properties, whereas some other oxides and transition metal compounds, such as VO and Ti_2O_3, transform from the metallic to an insulating state when the temperature is changed.

Oxides of Transition Metals

Among the transition metal compounds, the oxides occupy a particular place owing to their interesting optical, electrical, and magnetic properties [405]. The electron structure of the oxides of transition metals are currently the subject of extensive theoretical and experimental investigation. Nemoshkalenko and Aleshin have measured the X-ray photoelectron spectra of the valence and core electrons of NiO, MnO, and ZnO crystals [54]. Shay and Spicer [46] and Pollak [339] have measured the photoelectron spectra of NiO with a somewhat better accuracy. However, the interpretation of the experimental data is not satisfactorily convincing, particularly with regard to the energy position of 2*p* states of oxygen in the valence band. Pollak for example, suggested [339] that the peak at 11.5 eV is due to the reflection of oxygen 2*p* states in the X-ray photoelectron spectrum of the valence electrons of NiO. In our studies, in order to determine the energy location of the oxygen 2*p* states, we have used the data provided by X-ray spectroscopy.

The *K* X-ray emission spectrum of oxygen in NiO, as reported by Brytov *et al.* [407], is shown in Figure 156. It can be seen that the center of gravity of the oxygen 2*p* band is situated at a binding energy of 6.5 eV, although in the X-ray photoelectron spectrum there is no particular structure in this energy region. This result demonstrates that the photoionization cross section of nickel 3*d* electrons is much higher than the photoionization cross section of the oxygen 2*p* electrons. In Figure 156 the $L\alpha$ spectrum of nickel in NiO is also shown. All three spectra have been drawn on the same energy scale, by making use of the known values of the core-electron binding energies. The position of the maximum intensity in the $L\alpha$ spectrum of nickel coincides with the position of the maximum in the photoelectron spectrum. Therefore, the maximum in the photoelectron spectrum reflects the 3*d* states of the transition element. An interesting feature of the oxygen *K* X-ray emission spectrum is the bump in the energy region where the 3*d* states of nickel are localized. It is possible that this

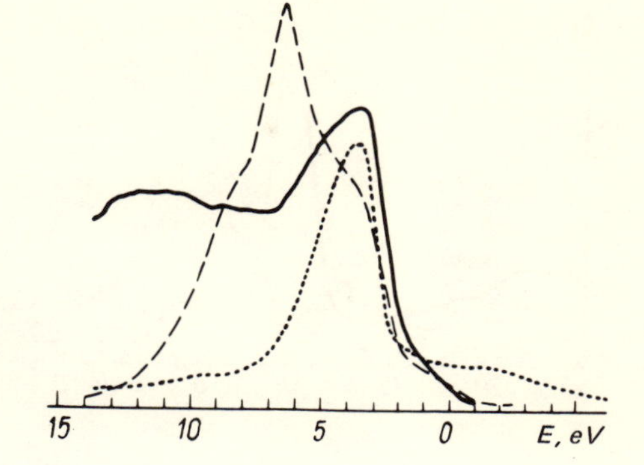

Figure 156: Valence band photoelectron spectrum (——); X-ray emission K spectrum (– – –) of oxygen; and Lα emission spectrum of nickel (· · ·) in the compound NiO.

bump reflects the contribution of the 3*d* states of the transition metal to the X-ray emission spectrum of oxygen. Also to be noted are the dip in the photoelectron spectrum on the high-energy side of the main maximum and the asymmetry of the main peak of the *L*α emission spectrum of nickel in NiO. The asymmetry in the distribution of 3*d* states in NiO may possibly be due to effects related to the splitting of the 3*d* orbitals into e_g and t_{2g} in the crystal field, and also to the fact that electrons with different spin orientation have different binding energies. The maximum at 11.5 eV may be explained in terms of the excitation of plasma oscillations, since Bakulin *et al.* [408] have shown that the magnitude of characteristic losses in NiO is of the order of 6 eV.

Theoretical calculations of the structure of the energy bands of the transition metal oxides are contradictory and do not agree with the experimental data. Switendick, for example, maintains that the 2*p* band of oxygen is situated above the band that contains mainly the 3*d* states of nickel [409], but the results obtained by Mattheiss [410] suggest that the 2*p* band of oxygen is situated below the 3*d* band of nickel, and that between these bands there exists a forbidden gap with a width of 4.5 eV. The distance between the centers of gravity of the bands (6.8 eV) is 4 eV greater than the value determined experimentally.

Figure 157 shows the X-ray photoelectron spectrum and the *K* emission spectrum of oxygen in MnO, obtained by Koster [411]. In the *K* emission spectrum a maximum is observed which represents a satellite due to multiple ioniza-

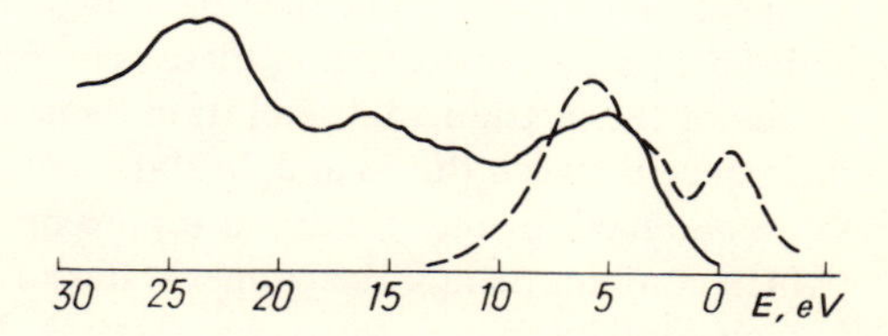

Figure 157: Valence band photoelectron spectrum (——) and X-ray K emission spectrum (– – –) of oxygen in the compound MnO.

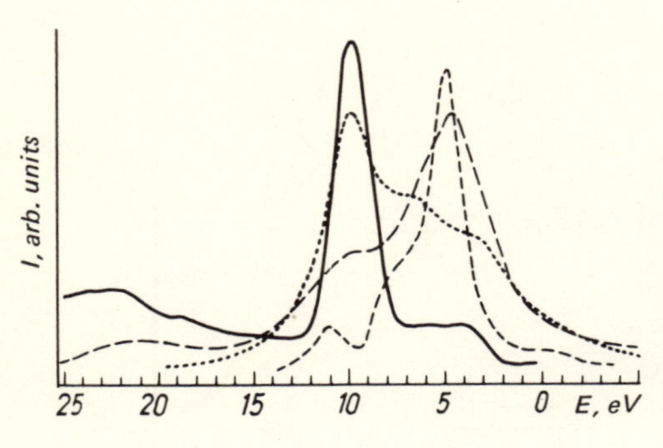

Figure 158: Valence band photoelectron spectrum and X-ray emission spectrum of compound ZnO: (——) photoelectron spectrum; (— —) X-ray K emission spectrum of oxygen; (– – –) X-ray K emission spectrum of zinc; (· · ·) X-ray emission Lα spectrum of zinc.

tion. The experimental results indicate that the distance between the 3*d* and 2*p* states is 1 eV, but the calculation yields a value of 7.4 eV.

The most complete information regarding the X-ray spectra of ZnO is shown in Figure 158. The *K* emission spectra of oxygen and zinc in ZnO are drawn on the same energy scale. The Zn *K* emission spectrum exhibits two satellites [412]. It is evident that in ZnO, unlike in atomic zinc, there are states of the *p* type, which are localized in the vicinity of the zinc nucleus. Contrary to the conclusion of Zyryanov and Nemnonov [413], it is to be noted that the maximum on the high-energy side of the binding energy peak in the *K* emission spectrum of oxygen is probably a satellite and not the result of the influence of the 3*d* states of zinc. This maximum is situated 1.2 eV from that corresponding to the 3*d* band. This energy difference is greater than the error in the localization of lines in the combined spectra, which is of the order of 0.5 eV.

In the comparison of the X-ray emission spectra and photoelectron spectra, we have used the values of the binding energy of oxygen 1*s* electrons (531.0, 531.0, 530.7 eV) and of the $2p_{3/2}$ electrons of the transition metals (855.8, 641.6, and 1021.2 eV) for NiO, MnO, and ZnO, respectively.

In conclusion, simultaneous interpretation of the X-ray emission spectra and photoelectron spectra offers the possibility of obtaining information about the spatial and energy localizations of electrons of various symmetry types, which is particularly valuable in those cases in which electrons with different types of symmetry have strongly different ionization cross sections.

Hüfner and Wertheim [414] have measured the X-ray photoelectron spectra of the valence band electrons of the oxides TiO_2, Cu_2O, and CoO, using an IEE-15 spectrometer. As can be seen from Figure 159, the bands containing the 2*s* and the 2*p* states of oxygen in these compounds are much wider than the magnitude of the instrumental resolution (which is approximately 1 eV). The energy difference between the 2*s* and 2*p* states of oxygen in these oxides is close to the value obtained for the free atoms (approximately 16 eV). In Cu_2O, the 3*d* band (which contains mainly the copper states) overlaps appreciably the 2*p* band (which contains mainly the oxygen states). The maximum due to the 3*d* elec-

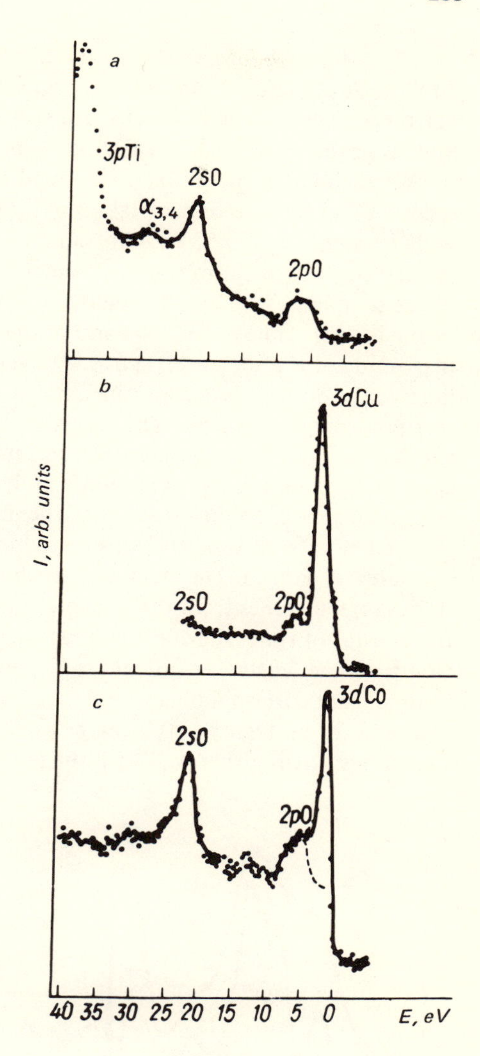

Figure 159: Photoelectron spectra of the compounds (a) TiO_2, (b) Cu_2O, and (c) CoO.

trons of copper in Cu_2O has a width at half-maximum of 1.5 eV, which is much lower than the value corresponding to pure copper. It should be noticed that this maximum is characterized by a sharp peak near the top of the valence band. The band-structure calculations performed by Dahl and Switendick for Cu_2O [415] give results for the $2p$ states of oxygen and the $3d$ states of copper which are in good agreement with the experimental data. The calculated value of the binding energy of oxygen $2s$ electrons, however, is not confirmed by experiment.

The electron configuration of the transition metal atom is $3d^0$ in TiO_2, $3d^{10}$ in Cu_2O, and $3d^7$ in CoO. The $3d$ states in CoO are characterized by a narrower peak than in MnO. The fine structure observed in the spectrum of MnO is probably due to the splitting of the final $3d^6$ state in the crystal field.

Wertheim *et al.* [416] have measured the valence band photoelectron spectra of an ReO_3 monocrystal, using an HP-5950A electron spectrometer with a resolution of 0.55 eV. The sample was oriented with the (110) plane parallel to the exit slit of the spectrometer. An indication of the cleanliness of the sample surface was the relatively low width and the symmetric shape of the oxygen 1*s* line (1.02 eV) and of the rhenium 4*f* line (0.78 eV). The magnitude of the spin-orbit splitting of the 4*f* states is 2.46 eV. Each of the lines of the doublet is accompanied by an extended satellite structure with a plasmon energy of 2.1 eV. The X-ray photoelectron spectrum of the ReO_3 valence band, shown in Figure 160, consists of two parts. The component of the spectrum having a width of approximately 2 eV and situated close to the Fermi level corresponds to the 5*d* states of rhenium. The structure of the spectrum situated between 3 and 10 eV is due to both the 5*d* states of rhenium and the 2*p* states of oxygen. The greatest contribution to the electron density of states in this region comes from the 2*p* states of oxygen [416–418]. The main peak in the density of states, having a binding energy of 3 eV, contains a small contribution from the 5*d* states of rhenium. However, in the photoelectron spectrum, in the region corresponding to this peak, a maximum of much lower intensity is observed. From Figure 160 it can be seen that the distribution of the density of *d* states agrees rather well with the structure observed in the X-ray photo-

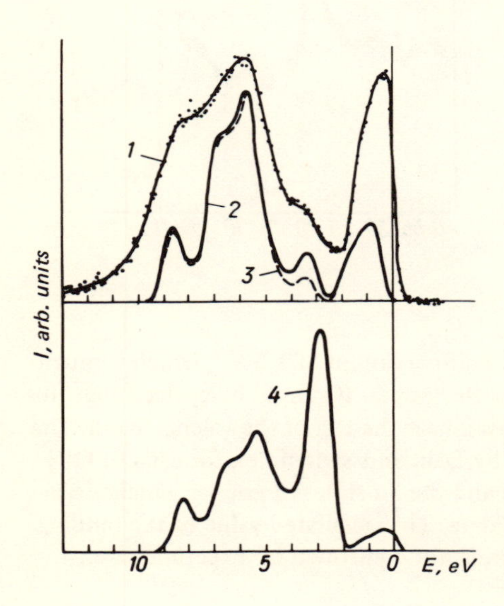

Figure 160: Valence band photoelectron spectrum of crystalline ReO_3: (1) photoelectron spectrum; (2) superposition of the partial density of d states and 0.025 partial density of 2p states in oxygen; (3) partial density of d states; (4) valence band electron density of states.

electron spectrum [119]. Better agreement with experiment is obtained if 2.5% of the density of *p* states is added to the density of *d* states and if the resulting curve is shifted 0.45 eV toward higher values of the binding energy. The shape of the maximum of the 5*d* states in the total density of states agrees well with the peak in the experimental photoelectron spectrum.

The band structure of ReO_3 calculated by Wertheim *et al.* [416–419] agrees well with the data on the optical properties, the dimensions of the sections, and the topology of the Fermi surface. Thus, the band related to the 2*p* states of oxygen appears in the X-ray photoelectron spectrum as a result of mixing with the *d* states of rhenium. The results obtained show that the ionization cross section of the 5*d* electrons of rhenium is 20 to 40 times greater than the photoionization cross section of the 2*p* electrons of oxygen. In general, the 2*p* electrons of oxygen make a small contribution to the X-ray photoelectron spectrum. This can be deduced by a comparison of the relative intensities of the photoelectron lines of 2*s* and 2*p* electrons in various compounds containing oxygen. In MgO, for example, in which the metal does not have filled *d* states, the intensity of the 2*p* lines is low.

Sulfides of Transition Metals

Hüfner and Wertheim [420] have measured the valence band X-ray photoelectron spectra of NiS using an IEE-15 electron spectrometer with the magnesium $K\alpha_{1,2}$ line as the photon source. The sample was prepared in the form of a thin film deposited on a holder of pure nickel. The nickel surface was heated up to 500°C and then allowed to react for a short time with hydrogen disulfide, after which the sample was quickly cooled down to the room temperature. At 264°C, the sulfide undergoes a transition from the metallic state to the semiconductor state [421]. The low-temperature phase is an antiferromegnetic semiconductor, and the high-temperature phase is a paramagnetic metal [421]. As can be seen from Figure 161, the X-ray photoelectron spectrum of NiS gives a rather clear representation of the structure of the density of electronic states in the valence band [422]. The theoretical and experimental results regarding the position and width of the *d* bands agree well, but they differ somewhat in the case of the *s* and *p* bands. Thus, experiment indicates that the 3*p* band of sulfur is much closer to the 3*d* band of nickel than is predicted by the theoretical calculations.

White and Mott have performed a detailed study of the electron structure of NiS [423]. One of the most important results of their experiment was that in NiS there exists a significant overlap of the *d* and *p* states. Therefore, the screening due to *p* electrons lowers the Coulomb energy of the valence electrons down to such a value that metallic conductivity becomes possible.

Oshawa *et al.* [424] have measured the X-ray photoelectron spectra of the

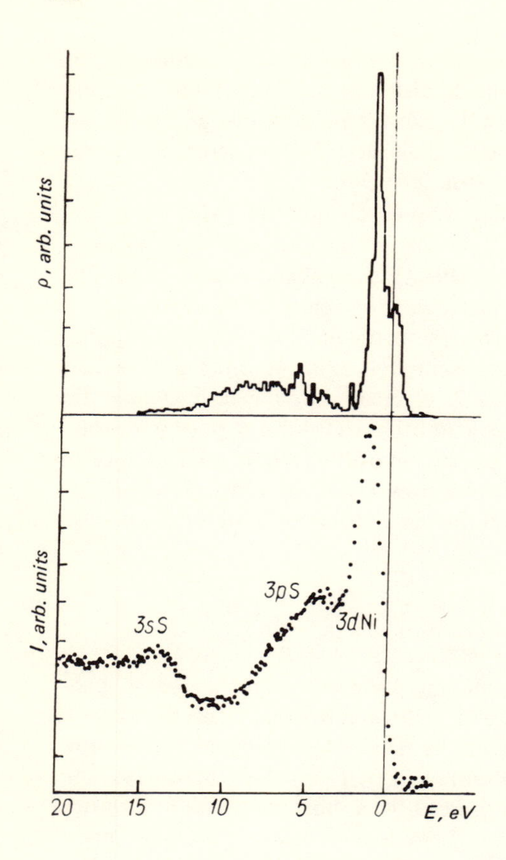

Figure 161: Valence band density of states (a) and photoelectron spectrum (b) of crystalline NiS.

disulfides of iron (FeS_2), cobalt (CoS_2), and nickel (NiS_2), using an AEI-ES 200 electron spectrometer. The electron structure of these compounds has been studied by a number of authors [425–427]. The highest component in the valence band of FeS_2 is the t_{2g} band, which contains the $3d$ states of iron. The band that contains the $3p$ states of sulfur is characterized by a large width (of the order of 7 eV). As a result of the hybridization of the states in this band, a significant mixing of $3d$ and $4s$ states is observed. As can be seen from the experimental results (Figure 162), in FeS_2 there exists a sharp maximum close to the Fermi level having a width at half-maximum of 1.3 eV. The width of this peak is dictated almost entirely by the spectrometer resolution. Studying the photoelectron spectra of FeS_2 at an exciting energy of 40 eV, Li *et al.* [427] established that the $3d$ band of iron extends over about 0.8 eV. The maximum at a binding energy of 4 eV corresponds to the $3p$ states of sulfur.

In CoS_2, besides the six t_{2g} electrons, there exists an additional electron that occupies the antibonding orbital e_g, which results in a metallike conductivity.

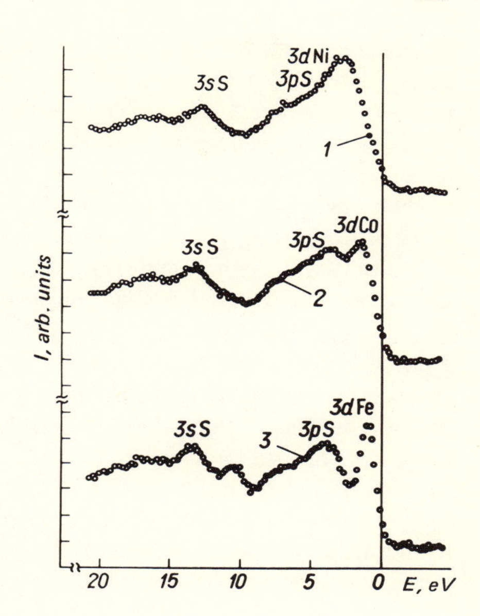

Figure 162: Valence band photoelectron spectra of some disulfides: (1) FeS_2, (2) CoS_2, and (3) NiS_2.

Therefore, the maximum in the photoelectron spectrum at a binding energy of approximately 1.5 eV corresponds to both t_{2g} and e_g electrons.

Since NiS_2 exhibits semiconductive properties, the antibonding e_g orbital is probably split by the Hubbard interaction in NiS_2, the $3d$ states of nickel overlap the $3p$ states of sulfur significantly. According to data of Li *et al.* [427], the t_{2g} maximum in NiS_2 is situated at a binding energy approximately 1.4 eV higher than the corresponding maximum in FeS_2. This energy difference reflects the fact that the $3d$ electron in the nickel atom has a higher binding energy than the $3d$ electron in the iron atom.

Halides of Transition Metals

Kono *et al.* [428, 429] have studied the X-ray photoelectron spectra of CuI, CuBr, and CuCl, while Sasisaka *et al.* [430] have studied compounds of transition metals (manganese, iron, cobalt, and nickel) in their bivalent state with chlorine and bromine.

Herman and McClure [431] have proposed that the valence band in CuCl, CuBr, and CuI is made up of two subbands separated by a forbidden band, the upper subband containing mainly d states and the lower subband mainly p states. It is considered that the upper subband is made up of the $3d$ states of the Cu^+ ion, and the lower subband by the np states of the halogen ion, where $n = 3, 4$, and 5 for CuCl, CuBr, and CuI, respectively.

The speciments were prepared [429] by evaporation onto a steel substrate, in a vacuum of 2–4 × 10^{-6} torr. In order to prepare the compound CuI, the substrate was heated up to 150°C. In CuCl, Novakov [432] observed satellite lines related to the $2p$ electrons of copper. It was shown, however, [429] that their occurrence is due to the absorption of water or oxygen. Frost *et al*. [433] have shown that the occurrence of an additional satellite structure is related to the presence of bivalent copper ions in the sample. In the previously quoted work of Kono *et al*. [429], no similar satellite structure was observed. This indicates that the compounds they studied were not oxidized and did not contain water. The X-ray photoelectron spectra, measured with a resolution of 1.0 eV using the magnesium $K\alpha_{1,2}$ line as the photon source, are shown in Figure 163. Each of the spectra is characterized by two maxima. For all of the specimens investigated, the most intensive maximum exhibits a bump on its low-energy side. Due to the rather poor resolution, the bandwidth determined from the X-ray photoelectron spectra in Figure 163 is large. However, in the case of CuCl and CuBr, it can be considered that the experimental X-ray photoelectron spectra confirm the existence of the theoretically predicted forbidden band between the two valence subbands. By using the method of Jansson [434] for the mathematical treatment of X-ray photoelectron spectra, it can be shown

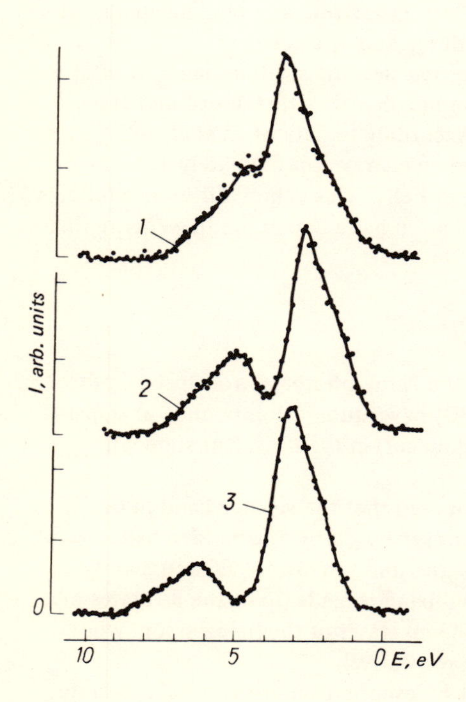

Figure 163: Valence band photoelectron spectra of some transition metal compounds: (1) CuCl; (2) CuBr; (3) CuI.

that, also in the case of CuI, two subbands of the valence band are separated by a forbidden gap.

Kono *et al.* [428, 429] also determined the magnitude of the relative intensity of the upper and lower valence bands:

$$R = \frac{\int dEI^U(E, \hbar\omega)}{\int dEI^L(E, \hbar\omega)} .$$

It is related to the magnitude of the moduli of the coefficients that determine the degree of hybridization of the p and d wave functions. The contribution of the d wave function to the electron wave function in the upper subband of the valence band amounts to 0.86 for CuCl, 0.69 for CuBr, and 0.77 for CuI, while in the lower subband it is 0.24, 0.52, and 0.39, respectively. The results obtained by the method of X-ray photoelectron spectroscopy also offer the possibility of interpreting absorption spectra in the ultraviolet region [429].

Goldman *et al.* [435] have studied the valence band photoelectron spectra of CuCl, CuBr, and CuI, in the ultraviolet and in the X-ray regions of incident radiation spectra. The X-ray photoelectron spectra, measured with an ESCA-3 spectrometer, are consistent with the spectra in Figure 163. Comparison of the X-ray photoelectron spectra with the spectra obtained at photon energies of 16–48 eV indicates that the photoionization cross sections of the p and d electrons depend strongly on the incident photon energy. In the ultraviolet region, the photoionization cross sections of p and d electrons are approximately equal, but in the X-ray region, the photoionization cross section of d electrons is greater than for p electrons.

Figure 164 shows the X-ray photoelectron spectra of the compounds of bivalent transition metals with halogens [430]. The spectra were recorded with a resolution of 1.0 eV, using an electron energy analyzer of the electrostatic condenser type. The measurements were made with a vacuum of 10^{-6} torr in the spectrometer. The $K\alpha_{1,2}$ emission line of magnesium was used as source of photons. The specimens were prepared by *in situ* evaporation onto an aluminum substrate heated to 200°C. The samples were kept at this temperature for half an hour and then cooled down slowly to room temperature. From Figure 164 it can be seen that the X-ray photoelectron spectra of the investigated specimens are similar for the chlorides and bromides of transition metals, with the exception of the compounds of iron.

The photoelectron spectra mainly reflect the splitting of 3d electrons, since the matrix elements of the probability of transition from d states to the conduction band are greater than those from np states. This is the explanation for the observed similarity of the photoelectron spectra.

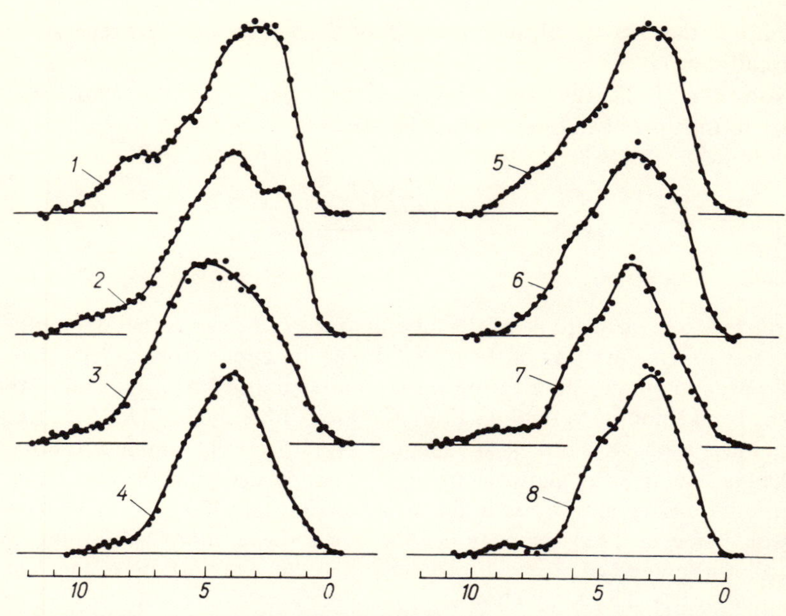

Figure 164: Valence band photoelectron spectra of some transition metals compounds with Cl and Br: (1) $NiCl_2$; (2) $CoCl_2$; (3) $FeCl_2$; (4) $MnCl_2$; (5) $NiBr_2$; (6) $CoBr_2$; (7) $FeBr_2$; (8) $MnBr_2$.

Manganese Borides

Nemoshkalenko *et al.* [436] have studied the binding energies of the $2p$ and $3p$ core electrons of manganese and the $1s$ core electrons of boron in borides, using an IEE-15 electron spectrometer. Powder samples were fastened onto adhesive tape. The spectra were calibrated with respect to the $1s$ line of carbon (284 eV).

The problem of how the charge is distributed between the metallic and nonmetallic atoms in the borides of the transition metals (see, for example, [437]) is not yet resolved; there exist new data that some authors interpret as an indica-

TABLE 47: Values of Core-Electron Binding Energies of Mn and B in Manganese Borides

Level	Pure elements	Mn_4B		Mn_2B	
		E_B	ΔE_B	E_B	ΔE_B
$2p_{1/2}$ Mn	652.0	654.7	2.7	653.8	1.8
$2p_{3/2}$ Mn	640.4	642.6	2.2	641.7	1.3
$3p$ Mn	47.4	49.7	2.3	49.1	1.7
$1s$ B	186.8	188.4	1.6	188.1	1.3

tion of electronic charge transfer from the metal atoms to the boron atoms [438], although other authors argue in favor of donor behavior by the boron atoms [439].

Ramqvist *et al.* [440, 441] have shown that, in the study of the problem of charge transfer in such compounds as carbides, nitrides, and borides, the simultaneous use of X-ray spectroscopy and electron spectroscopy is very useful.

As an extension of the previously mentioned studies of the X-ray spectra of Mn–B compounds [442], core-electron binding energies have been measured for the atoms of both components of the following borides: Mn_4B, Mn_2B, MnB, Mn_3B_4, and MnB_2. All borides of manganese are good metallic conductors (even better than the pure metal manganese [443]), and therefore the measured values of the core-electron binding energies are as correct as the choice of the Fermi level of the instrument as origin on the energy scale allows.

In all the samples investigated, the spectra of $1s$ electrons exhibit two lines, which probably indicates that the oxide B_2O_3 formed on their surface. However, the occurrence of peaks corresponding to the oxide did not introduce difficulties in interpretation since the energy difference between the boron $1s$ line in the oxide and that in borides is relatively large (approximately 4 eV). In the photoelectron spectra of manganese atoms, the measured $3p$, $2p_{3/2}$, and $2p_{1/2}$ lines are rather broad and possess a structure that may be interpreted in terms of the existence of an oxide layer, or of multiplet splitting [444].

The experimental results are shown in Table 47, which includes the binding energy values (E_B) of the $3p$, $2p_{3/2}$, and $2p_{1/2}$ core-electron levels of manganese as well as the binding energy values of the $1s$ electrons of boron in the borides. Also included are the corresponding values for the pure elements, as taken from the paper of Bearden and Burr [445]. The data on pure elements were used for the determination of the chemical shifts of the core levels (ΔE_B) in the formation of the compounds.

As can be seen from Table 47, all the measured values of the core-level binding energies of manganese in borides are greater than in the pure metal, which indicates that the manganese atoms in the compounds have a supplementary positive charge. The $1s$ level of boron in the various compounds is situated on either side of the maximum corresponding to pure boron (according to

MnB		Mn_3B_4		MnB_2		Precision of measurement
E_B	ΔE_B	E_B	ΔE_B	E_B	ΔE_B	
653.2	1.2	653.6	2.2	654.0	2.0	±0.2
641.2	0.8	641.6	1.2	642.1	1.7	±0.2
48.5	1.1	48.8	1.4	49.4	2.0	±0.1
187.7	0.9	187.8	1.0	188.4	1.6	±0.2–0.3

Bearden and Burr [445]). The shifts of the 1*s* line are, however, rather small. In general, it has been found that the shift of the 1*s* line of boron can reach up to approximately 9 eV, depending on the electronegativity of the second component in the compound. It should be noted, however, that with the exception of MnB and Mn_3B_4, the shift of the core levels of both components with respect to those of the pure elements has the same sign. According to our results, such a tendency is not observed in other compounds, either of ionic type (which is natural), or with mixed bonding. According to the existing results for boron compounds [447], for chalcogenides [448], and even for metallike compounds of carbon [440, 441], oxygen and nitrogen [440], the shift of the core levels of the two components is generally in opposite directions. Nevertheless, the results of Shabanova *et al.* [450] on the binding energies of iron $2p_{3/2}$ electrons and silicon 2*p* electrons in iron silicates show that the levels of both components are shifted in the same direction with respect to those of the pure elements. This provides an additional indication of the similarity between borides and silicates from the point of view of the chemical binding [437].

The chemical shifts of the core-electron binding energies of manganese in borides are not the same for different levels, as is demonstrated by the data in Table 47; more outer levels are characterized by smaller shifts. On the basis of a simple model it might be expected [440] that, under the influence of chemical binding, all the internal levels would suffer approximately the same shift. However, as more recent data have shown, this is not always the case: to a large extent it is true for ionic compounds [450] but less so for compounds with mixed ionic–covalent bonding [448].

The most obvious characteristic of the binding energy shift, in going from one boride to another, is the fact that for manganese and boron it occurs in the same direction; that is, if the binding energy of manganese core electrons increases, the binding energy of the boron 1*s* electrons also increases. Moreover, the dependence of core-level shift on the composition of the boride is almost linear, and in the region of the monoboride MnB it changes sign (Figure 165).

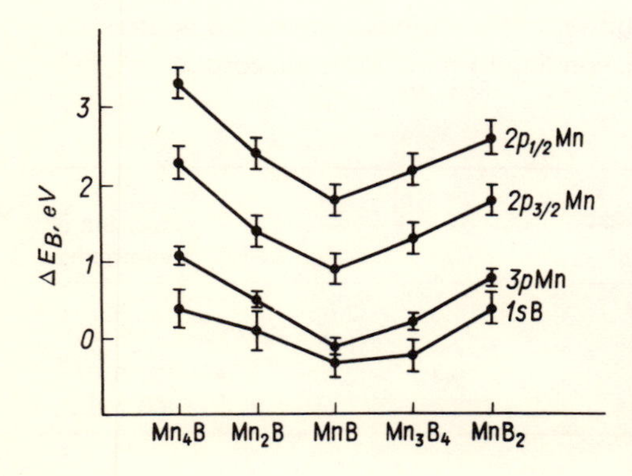

Figure 165: Mean value of the energy shift of the $2p_{1/2}$, $2p_{3/2}$, and 3p photoelectron lines of manganese, and of the 1s photoelectron lines of boron in manganese borides.

Therefore, the decrease of binding energy in passing from the lower borides of manganese (Mn_4B, Mn_2B) to the monoboride MnB, shows that the atoms of both components get some charge. In contrast, the increase of binding energy on passing from the monoboride MnB to the diboride MnB_2 indicates some decrease of the electron density, i.e., a loss of electron charge in the region of both manganese and boron atoms.

It is worth mentioning here that, in the region of MnB, a similar change of sign also occurs in the variation of the energy shift of X-ray emission lines in the *K* series of manganese and boron [442]. A change of sign at the concentration corresponding to MnB has also been observed in the variation of some other parameters of the X-ray emission spectra of manganese borides, such as the width and the asymmetry index of the *K* and *L* lines of the metal and of the *K* line of boron [442]. It can be said in conclusion that the position and shape of the valence band are modified in a similar way, by the same factors related to chemical binding as those that determine the regular change of energy of the core-electron levels.

What are these factors? In the system of manganese borides, the monoboride phase is characterized by a minimum distance between the manganese and boron atoms [437] and by a maximum in the hardness characteristics of the melting temperature and the microhardness [451]. For a series of metallike compounds of carbon, Ramqvist *et al.* [440] have established that the shifts of inner lines are proportional to the heat of formation, i.e., with the strength of the binding, and have concluded that such a correlation can be attributed to the ionic component of the binding. This result applied to the manganese borides, together with the fact that the variation of E_B with the procentual content in the components is essentially linear, indicates that the observed dependence (Figure 165) is mainly determined by the ionic component of the binding. At the same time, the influence of the covalent component is revealed by experimental observations such as the fact that the direction of shift of E_B is the same for both components, that E_B depends on the energy position of the level, and possibly also that the order of magnitude of the shift of the manganese levels is not low. From crystal chemistry one can deduce that in the borides covalent binding may exist between metal and boron atoms, as well as between boron atoms: starting from the monoboride phase, the occurrence of independent boron-boron bonds imparts to the boride compounds a particular structural individuality, which is characteristic of the pure element boron [437].

In conclusion, the analysis of experimental data points to the following result: when passing from the pure component elements to the lower borides (Mn_4B, Mn_2B), part of the electrons are removed from the spheres of both component atoms, in order to form a covalent bond; when passing from Mn_4B to MnB, the covalent component decreases or remains unchanged, while the ionic component increases and reaches a maximum for MnB (here, the shifts are at a minimum and, as an exception, have different signs); when passing from

MnB to MnB_2, the ionic component decreases, and the covalent one increases, probably on account of the boron-boron bond.

Since there exist controversial opinions concerning the direction of electron transfer and the magnitude of the transferred charge, it is interesting to make an attempt to extract these from the results obtained from the shift of the core levels. For this purpose, it is necessary to know the calibration of the energy shift of each element: for the $2p_{3/2}$ level of manganese, this has been established by Carver and Schweitzer [452], and for the $1s$ level of boron, by Hendrickson *et al.* [446]. In the first case, the atomic charges were calculated by taking into account the electronegativity according to Pauling, and in the second case, by using the method of molecular orbitals of Huckel and the CNDO method.† It was shown that the charge at the boron atoms, determined by the CNDO method, agrees rather well with the measured values of the energy of the boron $1s$ level [446].

The evaluation of effective charges in the borides of the transition metals, using such semiempirical calibrations, has evidently only an approximate character, since, as yet, it is difficult to say to what degree the linear correlation between the shift and the charge might be true for compounds with intermediate binding character. Moreover, the calculations based on molecular orbitals assume that the charge separation is rather high, which in borides is probably not the case. By using the semiempirical scale of the chemical shifts of the manganese $2p_{3/2}$ levels and boron $1s$ level [446], as well as the data in Table 47, it is found that when passing from Mn_2B to MnB, the charge increases by approximately 1 electron in the vicinity of manganese atoms, and only insignificantly (according to calculations by the CNDO method) in the vicinity of boron atoms. When the boron concentration is increased, the variation of the shift is reversed, and in MnB_2, the charge in the vicinity of the manganese atoms decreases by approximately 1.8 electrons, compared to that in MnB, while in the vicinity of the boron atoms it decreases by approximately 0.1 electrons. Of course, these estimations refer to the resultant charge, as built up by both component atoms. In compounds the total charge distribution may be very complicated, since the metal-nonmetal bonds involve contributions from the $3d$, $4s$, $4p$ orbitals of the metal atoms as well as the $2s$, $2p$ valence orbitals of the boron atoms. Therefore, it is difficult to correlate directly the charge variation with the variation of a given type of electron density. However, some information about such a correlation may be obtained from other experimental data. In particular, for the series Mn_2B-MnB-MnB_2, successively increasing values of the relative intensity of manganese $L\alpha$ lines have been observed, which may indicate an increasing density of d electrons. A similar conclusion can be drawn from measurements of the magnetic properties [439] and the electron-specific thermocapacity [453],

†In the CNDO method, differential overlapping is completely neglected.

namely, that in the series Mn_2B-MnB-MnB_2, the addition of each boron atom is accompanied by a corresponding binding of the *d* electrons of the metal with approximately 1.8 electrons of boron. In this process, the density of 3*d* electrons remains unchanged, or even increases. Therefore, the variation of the total charge, indicated by our experimental data (see Figure 165), is evidently caused by the simultaneous change of the density of the manganese 4*s*, *p* electrons and boron 2*s*, *p* electrons. It is possible that, up to the MnB phase, the charge cloud of valence electrons originating from both atoms is to a large extent shifted toward the manganese atoms, while beyond MnB, it is shifted toward the boron atoms due to the additional contribution of 4*s*, *p* electrons. The assumption that a number of electrons are transferred from the metal atoms in order to create the binding in borides with strong boron–boron bonds (MnB-MnB_2) has already been used in the interpretation of the X-ray emission spectra of these borides [442]. However, as is indicated by the increase of the binding energy of boron 1*s* electrons, the electrons that are removed from the vicinity of the metal atoms are probably not transferred to the boron atoms. It is possible that the formation of individual bonds between the boron atoms is accompanied by a decreasing localization of the 2*s*, *p* electrons of boron.

Carbides of Transition Metals

Ramqvist *et al.* [454] have studied the carbides of vanadium, niobium, and tantalum, using the electron spectrometer described in [1]. The binding energies were measured for the $2p_{3/2}$ electrons of pure vanadium and VC_X carbides (where $X = 0.876$ and 0.716), the $3d_{1/2}$ electrons of pure niobium and NbC_X carbides (where $X = 0.940$, 0.904, and 0.745) and of the $4f_{7/2}$ electrons of tantalum in Ta_XC (where $X = 0.990$, 0.963, 0.887, and 0.745). The values of the binding energies in the carbides are shifted toward greater energies compared with the values for the pure metals, which indicates a decrease of the electron density at the transition metals atoms. It was found that the energy shift of vanadium $2p_{3/2}$ electrons increases, while that of the niobium $3d_{1/2}$ electrons and the tantalum $4f_{7/2}$ electrons decreases when the carbon content in the carbides is increased. Since the binding energy of tantalum $4f_{7/2}$ electrons is not large (23 eV), special care should be exercised when interpreting the chemical shifts of this level in terms of electron charge transfer. The binding energy of the carbon 1*s* electrons decreases, which indicates that the electron density increases in the vicinity of the carbon atoms. However, the binding energy of carbon 1*s* electrons is almost the same for each carbide phase. Therefore, in these compounds, the carbon atoms are in the same chemical state. In the carbides of stoichiometric composition, the electron lines of the metal atoms are narrower than those in carbides of nonstoichiometric composition. Here, the broadening is caused by the fact that not all the transition metal atoms have the same

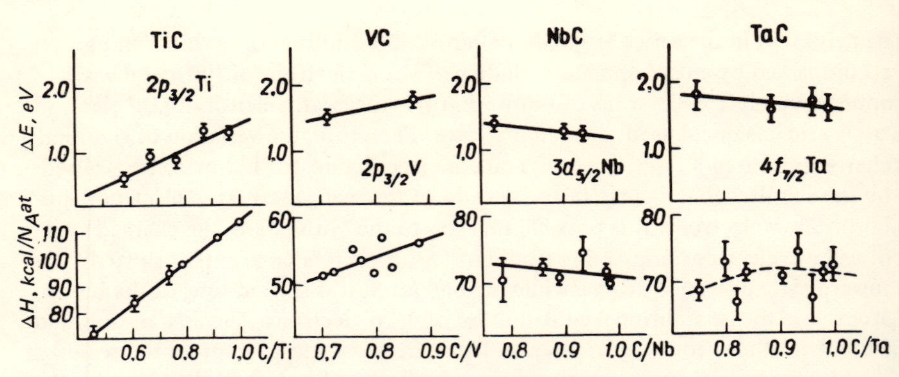

Figure 166: Energy shift of core-level photoelectron lines of transition metals and the heat of formation of transition metal carbides, in the region of homogeneity.

number of neighboring carbon atoms. The studies of titanium carbide by Ramqvist *et al.* [455] have shown that the above correlations are also valid for this compound.

Figure 166 shows the relationship between the chemical shifts of the core levels and the energy of formation of the investigated compounds, as given by Ramqvist *et al.* [441, 454]. This indicates that the variation in the energy of formation between the different compounds is related mainly to the variation in the ionic character of the transition metal–carbon bonding.

7

Characterization of Surfaces

The development in the knowledge of the electron structure of crystal surfaces has led to the solution of many problems in the physics of thin films and thin-film microelectronics. The electronic processes that take place at solid surfaces are of vital importance in such phenomena as chemisorption and heterogeneous catalysis. At present, solid surfaces may be characterized not only by vacuum ultraviolet spectroscopy, but also by means of X-ray photoelectron spectroscopy and Auger spectroscopy. Among the problems that may be investigated by X-ray photoelectron spectroscopy are the study of the energy states of the electrons situated near the crystal surface, the quantitative and qualitative analysis of the composition of the layer of material at the sample surface, the detection of impurities, and the determination of the valence states of atoms at the sample surface. For a number of reasons, Auger spectroscopy is characterized by a higher sensitivity, albeit combined with a lower resolution power, than X-ray photoelectron spectroscopy. Auger spectroscopy may be used for both quantitative and qualitative analysis, as well as for the identification of substances formed as a result of chemical reactions at the sample surface.

Auger Spectroscopy as a Method for Surface Characterization

As a result of the ionization of atoms by electron or photon bombardment, the newly created states may contain holes either in the valence band or in the core levels. Auger spectroscopy investigates the energy distribution of electrons emitted by the relaxation of these states. The Auger spectra can be measured with the same instruments as X-ray photoelectron spectra.

The Auger process consists in the nonradiative filling of holes by electron transitions from higher energy levels. The energy liberated in these transitions is transferred to another electron that is emitted from the sample. The energy of the emitted Auger electrons is determined by the binding energies of the electrons participating in the Auger process and the electron work function of the sample material [456]. In order to label the Auger transitions, the usual X-ray terminology is used. Thus, the KL_1L_2 Auger process represents the transition of an electron from the L_1 level to the previously ionized K level, together with the simultaneous emission of another electron from the L_2 level.

A competitive relaxation process, in which the atom containing a hole on an inner level undergoes transition to the ground state, is the emission of X-ray quanta. At low energies, the Auger-type relaxation process dominates [457]. At 4000 eV, the number of Auger electrons becomes comparable with the number of emitted X-ray photons. Restricting the discussion to low-energy transitions, Auger spectroscopy in this energy region is very sensitive to all of the elements, excepting hydrogen and helium, which do not have core electrons.

The kinetic energy of Auger electrons can be determined approximately from the values of X-ray energies [458, 459]. Thus, the energy of KL_1L_2 Auger electrons is equal to $E_{KL_1L_2}(Z) \approx E_K(Z) - E_{L_1}(Z) - E_{L_2}(Z)$. Here $E_{L_1}(Z)$ and $E_{L_2}(Z)$ represent the energies of the corresponding levels, when involved in the single ionization of atoms. Owing to the absence of the electron in the L_1 level, the binding energy of the electron on the L_2 level increases and approaches the binding energy of an electron on the same level in an atom with a higher atomic number. Therefore, a more correct expression for the energy of Auger electrons can be written as follows [460–462]:

$$E_{KL_1L_2}(Z) = E_K(Z) - E_{L_1}(Z) - E_{L_2}(Z + \Delta) - \varphi_{sp},$$

where Δ has a value between Z and $Z + 1$. The best agreement with experiment is obtained by chosing a value for Δ between $\frac{1}{2}$ and $\frac{3}{4}$. Since the binding energy of electrons is measured with respect to the Fermi level of the spectrometer construction material, it follows that the energy values of Auger electrons are lowered by the magnitude of the electron work function φ_{sp}.

In the interpretation of effects related to Auger processes, use has been made mainly of empirical data, since the theory of Auger processes is less developed than the theory of X-ray emission. It suffices to mention that the selection rules for Auger transitions in atoms are at present not well established. In elements with atomic numbers from 3 to 14, the KLL Auger transitions are excited most frequently. With increasing binding energy, the ionization cross section decreases rapidly. At the same time, an increase of the X-ray emission yields is observed, and for elements with $Z > 14$ the experimental investigation of KLL Auger spectra becomes difficult. For silicon ($Z = 14$) it is still possible to observe high-intensity LMM lines. In elements after zirconium ($Z = 40$), the MNN lines become more intense than the LMM lines. The MNN lines can be used for identification of elements with atomic numbers 40–79.

The energy extension of Auger-electron spectra depends mainly on the width of the energy levels involved in the Auger transitions. Owing to the short lifetime

(10^{-14}-10^{-17} sec) of the initial ionic state, the width of Auger lines is, at best, equal to several electron volts. The width of Auger lines is particularly high in those cases in which valence electrons are involved in the process. If one valence electron is involved in an Auger transition, the width of the Auger line will not be less than the width of the valence band of the crystal. Even larger values of Auger linewidths are obtained when the Auger transitions involve two valence electrons. The yield of Auger electrons is determined to a large extent by the cross section for photoionization of atoms in nonexcited states. The probability of Auger transitions increases from zero at an incident electron energy equal to the binding energy of electrons situated on the energy level E_B, and reaches a maximum value at incident energies of about $5E_B$ [463]. However, the dependence of Auger electron yield on the energy of incident (exciting) electrons does not have a form analogous to that of the energy dependence of the photoionization cross section of atoms in nonexcited states. The observed discrepancy between the two curves may be explained by the presence of scattered electrons, which may produce ejection of additional Auger electrons. The contribution of scattered electrons to Auger emission increases with increasing energy.

The problem of the energy dependence of the escape depth of electrons from the sample has been discussed in Chapter 1. Here, we only mention that in Auger spectroscopy, the electron escape depth is a complicated function of both the probability of Auger transitions and the energy of incident electrons. For one monolayer of oxygen atoms, the yield of Auger electrons is approximately equal to $10^{-4} KLL$ electrons for each incident electron [464].

Auger electron spectroscopy is mainly used for qualitative analysis. To solve many practical problems, a qualitative or a semiquantitative analysis is often sufficient. This is probably why rather few works have been devoted to the use of Auger spectroscopy in quantitative analysis. However, use of appropriate standards, knowledge of the values of electron escape depth for the investigated materials, and taking account of the effects of incident electron scattering and of sample surface topography makes it possible to use Auger electron spectroscopy as an efficient method for quantitative analysis.

The vacuum level necessary in Auger spectroscopy can vary over a wide range of several orders of magnitude, depending on the reactivity of the investigated surfaces. At pressures of about 10^{-9} torr, one monolayer of gas is adsorbed onto a crystal surface in about 10^3 sec, if the sticking coefficient is equal to unity. However, except for metal surfaces, the sticking coefficient is much lower than unity, so that in the Auger spectrometer a vacuum worse than 10^{-9} torr is permissible. Usually, for most practical problems, a very high vacuum is not necessary. Use of oil diffusion pumps, however, is not recommended since an ionizing beam of electrons promotes the cracking of hydrocarbons, and the presence of oil vapor inside the spectrometer results in the formation of a carbon contamination layer on the sample surface.

Auger spectroscopy can be used successfully in the study of surface chemical processes, for assessing surface cleanliness, and for determining the degree of coverage with adsorbed layers as well as the degree of purity of these adsorbed layers. It can also be used for detecting impurity migration on the sample surface.

Chemical effects in Auger spectroscopy can be grouped into three categories: (a) energy shifts of Auger lines within the spectrum due to core-level shifts resulting from changes of the valence state of atoms; (b) changes of the line shapes, due to the redistribution of electrons in the valence band; and (c) changes of the line shape on the low-energy side of Auger lines due to changes in the mechanism of energy losses in the matrix. A reliable interpretation of chemical shifts in Auger-electron spectroscopy is not easy, since account should be taken of the energy changes in three levels. For example, the chemical shift in KL_1L_2 transitions is determined by the expression:

$$\Delta E_{KL_1L_2} = E_K - E_{L_1} - E_{L_2} - (E_K + \Delta E_K - E_{L_1} - \Delta E_{L_1} - \\ - E_{L_2} - \Delta E_{L_2}) = - \Delta E_K + \Delta E_{L_1} + \Delta E_{L_2},$$

where ΔE_K, ΔE_{L_1}, Δ_{L_2} represent the shifts of the K, L_1, and L_2 levels. These shifts are in general not equal. The situation is even more complicated when valence electrons take part in the Auger transitions. Besides of the energy shifts, the shape of Auger lines can also be affected by the charge redistribution in the valence band occurring during the process of oxidation of the sample. Although the interpretation of chemical shifts in Auger spectroscopy is difficult, the observed shifts allow a determination of the chemical environment of atoms in compounds.

In quantitative analysis, the following formula may be used as a first approximation [459]:

$$C_X = C_S \frac{I_X}{I_S},$$

where I_X and I_S represent the Auger-line intensities of the sample under investigation and of another material used as a standard, respectively. This formula is valid only if the Auger excitation probability and the Auger electron yield are the same for the sample and the standard. Otherwise, large errors may appear. In quantitative analysis, account should be taken of the presence of scattered electrons, which may ionize the atoms and therefore give rise to additional Auger electrons. The contribution of scattered electrons to the Auger spectrum may be decreased by chosing the excitation energy only a little higher than the ionization energy of the sample. Quantitative analysis is especially difficult when the sample composition varies within a depth smaller than the electron escape depth of the sample.

X-ray photoelectron spectroscopy is usually considered as a method for the study of the electron structure of substances, whereas Auger spectroscopy is a useful method for the rapid identification of the chemical composition of the sample. The electron analyzers used in Auger spectroscopy are characterized by a higher sensitivity but lower resolution than the electron analyzers used in X-ray photoelectron spectroscopy. In fact, in Auger spectroscopy there is no point in increasing the instrumental resolution, because of the large natural widths of the Auger lines. Consequently, the speeds of spectra recording in Auger spectroscopy and in X-ray photoelectron spectroscopy differ by a factor of approximately 10^2–10^3. It should also be noted that the ionization cross section with electrons is usually higher than the ionization cross section with photons. Moreover, the flux density of photons and that of the electrons used in the excitation of Auger spectra are different. The electron beam may be focused onto a small area. However, bombardment of the sample surface by high-energy electrons may result in an increase of its temperature. Heating of the sample surface may change its adsorption properties and may cause desorption, decomposition, or diffusion of adsorbed layers. If the diffusion occurs into the bulk of the sample, it may hardly be distinguished from desorption. Diffusion is more probable in insulators. All of these processes can, however, be monitored by the method of Auger spectroscopy itself. The simplest way to avoid unwanted phenomena is to lower the electron current; alternatively, a comparison can be made of measurements performed on surface regions submitted to the action of high-energy electrons with those from previously non-exposed regions.

As a rule, Auger spectra are studied by using specially constructed Auger spectrometers. The energy derivative $dI(E)/dE$ of the intensity of secondary electrons $I(E)$ is recorded, since it is sensitive to minute changes in $I(E)$.

Adsorption and Oxidation Processes at Metal Surfaces

The simplest possible case of adsorption is the adsorption at the surface, in which the adsorbate atoms do not penetrate into the substrate, while the surface is completely or partially covered with the adsorbate. This means that one surface atom corresponds to one or less adsorbate atom. This type of adsorption is termed type A. A more complicated type of adsorption occurs when the adsorbed atoms penetrate into the substrate, as, for example, in oxidation. These forms of adsorption can be grouped into two different classes, termed B and C. In cases of type B, part of the adsorbed substance penetrates into the substrate under the action of temperature or pressure. Type-C adsorption is characterized by the fact that the adsorbed substance penetrates into the sample before forming a monolayer of material on its surface.

By taking into account the magnitude of the mean free path of electrons in crystals, given in Chapter 1, one can consider that at photoelectron energies of 1400 eV, only 15% of the corresponding photoelectron line represents the contribution of atoms situated at the sample surface, but at energies of 100–200 eV, this contribution amounts to 50% [465]. X-ray photoelectron spectroscopy allows the observation of different adsorption states by a study of the chemical shifts of atomic core levels. This makes possible a determination of the type of chemical bonding between the atoms of the adsorbed substance and the atoms of the substrate. The simplest case of oxygen adsorption by metals, that is, a type-A process, is easiest to study since the chemical shift (the difference between the position of the metal line and the position of the line corresponding to the same metal in oxides) for solid samples is well known. This chemical shift may take rather high values. For example, the 4*f* photoelectron line of tungsten is shifted by about 4 eV when comparing pure tungsten with tungsten in WO_3 oxide. In the following processes of adsorption of one monolayer of CO and O_2 on Mo, CO on W, N_2 and NO on W, CO on Pt, CO on Ni, and Xe or O_2 on W, no energy shifts in the core-level photoelectron lines of the substrate occur. Figure 167 illustrates the photoelectron spectra of platinum. It can be seen that no shift of the 4*f* line of platinum is caused by oxygen adsorption.

A B-type adsorption of oxygen on molybdenum can take place if it is induced by heating in a gas atmosphere at a pressure of 1 torr. Figure 168 shows the change of the molybdenum 3*d* line that accompanies oxygen adsorption. The maximum corresponding to oxygen in the photoelectron spectrum becomes more intense than the metal line. Its position is characteristic for MoO_3. Unfortunately, since the resolution power of the spectrometer used by Brundle in this experiment [465] was rather low, he was not able to establish whether the oxidation of molybdenum to MoO_3 takes place via the intermediate

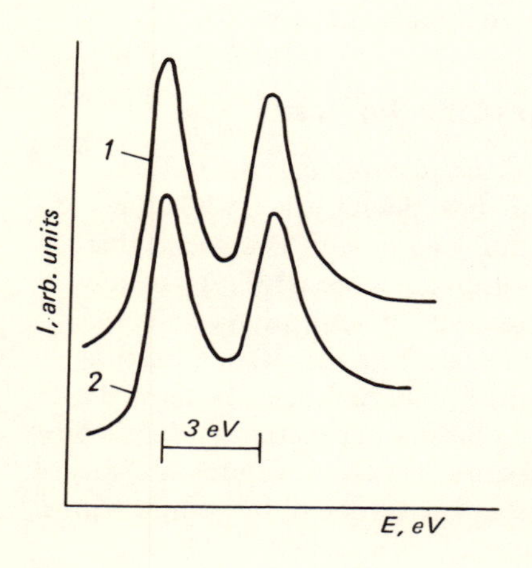

Figure 167. 4f photoelectron spectra of platinum: (1) platinum deposited on a substrate; (2) platinum covered with an adsorbed layer of carbon oxide.

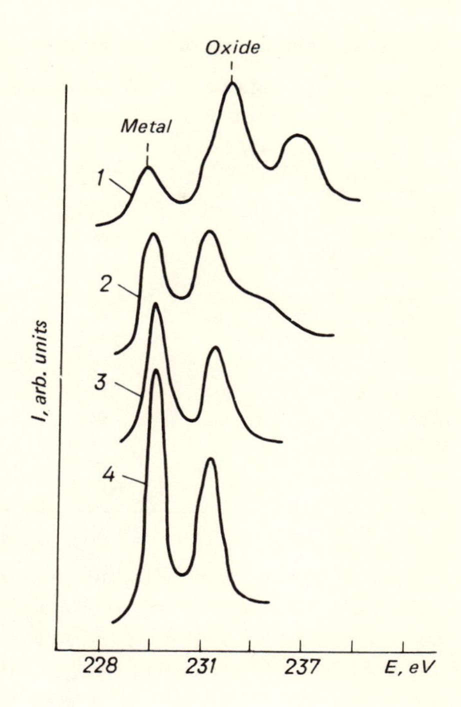

Figure 168. 3d photoelectron spectra of molybdenum in oxygen adsorption processes: (1) molybdenum with 7 monolayers of oxygen adsorbed on its surface; (2) molybdenum with 3 monolayers of oxygen adsorbed on its surface; (3) molybdenum with one monolayer of oxygen adsorbed on its surface; (4) pure molybdenum.

oxidation state MoO_2. It should also be noted that to within ±0.2 eV the position of the 1*s* line of oxygen remained unchanged during the adsorption of oxygen on molybdenum and during its oxidation. The adsorption of oxygen on nickel, vanadium, iron, lead, sodium, magnesium, and aluminum is a C-type process.

In analyzing the line shifts, it should be established with certainty that the new lines do not arise from satellites or multiplet splitting. In the case of oxygen and carbon, the interpretation of the observed shifts in terms of the chemical binding is very important. One way of interpreting line-shift data is purely empirical. It requires knowledge of the values of binding energies obtained for various possible chemical conditions. It is known that the weakest type of interaction between the adsorbate and the sample surface—i.e., condensation—causes a small shift of the core-level lines of the adsorbate toward lower binding-energy values, while the strong type of adsorption—type B and type C—causes a larger shift to lower binding-energy values.

As we have mentioned in Chapter 1, the cleaning of sample surfaces can be accomplished rather efficiently by bombardment with argon ions. Unfortunately, this gives rise to a series of complicated processes that need careful investigation. Among such effects and processes one could mention: a different

TABLE 48. Free Energy of Oxide Formation and the Possibility of Their Reduction by Bombardment with Argon Ions

Sample	Free energy ΔG_f° (kcal/mol)	Reduction	Sample	Free energy ΔG_f° (kcal/mol)	Reduction
Au_2O_3	−39	Yes	$Ni(OH)_2$	108	No
Ag_2O	2.6	Yes	WO_2	118	Yes
CuO	30	Yes	MoO_2	119	No
Cu_2O	35	Yes	SnO_2	124	No
IrO_2	40	Yes	MoO_3	162	Yes
PbO	45	Yes	Fe_2O_3	177	Yes
NiO	52	Yes	WO_3	182	Yes
PbO_2	52	Yes	SiO_2	192	No
CdO	54	Yes	Cr_2O_3	250	No
FeO	58	Yes	Ti_2O_3	346	No
RuO_2	60	Yes	Al_2O_3	377	No
ZnO	76	No	Ta_2O_5	471	No

quality of surface cleaning for different impurity elements, diffusion induced by ion bombardment, chemical interaction between the atoms on the sample surface and the incident ions, and strong heating of the sample surface due to the interaction between the material at the sample surface and the high-energy ions. Kim *et al.* [466] studied the removal of oxides from the surface of a number of metals by bombardment with argon ions of 400 eV energy. The metal surfaces had been oxidized by heating them up to high temperatures in air. As is seen from Table 48, many oxides have the tendency to undergo a reduction from higher oxidation states to pure metals or to lower oxidation states. All the oxides having a free energy of formation $\Delta G_f^\circ < 60$ kcal/mol were reduced to pure metals under bombardment with argon ions. A number of oxides, such as Ta_2O_5, Al_2O_3, SiO_2, do not have stable intermediary oxidation states and are, therefore, stable when bombarded with argon ions. The observed correlation between the reactivity and the free energy of formation is consistent with the idea of the existence of a quasiequilibrium developed during the cleaning process between the oxide and its components.

Let us consider the interaction between argon ions and a palladium surface. Layers of palladium oxide, several microns thick, were obtained by heating in air at 600°C for several hours [466]. After ion bombardment, reduction of the oxide to the pure metallic state was observed. The chemical shift of the $3d_{5/2}$ core level of palladium in PdO is large (as referred to pure palladium), and this fact allows the determination of the approximate thickness of metallic palladium after different bombardment times. As can be seen from Figure 169, the thickness of the metal layer increases with the bombardment time and with the ion current. The different slope of the curves for different kinetic energies of

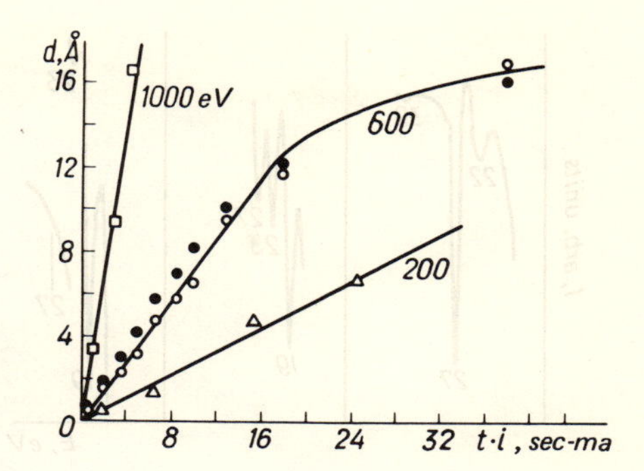

Figure 169. Thickness of palladium layer on PdO versus sputtering time and sputtering current, during bombardment with argon ions of various kinetic energies.

the argon ions is mainly due to the change of area of the ion-beam cross section with the increasing kinetic energy of the ions. After correcting for the variation of cross section, it is found that, for films of thickness under 10 Å, the rate of reduction of oxides to pure metals is constant, to within about 20%. From the radiation dose and the thickness of the palladium film formed, it was calculated that approximately two argon ions are necessary to reduce one palladium atom.

In Auger spectra, practically no chemical shift of the electron lines is observed in the case of A-type adsorption. Instead, the widths of Auger lines are more sensitive to changes of the chemical environment of the given atom. In a given situation, the X-ray photoelectron spectra may not show any observable change, although in the Auger spectra the lines are significantly broadened. Thus, after adsorption of one monolayer of carbon monoxide or oxygen onto pure molybdenum, the photoelectron spectrum of molybdenum 3*d* electrons does not change, but the *MNN* Auger line at 180 eV broadens by 1 eV.

Such changes in the shape of Auger spectra are often used empirically for the detection of changes in the chemical binding of the adsorbant. Thus, a change in the shape of the carbon *KLL* Auger spectrum indicates a change in the chemical state of carbon atoms at the surface [467]. Also in a number of other cases, use of Auger spectroscopy in the study of surface states is preferable to the use of X-ray photoelectron spectroscopy. Let us consider as an example the adsorption of oxygen on metallic sodium [465]. In sodium, the high-energy peak in the *KLL* Auger spectrum has a kinetic energy of 965 eV, and the low-energy peak in the $L_{2,3}VV$ Auger spectrum has a kinetic energy of 27 eV. The kinetic energy of 1*s* photoelectrons is 414 eV. Upon oxidation, the peak in the *KLL* Auger spectrum of sodium is shifted by about 4 eV, which is significantly greater than the 0.4 eV shift of the 1*s* photoelectron line. The shift of the Auger line is probably due to relaxation effects. After the formation of $\frac{1}{4}$ of a monolayer of oxygen on the sodium surface, the $L_{2,3}VV$ Auger peak at 27 eV (Figure 170) decreases drastically and almost disappears. At the same time two new peaks with significant intensities appear at 19 eV and 23 eV. A

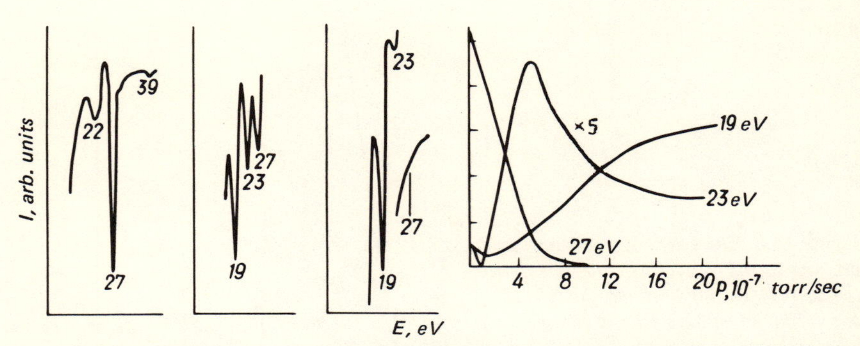

Figure 170. Variation of the structure and intensity of low-energy Auger lines of Na during the adsorption of oxygen.

detailed investigation of possible mechanisms responsible for the occurrence of these peaks shows that they are due to cross transitions: the peak at 19 eV corresponds to the transition $(Na^+L_{2,3})(O\text{-}V)(O\text{-}V)$, and the peak at 23 eV to the transition $(Na^+L_{2,3})(Na\,V)(O\text{-}L_{2,3})$.

It should be noted that in this case a new situation has arisen: the new peaks do not appear as a result of chemical shifts of the electron lines of the substrate or adsorbate, but as a result of the creation of new reaction channels, from the interaction between the substrate and the adsorbate. These lines also appear to be more sensitive to small amounts of adsorbate than the usual transitions ($OKLL$), since they are situated in the energy region in which the electron mean free path is small. However, it is difficult to investigate this energy region, because of the high intensity of the background of scattered electrons. Complicated but rather consistent effects have also been observed in the low-energy Auger spectroscopy of some other metals [467]. However, care should be taken when interpreting Auger spectra in terms of changes in the character of chemical binding in the adsorption process, since the Auger transition occurs in a system that is already once ionized, so that after emission of the Auger electron the atom is left in a twice-ionized state. Therefore, the possibility of describing such atomic states in crystals on the basis of conventional models of chemical binding should be regarded with caution.

Use of X-Ray Photoelectron Spectroscopy in Quantitative Analysis

The mean free path of photons with energies of the order of 1 keV is several orders of magnitude greater than that of emitted electrons. Therefore, the escape depth of electrons is mainly determined by the probability of emitted electrons to escape from the crystal without undergoing inelastic collisions. The angular distribution of photoelectrons, as described by formula (50) should also be taken into account. Manson [468] has calculated the dependence of β on the atomic number of the element investigated. It was found that, at high photoelectron energies, β reaches some limiting value at the given angular momentum, and the corrections for relative intensities, measured at a fixed value of the angle θ, are small. Thus, the ratio of integral intensities to those measured at 90° is 10% greater for p orbitals than for s orbitals. In solids, the angular dependence is less significant than in free atoms and molecules, because of effects related to elastic scattering processes. In order to avoid the influence of the angular dependence, the measurements should be performed at an angle of 54.7° at which the intensity of photoelectrons does not depend on the magnitude of β. From the analysis in Chapter 1, it follows that the intensity of photoelectrons from solid samples may be expressed by the formula

$$I = F\sigma N k \Lambda.$$

The relative intensity of two different photoelectron lines of the same sample is given by the relation

$$\frac{I_1}{I_2} = \frac{\sigma_1 N_1 \Lambda_1}{\sigma_2 N_2 \Lambda_2}.$$

From the discussion in Chapter 1 it follows that the magnitude of the electron mean free path Λ is a monotonous function of its kinetic energy. In the energy region 100–1500 eV, it can be considered that $\Lambda \sim E^{1/2}$. Carter, Schweitzer, and Carlson [469] chose carbon as the standard element:

$$\frac{I_{Z,nl}}{I_{C,1s}} = \frac{\sigma_{Z,nl} \sqrt{h\nu - E_B(nl)}}{\sigma_{C,1s} \sqrt{h\nu - E_B^C(1s)}}, \tag{122}$$

where $I_{Z,nl}$ represents the integral intensity of the photoelectron peak corresponding to the nl subshell of the element with atomic number Z; $h\nu$ represents the energy of the X rays; and $E_B(nl)$ is the binding energy of atomic electrons in the nl state. The photoionization cross section σ_Z in this formula has been calculated by Scofield [470] using relativistic Hartree–Fock–Slater functions. The values obtained were compared with experimental data for some simple molecules in the gaseous state [469]. Good agreement was observed between experiment and theory.

Equation (122) has been used to calculate the relative intensities of photoelectrons (Figure 171) for some specially chosen subshells of all of the elements in the periodic system, namely, those subshells that are most often used in X-ray

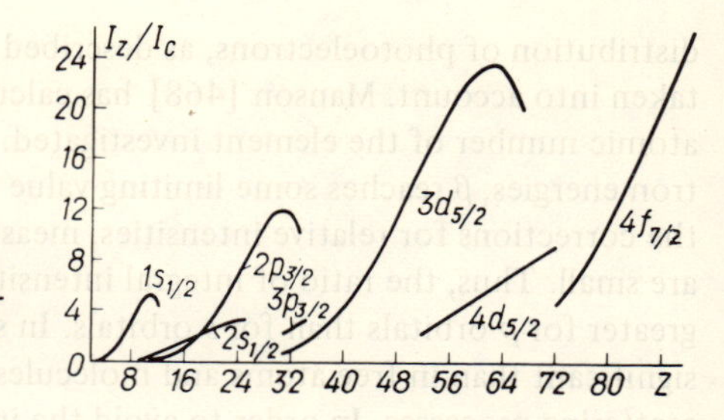

Figure 171. Intensity of some photoelectron lines versus atomic number of the elements. The values are given relative to the 1s line of carbon.

photoelectron spectroscopy. It is usual to choose subshells with high orbital quantum number at the given principal quantum number, since a vacancy in these subshells cannot be filled via Coster–Kronig transitions, which lower the lifetime of the ionized states and therefore broaden the photoelectron lines. In Table 49 are given the relative intensities of photoelectron lines for a number of solid samples, as well as the ratios calculated from equation (122). Consistency between experiment and theory is rather good (being to within about 10%). In certain circumstances, this consistency can even be improved. Thus, it was observed that the intensity ratio of photoelectron lines of the same type of atom in chemically different molecules does not always correspond to the stoichiometric ratio. This is probably due to the different probability of shake-up and shake-off processes for different atoms in molecules.

The model proposed by Wagner [471] for the quantitative analysis of the surface composition of solids assumes a uniform distribution of the investigated species over the sample. Any deviation from the expected behavior of relative intensities for homogeneous materials may help in the characterization of the

TABLE 49. Comparison between the Relative Intensities of Photoelectron Lines for Solids

Ratio	Experiment [471]	Experiment [469]	Experiment [104]	Theory [469, 470]
1*s*C/1*s*F	0.24	0.29	0.24	0.277
1*s*O/1*s*Na	0.61	0.53	0.35	0.522
1*s*Na/1*s*F	2.09	1.44	1.89	1.32
$2p_{3/2}$Si/1*s*F	0.17	0.23	0.15	0.161
$2p_{3/2}$P/1*s*Na	0.26	0.18	0.12	0.167
$2p_{3/2}$S/1*s*Na	0.33	0.30	0.18	0.232
$2p_{3/2}$Cl/1*s*Na	0.46	0.43	0.25	0.312
$2p_{3/2}$K/1*s*F	0.85	1.03	0.83	0.723
$2p_{3/2}$Ca/1*s*F	1.01	1.06	0.98	0.903
$4f_{7/2}$Pb/1*s*F	4.10	4.12	–	3.74
2*s*Na/1*s*Na	0.065	0.145	0.077	0.0919

nature of surface layers of a sample. For example, if a homogeneous material contains two elements of known concentration, situated under an overlayer, the observed intensity is determined by the expression

$$\frac{I_1}{I_2} = \frac{I'_1}{I'_2} e^{d\left(\frac{1}{\Lambda'_2} - \frac{1}{\Lambda'_1}\right)},$$

where I'_1/I'_2 represents the intensity ratio of the homogeneous material, as determined by equation (122). The values of electron mean free paths, Λ'_1 and Λ'_2, correspond to electrons with kinetic energies E_1 and E_2. If the magnitude of $1/\Lambda'_2 - 1/\Lambda'_1$ is known, then from the expression given above the thickness d of the overlayer can be calculated. This method may be used in those cases in which the thickness of the overlayer is greater than the lowest of the mean free paths Λ'_1 and Λ'_2. It is also desirable that the energy of the photoelectrons be significantly different.

In the study of thin films, the characteristics of the elements situated inside the impurity layer can also be measured.

Angular Distribution of X-Ray Photoelectrons with Regard to Quantitative Analysis

As is known, X-ray photoelectron spectroscopy is extremely sensitive to the surface state of the investigated sample. This has been demonstrated in a number of works [472–474] in which it has been shown that the method permits the detection of amounts of surface materials down to a limit of 1/50 of a monolayer. In order to separate from the total intensity the contribution of photoelectrons emitted from the very top layer of the sample surface, use can be made of the dependence of photoelectron intensity on the emission angle θ (with respect to the surface) or the incidence angle Φ of the X rays. In the case of $4f$ electrons of gold excited by $K\alpha_{1,2}$ radiation of aluminum, the electron mean free path amounts to 22 Å. The number of atomic layers corresponding to this value of the mean free path is 9 for $\theta = 90°$ and 2 for $\theta = 10°$. Therefore, in the second case, it is the electron structure of the surface that is observed. The effects related to the angular dependence of photoelectron intensity were first observed by Fadley and Bergström [475]. They measured the X-ray photoelectron emission of a gold crystal with its surface covered by a thin layer containing carbon. The relative intensities of the surface carbon atoms increased by approximately one order of magnitude at small electron emission angles. A similar effect has also been observed in the study of molybdenum foils covered with cesium monolayers [473].

A second effect is related to the change of orientation of the sample with respect to the source of X rays. Henke [476] has shown that if the gold sample is investigated at low X-ray incidence angles Φ, the intensity of photoelectrons increases, and the depth of X-ray penetration into the sample decreases to approximately 20 Å. It should be noted that the variation of X-ray incidence angle Φ affects the photoelectron intensity to a much smaller extent than the variation of the angle θ. Nevertheless, the effect can also be used in some cases to characterize the sample surface state.

An angular dependence of the fine structure of photoelectron spectra is also possible. This arises from the nature of the geometrical distribution of atoms in the sample. This effect was observed in the photoelectron spectra of single crystals of NaCl [477] and gold [478]. In such cases, electron-diffraction (channeling) processes occur, which result in an oscillatory dependence of the angular distribution on the angle of electron emission from the sample and a preferential emission of photoelectrons along certain directions of the crystal. For simplicity, in the following we will only consider the case of polycrystalline samples with plane surfaces. Fadley *et al.* [479, 480] studied the angular dependence of photoelectrons in terms of a simple model.

The geometrical arrangement of the sample, the X-ray source, and the analyzer exit slit is shown in Figure 172. The beam of X rays of intensity I_0 falls onto the sample at an angle Φ. Inside the sample, due to refraction, it continues at an angle $\Phi' \neq \Phi$, though for X rays with an energy of the order of 1 keV the refraction effects are important only for incidence angles lower than a few degrees. It is also assumed that the electrons emitted from the sample at an angle θ undergo neither reflection at the surface nor diffraction processes. In

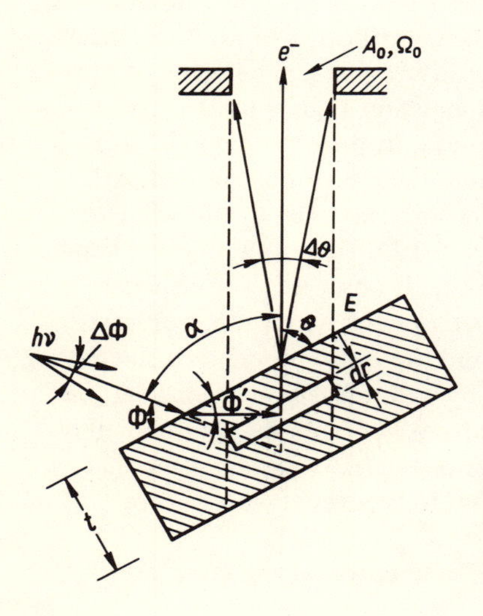

Figure 172. Principle of the experimental arrangement for the study of the angular dependence of the photoelectron current intensity in solid state samples.

most experiments the angle α between the X-ray beam and the beam of photoelectrons is constant, so that a rotation of the sample around an axis situated on the surface and perpendicular to the plane of the figure results in a simultaneous change of the angles θ and Φ. The angle Φ may be treated as a variable expressed by $\Phi = \pi - \alpha - \theta$. Figure 172 also shows the maximum possible uncertainty $\Delta\theta, \Delta\Phi$ in the values of the angles θ and Φ at the given geometry of the experiment. The problem is to determine the number of photoelectrons emitted in one second from the *nl* subshells of the atoms contained in a small volume of thickness *dr* at a distance *r* from the sample surface. The basic assumptions of the model are the following:

1. The sample surface is assumed to be atomically smooth.
2. The sample is amorphous or polycrystalline (in order to justify the neglect of channeling effects).
3. At the sample, the beam of X rays is homogeneous.
4. Reflection and refraction of X rays at the surface is described by the complex dielectric constant $n = 1 - \delta - i\beta$, where, in the X-ray region, δ and β are much smaller than 1 (e.g., 10^{-4}).
5. The X rays and the current of photoelectrons are damped exponentially with the distance traveled inside the sample. Some typical values for the damping lengths are of the order of 1000–10000 Å for X rays and 10–100 Å for photoelectrons.
6. The damping length of X rays is considered to be independent of the incidence angle Φ.
7. The length of the electron mean free path does not depend on the depth from which the electron is emitted or the emission angle θ.
8. It is considered that the effects of electron elastic scattering do not have any influence on the angular dependence of photoelectrons.

Thus, the intensity of the photoelectron current can be written as the product of the flux of X rays at the depth *r*, the number of atoms of the given type inside the unit volume of the sample, the probability of photoelectron emission at angle Ω_0, a factor describing the damping of X rays, the intensity of the electron current, and an instrumental factor:

$$dI = J_0 A_0 \Omega_0 D \rho \frac{d\sigma_{nl}}{d\Omega} (1 - R) \frac{\sin \Phi}{\sin \Phi' \sin \theta} \times$$
$$\times \exp\left[- r \left\{ \frac{1}{\Lambda_x \sin \Phi'} + \frac{1}{\Lambda_e \sin \theta} \right\} \right] dr, \qquad (123)$$

where R is the coefficient of X-ray reflection; ρ is the density of the sample material; $A_0/(\sin \theta)$ is the effective area of the sample; $d\sigma_{nl}/d\Omega$ is the differential cross section for photoionization of electrons from the *nl* subshell; $r/(\sin \Phi')$

and $r/(\sin\theta)$ are the distances traveled by the X rays and the electrons inside the sample; and D is a function describing the detection efficiency of the electron spectrometer at the given photoelectron energy E. By integrating equation (123) over r from 0 to t, where t is the thickness of the sample surface layer, the following expression is obtained for the photoelectron intensity:

$$I_t = I_0(1-R)\sin\Phi\,\frac{1-\exp\left[-t\left\{\dfrac{1}{\Lambda_x\sin\Phi}+\dfrac{1}{\Lambda_e\sin\theta}\right\}\right]}{\sin\Phi'+(\Lambda_e/\Lambda_x)\sin\theta}, \tag{124}$$

where

$$I_0 = J_0 A_0 \Omega_0 D\rho\,\frac{d\sigma_{nl}}{d\Omega}\,\Lambda_e(E).$$

For a sample with infinite thickness,

$$I_\infty = I_0(1-R)\,\frac{\sin\Phi}{\sin\Phi'+(\Lambda_e/\Lambda_x)\sin\theta}\,. \tag{125}$$

Since the angle Φ is much higher than the critical angle, $\Phi_c = (2\delta)^{1/2}$, which in typical cases amounts to several degrees (for gold, for example, $\Phi_c = 2.1°$ when using the $K\alpha_{1,2}$ radiation of aluminum), it follows that the effects related to reflection and refraction can be neglected, so that it can be assumed that $R = 0$, and $\Phi = \Phi'$ [476]. Moreover, since $\Lambda_e/\Lambda_x = 100$, the second term in the denominator in formulas (124) and (125) can be neglected, thus giving:

$$I_t = I_0[1-\exp(-t/\Lambda_e\sin\theta)], \quad I_\infty = I_0. \tag{126}$$

Thus, for a bulk sample the intensity of the photoelectron beam does not change with the angle θ.

Let us now consider the following case. The substrate of infinite thickness is covered with a thin adsorbate layer of thickness t'. The intensity of photoelectrons emitted from the surface layer and from the substrate will then be given by the following formulas:

$$I_{t'} = I_0'\left[1-\exp\left(-\frac{t'}{\Lambda_e'(E')\sin\theta}\right)\right], \tag{127}$$

$$I_{\infty,t'} = I_0\exp\left(-\frac{t'}{\Lambda_e(E)\sin\theta}\right). \tag{128}$$

In general, the kinetic energy E of the electron emitted from the nl subshell is different from the energy E' of the electron emitted from the $n'l'$ subshell.

Therefore, $\Lambda_e'(E) \neq \Lambda_e'(E')$, and the ratio of intensities of photoelectrons emitted from the sample surface layer and from the substrate is given by the following expression:

$$\frac{I_{t'}}{I_{\infty,t'}} = \frac{I_0'}{I_0} \exp\left(\frac{t'}{\Lambda_e(E)\sin\theta}\right)\left[1 - \exp\left(-\frac{t'}{\Lambda_e'(E')\sin\theta}\right)\right], \quad (129)$$

and is represented by a function rapidly increasing with θ at low θ values. If the energies E and E' are close to each other, the values of the electron mean free paths will also be close to each other: $\Lambda_e'(E) \approx \Lambda_e'(E')$.

Within a range of small t' values, there exists a region of θ values for which $t' \ll \Lambda_e'(E)\sin\theta$. In this case

$$I_{t'} \sim \frac{t'}{\sin\theta}; \quad (130)$$

and therefore

$$\frac{I_{t'}}{I_{\infty,t'}} \sim \left[1 + \frac{t'}{\Lambda_e(E')\sin\theta}\right] \frac{t'}{\Lambda_e'(E')\sin\theta},$$

or, to a first approximation,

$$\frac{I_{t'}}{I_{\infty,t'}} \sim \frac{t'}{\Lambda_e'(E')\sin\theta}. \quad (131)$$

An equation having a form intermediate between (129) and (131) can be written as follows:

$$\frac{I_{t'}}{I_{\infty,t'}} \sim \exp\left(\frac{t'}{\Lambda_e'(E)\sin\theta}\right)\left[\frac{t'}{\Lambda_e'(E')\sin\theta}\right].$$

For $\Phi \leqslant \Phi_c$, Φ' is in general different from Φ, and the reflection coefficient R is nonzero. For $\Phi \approx \Phi_c$, $\Phi' \ll \Phi$. Therefore, taking R small for these angles, formula (125) indicates–due to the factor $(\sin\Phi)/(\sin\Phi')$–that the intensity I_∞ for massive samples increases with θ. For gold, using the $K\alpha_{1,2}$ radiation of aluminum, $R = 0.18$ and $\Phi' = 0.46\Phi$. This results in an increase of I_∞ by a factor of approximately 1.2 as compared to the value I_∞ corresponding to $\Phi > \Phi_c$, this result being confirmed experimentally.

Henke [476] has calculated the intensity of a photoelectron beam assuming that the radiation incident on the surface of a sample with dielectric constant n is a plane wave. The result of this calculation can be written as follows:

$$I_\infty = I_0 F(\theta, \alpha, \delta, \beta, \lambda_x, \lambda_e),$$

where F is a universal function describing the deviation of the intensity of the photoelectron beam from the constant value $I_\infty = I_0$, and λ_x is the wavelength of X rays. It is convenient to express F as a function of three dimensionless variables:

$$X = \frac{\Phi}{\Phi_c}, \quad Y = \frac{\beta}{\delta}, \quad Z = \frac{\Lambda_e \sin\beta}{\Lambda_x \Phi_c},$$

which gives the following dependence:

$$I_\infty = I_0 F(X, Y, Z). \tag{132}$$

The function F, for values Y and Z close to those for gold, and for the case of $K\alpha_{1,2}$ radiation of aluminum, is illustrated in Figure 173. It predicts a sharp increase of the intensity I_∞ for angles $\Phi \approx \Phi_c$ ($\theta = 180°\text{-}\alpha\text{-}\Phi_c$). A similar behavior of this function is found for any possible pair of values (Y, Z), although its relative amplitudes are strongly dependent on the values of these variables. Such an increase in the value of the intensity I_∞ was observed in the investigations of Henke [476] and Fadley and Bergström [478]. Knowing the values of δ, β, and Λ_x, the value of z and, consequently, Λ_e can be found from the increase of the intensity I_∞ near the angle Φ_c. Another important effect appearing at low Φ is the decrease of the penetration depth of X rays as a consequence of the small values of the angle Φ' [476]. Thus, for example, for the $K\alpha_{1,2}$ radiation of aluminum, the penetration depth may have a value of the order of 20 Å.

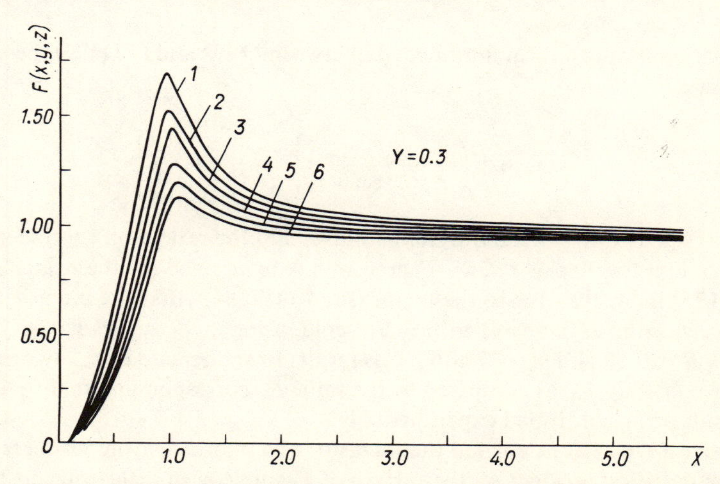

Figure 173. Shape of F(X, Y, Z) function versus the variable X at low incidence angles of the exciting X rays on the sample surface. The curves 1–6 are drawn for values of Z equal to 0.05, 0.10, 0.15, 0.20, 0.25, and 0.30, respectively.

The basic types of dependence of photoelectron intensity on the angle Φ in the above model are shown in Figure 174. Here, the dependence is illustrated for several cases of massive samples covered by a surface layer of thickness t'. The values of the angles α and θ_{min} were chosen arbitrarily, since these values are different in different experiments. The angle $\theta = 180°\text{-}\alpha$ corresponds to the angle $\Phi = 0$. In Figure 174, the angle Φ increases towards the left-hand side. Therefore, all photoelectron intensities for which $\theta < \theta_{min}$ and $\theta > 180°\text{-}\alpha$ must be equal to zero. The value Φ_c is reached at an angle slightly lower than $180°\text{-}\alpha$, and consequently the peaks that describe the particularities of X-ray refraction, analogous to that shown in Figure 173, should also appear in Figure 174 at the same value of the angle. In order to obtain the distribution of photoelectrons for angles of $\theta < \theta_{min}$, the angular distribution for $\theta > \theta_{min}$ must be multiplied by $\sin\theta$ [480]. In Figure 174, *a* represents the angular distribution of photoelectron intensity from a massive sample without an adsorbed surface layer. In this case, the intensity is described by the equations (126) and

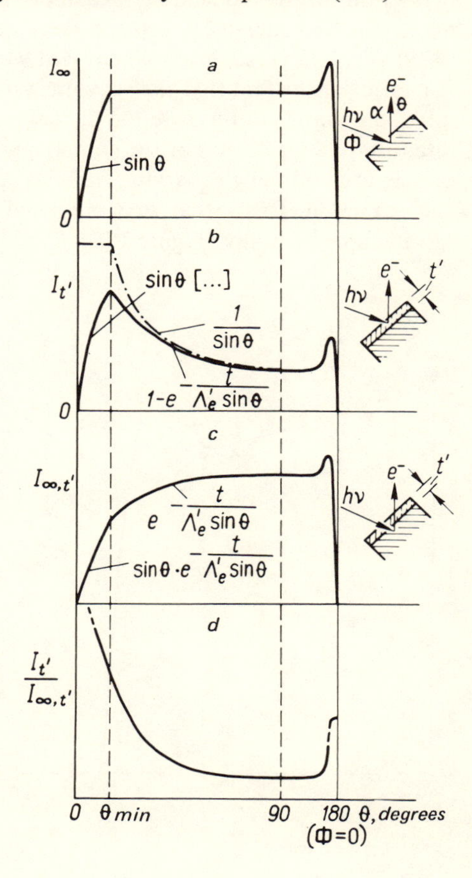

Figure 174. Theoretically predicted angular dependence of the photoelectron current intensity for (a) a bulk sample; (b) a surface layer of thickness t'; and (c) a bulk sample covered by an adsorbed layer of thickness t'. Also shown is the angular dependence of the photoelectron current intensity ratio corresponding to a surface layer of thickness t' and to the bulk (d).

(132) for angles $\theta > \theta_{\min}$. In the same figure, b represents the angular distribution of photoelectron intensity from an adsorbed surface layer of thickness t'. The continuous line represents the correct dependence as given by equation (127), and the dotted line represents the approximate dependence, valid for small values of t' [equation (130)]. In Figure 174, c represents the angular dependence of photoelectrons from the substrate, as given by equation (128). For small values of the thickness t' of the surface layer, the intensity distribution of photoelectrons from the substrate is only slightly different from the continuous line, since $t' \ll \Lambda_e' \sin\theta$. In Figure 174, d represents the ratio of photoelectron intensities from the surface layer and the substrate. Some increase in the intensity is observed for $\theta = 0$ and $\theta = 180°\text{-}\alpha$.

The relations obtained above are useful in the interpretation of experimental results. In spite of the fact that the amount of experimental data on the angular dependence of photoelectrons is sparse, it can be stated that the basic predictions of the model described above is qualitatively consistent with experiment. The first such measurements were performed by Fadley and Bergström [475] on monocrystalline gold covered with a surface layer of carbon formed as a result of the fact that the vacuum system of the spectrometer included oil diffusion pumps. In this case the increase of the carbon 1s photoelectron line intensity relative to that of the gold 4f photoelectron line was of the same order of magnitude for angles varying from 90° to 20°. The peaks due to reflection and refraction effects, that were predicted by equation (132), were in fact observed experimentally (Figure 175).

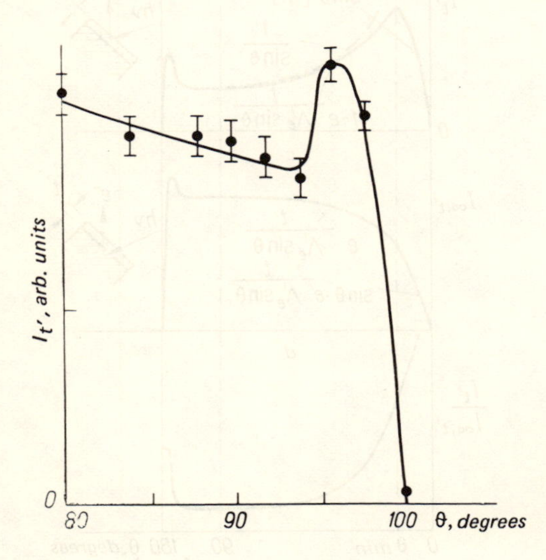

Figure 175. Angular distribution of $3d_{5/2}$ photoelectrons of (a) molybdenum and (b) a cesium monolayer deposited on the surface of a molybdenum sample.

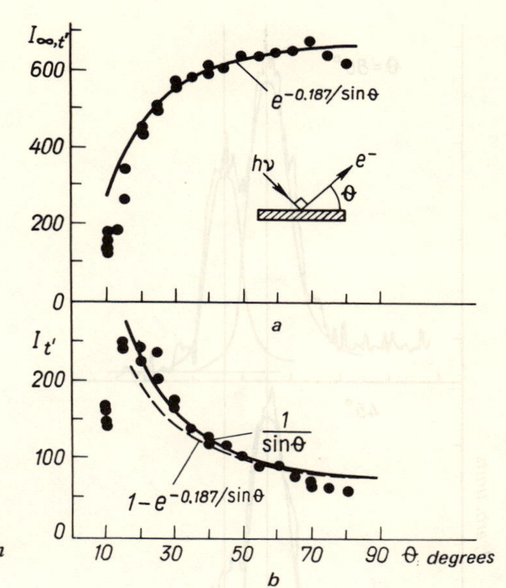

Figure 176. Angular distribution of $3d_{5/2}$ photoelectrons of (a) molybdenum and (b) a cesium monolayer deposited on the surface of a molybdenum sample.

Fraser *et al.* [473] studied the angular dependence of photoelectrons emitted by molybdenum foils polished mechanically, heated in an argon atmosphere, and covered by a monolayer of cesium in a vacuum of 10^{-10} torr. In this case, $\alpha = 90°$, $\Delta\theta$ amounted to several degrees and $\Delta\Phi$ was relatively high. Figure 176 shows the angular distributions of the $3d_{5/2}$ photoelectrons of molybdenum and cesium. Equations (127) and (128) describe the experimental data quite satisfactorily. For small thicknesses, t', the function $1/(\sin\theta)$ can also be used to describe the observed angular distribution. Fadley *et al.* [479] have measured the angular distribution of photoelectrons from aluminum covered by aluminum oxide and carbon. In this experiment, the samples were mechanically and chemically polished aluminum disks. The measurements were performed at $\alpha = 90°$, $\Delta\theta = 3°$, $\Delta\Phi = 10$–$15°$. By using large values of Φ, effects related to reflection and diffraction could be avoided. For this sample, it is easy to distinguish between aluminum and aluminum oxide by studying the angular dependence of the photoelectron intensity of the $2p$ electrons of aluminum (Figure 177). From the results, the increase in the intensity of the signal from the oxide at relatively low values of the angle θ is clearly observed.

In a careful comparison between theory and experiment, it is necessary to take into account the nonuniformity of the X radiation, variations in the thickness of the adsorbed layer over the surface of the sample, as well as effects related to the surface roughness [480].

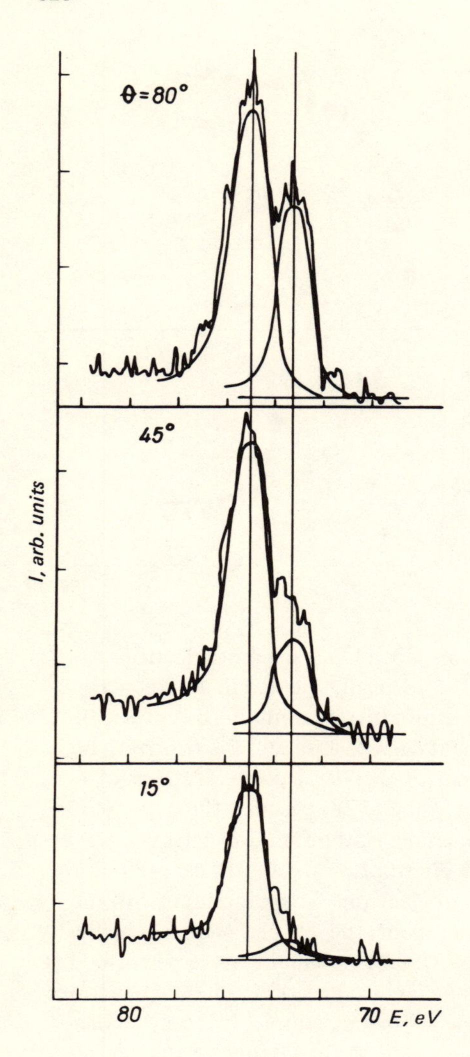

Figure 177. Angular distribution of 2p photoelectrons of aluminum covered by an oxide surface layer.

Use of X-Ray Photoelectron Spectroscopy in the Study of Catalysts

Brinen [481] has discussed the problems related to the use of X-ray photoelectron spectroscopy in the study of catalysis. X-ray photoelectron spectroscopy can be used to study changes in the chemical substances adsorbed on the surface of catalysts, the effects of oxidation and reduction reactions occurring

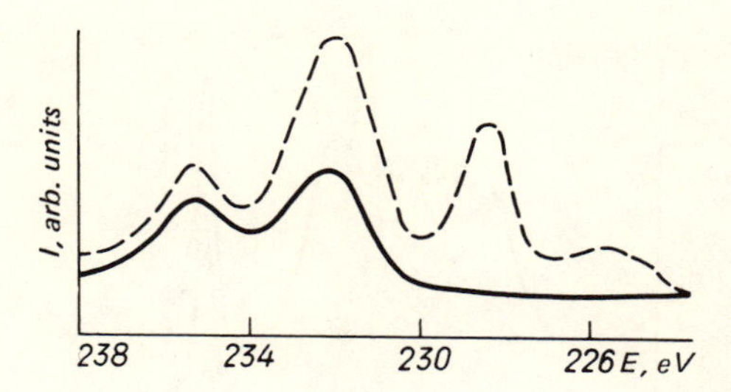

Figure 178. Photoelectron current intensity of the catalyst before (——) and after (– – –) use.

at metal surfaces, and changes at the catalyst surface after catalysis. Figure 178 shows the photoelectron spectra of molybdenum $3d$ electrons in fresh and used catalysts. It is seen that the photoelectron spectra can give information on the changes suffered by the catalyst. In the spectrum of the used catalyst there appear features characteristic of the oxide and sulfide of molybdenum.

The method of X-ray photoelectron spectroscopy also makes it possible to obtain information about differences in the catalytic capacity of various catalyts. Figure 179 shows some results for a catalyst containing rhodium in a carbon matrix. It is seen how the catalytic activity is correlated to changes in the chemical composition of the catalyst's surface layer. The large ratio between the metal oxide and the pure metal can be explained in this case as being due to high catalytic activity. If this ratio is low, the catalyst is bad.

The method of X-ray photoelectron spectroscopy can also be used to determine differences between the surface conductivities of catalysts. The full line in the photoelectron spectrum shown in Figure 180 contains four lines—two sharp and two broad. The dotted line represents the same region of the spectrum, recorded using a source of low-energy electrons to compensate for the charging of the sample surface. The charge is neutralized, and therefore the spectrum exhibits only two lines.

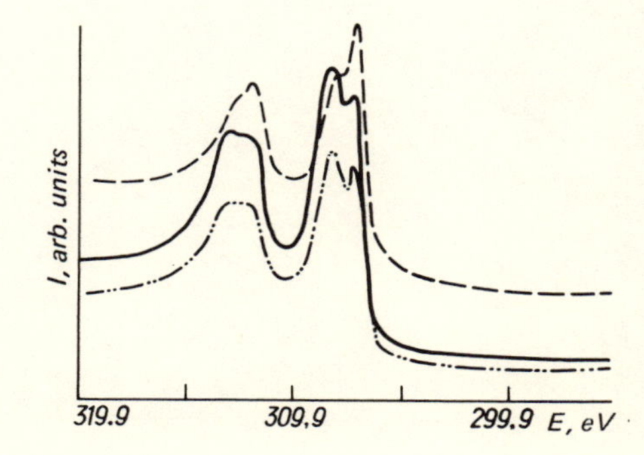

Figure 179. Current intensity of 3d photoelectrons in catalysts containing rhodium: high activity catalyst (— · · —) and (——); low activity catalyst (– – –).

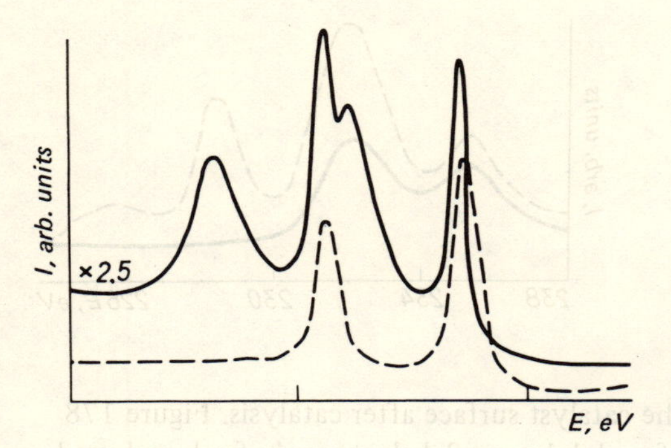

Figure 180. Photoelectron spectra of a catalyst having on its surface some charged and uncharged sites (——), and of a catalyst with an uncharged surface (– – –).

Using X-ray photoelectron spectroscopy, it is possible to detect the formation of new chemical compounds on the surface, the adsorption of ions at the surface, as well as differences between the charge of various parts of the surface. It is also possible to detect undesirable components among the reaction products.

8

Structure of the Photoelectron Spectra of Core Electrons

As we have already pointed out, the X-ray photoelectron spectra of core electrons contain important information about the electron structure of crystals. The chemical shifts of photoelectron lines observed when passing from one compound to another allow the determination of the magnitude of effective atomic charges, as well as the redistribution of valence electrons and the polarity of bonds between atoms of different types. The study of satellite lines observed in X-ray photoelectron spectra provides additional information about the electron structure of crystals and the characteristic features of the photoionization process. If the given compound contains atoms of the same type but in different valence states, the study of the energy position of the satellite lines yields information about the chemical state of a given atom, and their intensity indicates approximately the amount of atoms in different chemical states.

In the simplest case, the sample surface is covered with an oxide layer. The photoelectron spectrum then contains, besides the main line, an additional line situated at a higher binding energy, its position being characteristic for the particular oxide formed on the sample surface. In other cases, such as in catalysis, for example, further additional lines may appear situated at both greater and lower binding energies with respect to the main line. Knowing the position of these lines, it is in general possible to determine which particular compounds have formed on the catalyst surface. More difficult situations arise when the main photoelectron line is not split but only broadened. It is then not possible to determine the type of chemical compound formed on the surface. In some cases, it is important to evaluate the change in the magnitude of spin–orbit splitting, since it provides information about the crystal field.

In a number of cases, however, the occurrence of satellites of the photoelectron lines is not directly related to the change in the chemical state of atoms in crystals. If a whole series of approximately equally spaced satellite lines appears beside the main line, it can be assumed that they are due to the excitation of plasmons by the photoelectrons, on their way out of the sample. In other cases, the satellite lines are due to multiplet effects in atoms having unfilled d and f shells. Satellite lines resulting from multiplet splitting effects should be distinguished from satellites resulting from shake-up or shake-off type processes. Usually, the satellite structure is produced by the simultaneous contribution of complicated multiplet interactions, the effects of shake-up or shake-off processes and plasmon excitations.

In the most careful experiments, it has been observed that the photoelectron lines of many pure metals and alloys are not exactly symmetric. The observed deviation from symmetry is due to many-electron effects arising from the interaction of the hole on the core level with the electrons in the conduction band. The width of photoelectron lines provides information about the magnitude of radiative and nonradiative decay of hole states.

In the present chapter, particular attention will be paid to problems related to the multiplet splitting of photoelectron lines, the satellites of the shake-up and shake-off type, and the asymmetry of photoelectron lines of atoms containing unfilled d and f shells. We will not discuss electron lines resulting from plasma excitations, since at present only a relatively small amount of information on this problem has been obtained using the method of X-ray photoelectron spectroscopy, and this concerns only a restricted number of samples.

Multiplet Splitting of Photoelectron Lines

In Chapter 2, we reviewed basic theoretical ideas about the multiplet splitting of photoelectron lines. Here, we will discuss in more detail the experimental results and their agreement with theory.

The multiplet splitting of s states has been studied most extensively for the $3s$ states of transition metals, and for the $4s$ and $5s$ states of rare-earth metals. Investigation of the multiplet splittings of $2s$ states has been made only in a few cases.

The multiplet splitting of s states in transition metals has been studied by Fadley *et al.* [482, 483], Hüfner and Wertheim [484], Carver *et al.* [485, 487], Kowalczyk *et al.* [486] and McFeely *et al.* [488]. The experimental results obtained by these authors are given in Table 50. For the transition metals of the first transition period, the experimental values of the splitting are twice as large as the values calculated by formulas (35) and (36). This discrepancy can be explained by the fact that the d electrons in the compounds of transition metals are less localized because of their participation in the formation of chemical bonds. This results in a decrease of the overlap integrals (36) of s and d electrons. The action of the crystal field on the ligand part causes a redistribution

TABLE 50. Multiplet Splitting of 2s and 3s States of Transition Metals and Their Compounds

Compound	$\Delta [E_B(2s)]$	$\Delta [E_B(3s)]$
CrF_3	–	4.2 [485]
$CrCl_3$	–	3.9 [485]
Cr_2O_3	–	4.1 [485]
Cr_2S_3	–	3.2 [485]
MnF_3	–	5.6 [487]
MnF_2	5.9 [484]	6.0 [486]; 6.5 [484]
MnO	5.6 [484]	5.7 [482]; 6.1 [484]
MnS	–	5.4 [484]
α – Mn	–	4.1 [488]
FeF_3	–	7.0 [483]
FeF_2	4.3 [484]	5.9 [484]
Fe (metallic)	–	3.5 [484]
CoF_3	–	5.3 [485]
CoF_2	–	5.1 [484]
CO (metallic)	–	2.1 [484]
NiF_2	–	3.1 [484]
Ni (metallic)	–	1.8 [484]

of 3*d* electrons over the energy states, and this may lead to a decrease of the number of unpaired electrons. It is also important to take into account electron correlations inside the atomic shell. Inclusion of these correlations results in an additional decrease of the magnitude of the splitting. The effect is more pronounced for states with lower spin values. By taking into account the electron correlations with the method of configuration interaction, for example, it is possible to obtain a good agreement between the calculated values of splitting energies and the experimental data.

It should be noted that the behavior of the intensity ratio of the photoelectron peaks, corresponding to spin values $S + \frac{1}{2}$ and $S - \frac{1}{2}$, is different from the simple one predicted by formulas (34) and (38). Thus, for the compounds MnF_2 and MnO this ratio is 2.0 : 1.0 [483], which is different from the value 1.4 : 1.0 obtained on the basis of the one-electron model of atomic transitions.

As Table 50 shows, when the number of unpaired 3*d* electrons increases, the magnitude of the splitting generally increases. It also depends on the nature of the ligand: the greater the difference in electronegativity, the larger the magnitude of the splitting. As an illustration, Figure 181 shows the photoelectron spectra of the 2*s* and 3*s* electrons of manganese in the compound MnF_2.

McFeely *et al.* [488] studied the multiplet splitting of 3*s* photoelectron lines of α-manganese in the paramagnetic state, using an HP-5950A electron spectrometer with the $K\alpha_{1,2}$ line of aluminum as excitation radiation. The experiment was performed with a vacuum of 6×10^{-11} torr. Figure 182 shows

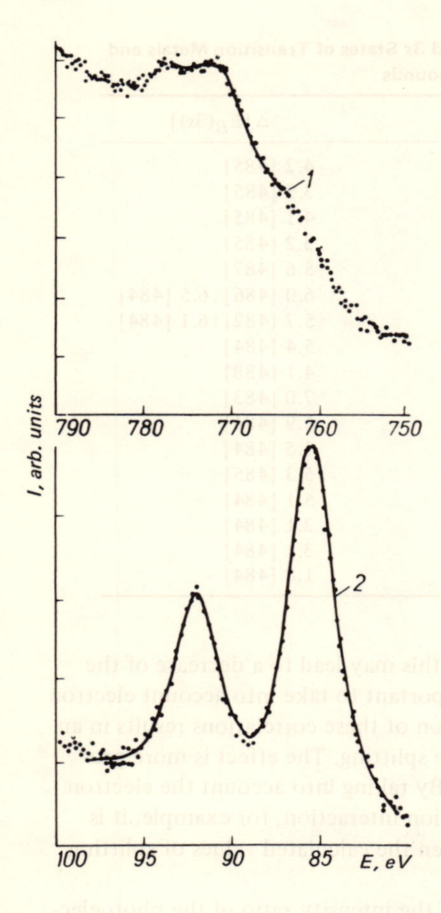

Figure 181. (1) 2s and (2) 3s photoelectron spectra of manganese in the compound MnF_2.

the results obtained by these authors on the multiplet splitting of 3*s* states in α-manganese together with the results of Fadley *et al.* [483] on MnO_2, Carver *et al.* [487] on MnF_3, and Kowalczyk *et al.* [486] on MnF_2. As the figure shows, the experimental points lie on a straight line. Assuming that in these compounds the exchange integral (36) is constant, and making use of formula (35), one obtains

$$\Delta\,[E_B\,(3s)] \cong \text{const}\,(2S + 1).$$

However, the experiment indicates that the following relation is valid:

$$\Delta\,[E_B\,(3s)] = 1.0\,(2S + 1) + 0.6.$$

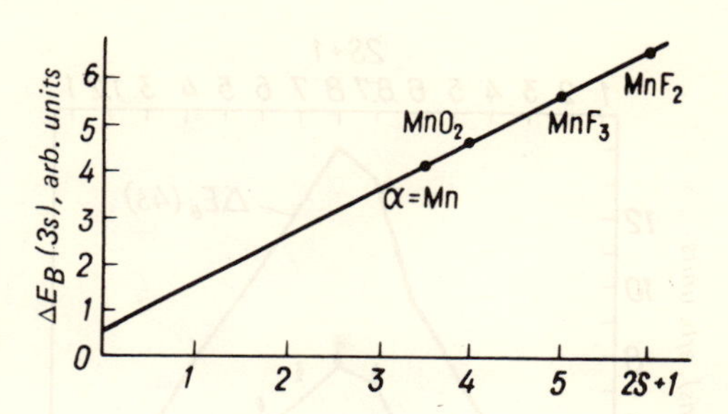

Figure 182. Multiplet splitting of 3s photoelectron lines in α-manganese and in some manganese compounds.

The occurrence of a nonzero term on the right-hand side of this expression reveals the approximate character of the model used. The exchange integral is probably variable. Moreover, it is evident that the $3s$–$4s$ exchange integral should also be taken into account. The experimental value of the splitting $\Delta [E_B(3s)] = 4.08$ eV for α-manganese corresponds to the value $S = 1.25$, which is consistent with the magnitude of the magnetic moment $\mu = 2.5\ \mu_B$.

Experimental results regarding the multiplet splitting of $4s$ and $5s$ states in rare-earth metals [489–492] are given in Table 51. As this table shows, the magnitude of the splitting for some rare-earth metals is close to values obtained for some trivalent compounds of the transition metals with fluorine (trifluorates) [492]. Such a close agreement demonstrates that the $4f$ electrons in rare-earth metals are not very sensitive to changes in the distribution of the valence electrons. McFeely *et al.* [491] have estimated the magnitude of multiplet splittings with formula (35) using atomic functions for the s and f orbitals. As Figure 183

TABLE 51. Multiplet Splitting of 4s and 5s States of Rare-Earth Metals and of Trivalent Compounds of Rare-Earth Metals with Fluorine

Element	$2S + 1$	$\Delta [E_B(4s)]$	$\Delta [E_B(4s)]^a$	$\Delta [E_B(5s)]$	$\Delta [E_B(5s)]$
Ce	2	1.4	–	1.0	–
Pr	3	2.0	–	1.4	1.2
Nd	4	2.7	–	1.6	1.7
Pm	5	–	–	–	–
Sm	6	5.4	–	2.9	2.8
Eu	8.7	7.4	–	3.8	–
Gd	8	7.8	8.2	3.6	3.7
Tb	1	–	7.5	3.2	3.4
Dy	6	–	6.4	2.8	2.7
Ho	5	–	5.6	2.4	–
Er	4	–	4.3	–	–
Tm	3	–	3.4	–	–

[a]Values of multiplet splitting energies for trifluorides are taken from Ref. [492].

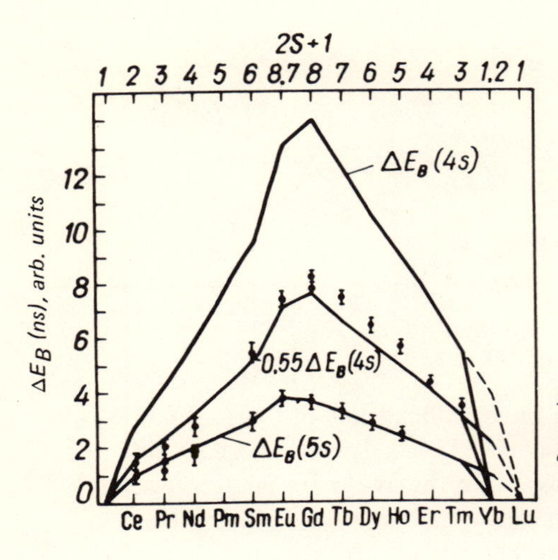

Figure 183. Experimental and theoretical values of the multiplet splitting of the 4s and 5s photoelectron lines of rare-earth metals: (•) metals and trifluorides [491, 492], (–) theoretical values [491].

shows, the calculated values of the multiplet splitting of the 5*s* states are rather close to the experimental ones, but for the 4*s* states, the theoretical values are about 80% greater than the experimental.

Satellite Structure of 2*p* Photoelectron Lines

The satellites of 2*p* photoelectron lines that appear in compounds of transition metals have been studied by Rosencwaig *et al.* [493], Wallbank *et al.* [494], Carlson *et al.* [495], and Novakov [496]. Completely different interpretations of the nature of these satellites have been proposed. Are they the result of multiplet splitting arising from the interaction between, on the one hand, the spin and orbital moments of the unfilled *p* shell (as a result of the removal of one electron from this shell) with, on the other, the spin and orbital momenta of the electrons in the unfilled 3*d* valence shells? Alternatively, are they the result of processes of the shake-up and shake-off type? In order to elucidate this question, let us first consider some existing experimental facts.

1. The satellites exist only in the spectra of the transition metal compounds. They are absent in the spectra of the pure metals.
2. The satellites of the 3*p* photoelectron lines are not observed with an intensity comparable to that of the satellites of the 2*p* lines.
3. For a given metal, the energy distance between the satellite and the main photoelectron line increases as the ligand becomes less electronegative.
4. Only one of the 2*p* photoelectron lines has a satellite line. There exist compounds for which each main line has two or more satellite lines.
5. The satellites are most intense in paramagnetic compounds.

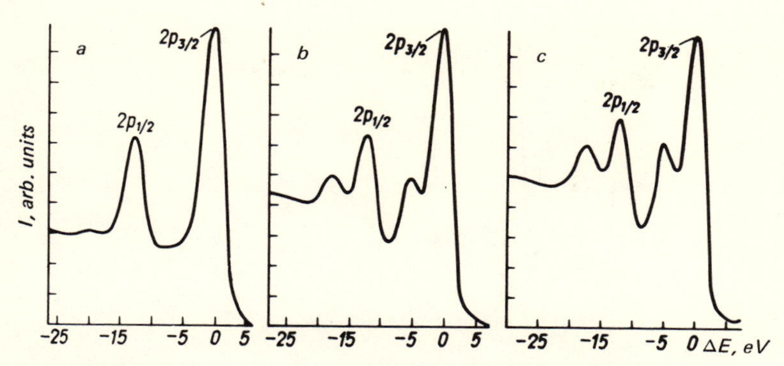

Figure 184. 2p photoelectron spectra of manganese in the compounds (a) MnF_2; (b) $MnCl_2$; and (c) $MnBr_2$.

6. The satellite structure is absent in compounds in which the $3d$ shell is completely occupied.
7. Satellites exist only in compounds of transition metals, in which the $3d$ shell is partially occupied.

To give an idea of the shape of a satellite line and its energy position, Figure 184 shows the X-ray photoelectron spectra of $2p$ electrons in the compounds MnF_2, $MnCl_2$, and $MnBr_2$, and Figure 185 shows the $2p$ spectra for Sc(acetylacetonate)$_3$, ScF_3, and TiF_4 in which the $3d$ shell according to the chemical formula is partially empty.

Yin *et al.* [497] have stated that no satellites exist in diamagnetic compounds. Carlson *et al.* [495] have shown, however, that diamagnetic compounds

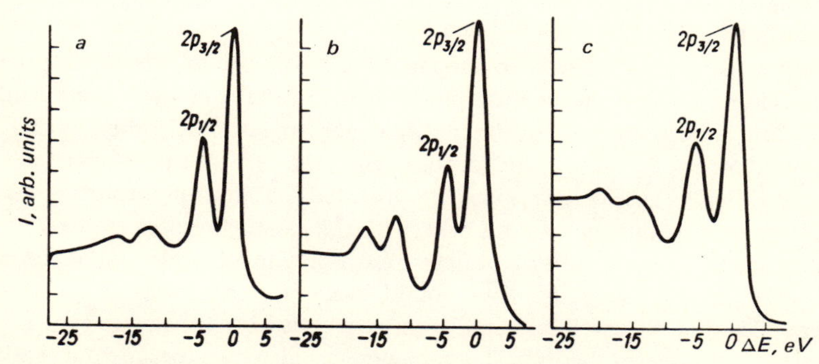

Figure 185. 2p photoelectron spectra of scandium and titanium in compounds. (a) Sc(acetilacetonate)$_3$; (b) ScF_3; (c) TiF_4.

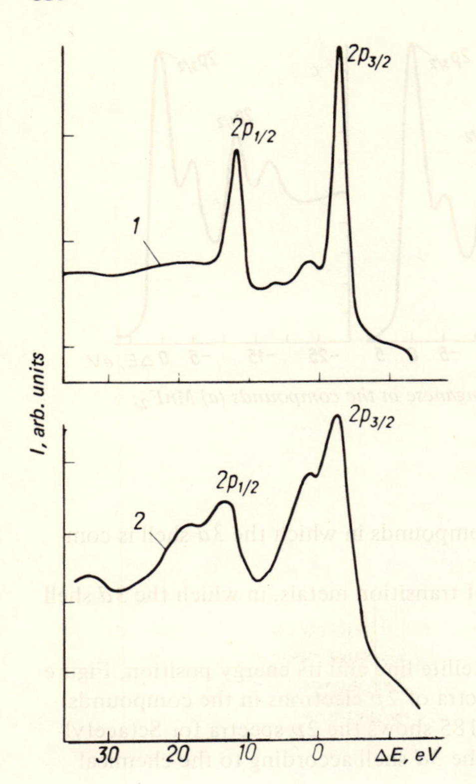

Figure 186. 2p photoelectron spectra of cobalt in the compounds. (1) Co(acetilacetonate)$_3$; (2) Co(acetilacetonate)$_2$.

can also have satellites, but that their intensity is much lower than those of paramagnetic compounds. For example, the paramagnetic compound Co(acetylacetonate)$_2$ is characterized by intense satellite peaks, while in the diamagnetic compound Co(acetylacetonate)$_3$ satellites exist, but their intensity is considerably lower (Figure 186).

Gupta and Sen [498] have investigated the possibility of interpreting these satellites as resulting from the multiplet splitting of the core levels. Very careful calculations were performed of the multiplet splitting in the $2p$ shell of the compound MnF_2, by taking into account spin–orbit effects and effects due to the crystal field. These calculations demonstrated that multiplet splitting plays an important role, but that, in this case, with the existing resolving power of electron spectrometers, multiplet effects result only in a broadening of the core $2p$ lines and an increase of the background level between them.

The idea that the satellites result from processes of the shake-up type is supported not only by theoretical arguments, but also by experimental results. The excitations of the shake-up type does not depend on the core atomic state in which the vacancy exists [499]. Also the intensity of the corresponding photoelectron lines exhibits an insignificant dependence on the state of the core hole.

Carlson *et al.* [495] have studied the satellites appearing in the 1*s* electron spectra of $FeCl_3$, $FeBr_3$, $K_3Fe(CN)_6$, and $K_4Fe(CN)_6$. In this case, the photoelectron lines had a lower resolution than the lines generated by the 2*p* shell using the $K\alpha_{1,2}$ radiation of aluminum or manganese as excitation source. This is explained by the fact that the $K\alpha_{1,2}$ line of copper used by Carlson and coworkers as radiation source is broader. Moreover, the natural width of the iron 1*s* level is greater than that of the 2*p* level. High-intensity satellites were observed in the spectra of the compounds $FeCl_3$ and $FeBr_3$. They are situated at a distance of about 4.7 eV from the main line, and their intensity amounts to 0.7 of the intensity of the main line in the case of $FeBr_3$, and 0.5 in the case of $FeCl_3$. This is consistent with the satellite structure observed in the 2*p* photoelectron spectra of iron in the same compounds. In the cyanide compounds, no satellites were observed for the 1*s* electrons. In these compounds, the 2*p* photoelectron lines of iron do not have satellites. The theoretical value [499] for the magnitude of the multiplet splitting of iron 1*s* photoelectron lines is very small, of the order of 0.1 eV. Thus, it is possible to conclude, for the compounds $FeCl_3$ and $FeBr_3$ at least, that the satellite structure in the 2*p* photoelectron spectra is a result of shake-up processes and not of multiplet splitting. It is possible that the satellite structure in the 2*p* photoelectron spectra resulting from multiplet splitting cannot be observed in this case because the spectrum of multiplet splitting is rather complicated and does not contain peaks clearly separated in intensity.

Rosencwaig *et al.* [493] have interpreted the satellites in the 2*p* photoelectron spectra as being the result of the excitation of a second electron from the 3*d* states of the valence band to the 4*s* states. As a support of this point of view, the authors compared the values of the energy distances between the satellites and the main photoelectron lines with the energy values of the optical transitions of electrons from the 3*d* to the 4*s* states of the free ions. The same mechanism of satellite formation has also been discussed by Demekhin *et al.* [500] within the framework of the many-electron theory. It is impossible, however, to understand why satellites are absent in compounds in which the 3*d* shell is completely filled. It is also difficult to understand the existence of a strong dependence of the intensity of the satellites on the type of ligand. In spite of the fact that there exists some correlation between the values of the corresponding energies, it is at present accepted that it is not this mechanism that is responsible for the occurrence of satellites in the spectra of 2*p* photoelectrons.

In our opinion, a more justified interpretation is that which states that the satellites related to transitions from the 2*p* shell result from the transition of electrons from the ligand to the 3*d* states. These transitions will be of the mono-

TABLE 52. Satellite Structure in the 2*p* Photoelectron Spectra of Transition Metals in Compounds

Compound	Number of *d* electrons	$\Delta[E_B(2p)]$		Compound	Number of *d* electrons	$\Delta[E_B(2p)]$	
		$2p_{1/2}$	$2p_{3/2}$			$2p_{1/2}$	$2p_{3/2}$
ScF_3	0	12.2 [495]		$FeCl_2$	6	5.7 [494]	4.3
Sc_2O_3	0	10.8 [494]	11.3	$FeBr_2$	6	4.0 [495]	
TiF_3	1	13.0 [495]		$FeCl_3$	5	4.4 [495]	
TiO_2	0	12.6 [494]	13.3	$FeBr_3$	5	4.7 [495]	
TiF_4	0	14.7 [495]		Fe_2O_3	5	8.2 [494]	8.0
VF_3	2	12.0 [494]		$CoCl_2$	7	5.1 [495]	
CrF_3	3	11.8 [494]		$CoBr_2$	7	5.6 [495]	
$CrCl_3$	3	10.3 [495]		CoI_2	7	3.9 [495]	
$CrBr_3$	3	11.0 [495]		NiF_2	8	6.1 [495]	
CrF_2	4	13 [494]		$NiCl_2$	8	5.6 [495]	
MnF_2	5	6.5 [495]		$NiBr_2$	8	5.4 [495]	
$MnCl_2$	5	5.1 [495]		NiI_2	8	6.1 [495]	
FeF_2	6	5.6 [494]	5.4				

pole type, conserving the parity and the symmetry type. For these transitions to be possible, it is necessary that the 3*d* shell not be completely filled. Such a model was proposed by Carlson *et al.* [495]. According to this model diamagnetic compounds can also have a satellite structure related to transitions from the 2*p* shell.

Larson and Connolly [501] have used the X_α method to perform calculations for a number of cluster groupings: TiF_6^{3-}, CrF_6^{3-}, $CrBr_6^{3-}$, MnF_6^{4-}, $MnCl_6^-$, and others. It was shown that the energies of electron transitions inside the limits of metal ions are less consistent with experiment then the energies of electron transitions from the ligands to the metal. The research in this field should, however, be extended, and in particular calculations should be performed, not only of excitation energies, but also of transition probabilities.

The energy values characteristic for the satellite structures of transition metal compounds, related to transitions from the 2*p* shell, are shown in Table 52.

Width and Asymmetry of Core-Level Photoelectron Lines

The width of hole states is determined by the sum of the width of the radiative transition, the Auger transition and the Coster–Kronig transition:

$$\Gamma = \Gamma_R + \Gamma_A + \Gamma_{CK}.$$

Thus, for the K level, the radiative width of the level is proportional to Z^4, and the Auger width is proportional to Z. The radiative and nonradiative widths of the K level become equal at $Z \approx 29$–30. For elements with $Z < 17$, the width of the K level is in general determined by the Auger transitions, mainly KLL transitions. The main contribution to the width of the L_1 level comes from the Coster–Kronig transitions. For the elements with atomic number $20 \geqslant Z \geqslant 13$, the width Γ_{CK} is more than 20 times higher than the field Γ_R. For the elements with $Z \leqslant 32$, the width of the L_2 and L_3 levels is almost completely determined by the Auger transitions [502] and is practically the same. A more detailed discussion of the contributions of the radiative and nonradiative transitions to the width of hole states may be found in the review article of Bambynen *et al.* [503].

The processes that determine the linewidth in compounds can be more complicated than in atoms. Friedman *et al.* [504] and Shaw and Thomas [505] have shown that the lifetime of core states in solids depends on the type of chemical compound in which the given atom is included. This effect has been observed in the study of the X-ray photoelectron spectra of a group of compounds in which the metal atom was characterized by a higher degree of oxidation, and the hole state by a lower decay width. The effect is due to the reduced number of valence electrons that can fill the hole. In several cases, however, the opposite situation has been observed. Citrin [506] has shown that when the width of the hole states in chemical compounds is to be determined, it is necessary to take into account not only the intraatomic processes, but also the interatomic ones. The probability of Auger processes can be expressed in terms of the matrix element

$$\left\langle \varphi_A(1)\,\Psi(2) \left| \frac{1}{|r_1 - r_2|} \right| \Phi_{A,B}(1)\,\Phi'_{A,B}(2) \right\rangle,$$

where $\varphi_A(1)$ is the initial hole state of atom A; $\Psi(2)$ is the state of Auger electrons (in the continuum); and $\Phi(1)$ and $\Phi'(2)$ are the initial bound states of the electrons that fill the hole and that are emitted out of the crystal, respectively. Significant differences in the width of hole states can appear in cases when these wave functions are localized at atom A (the atom with the hole) or at atom B (the nearest ligand to the atom with hole). Therefore, the following types of Auger processes are possible: AA, BB, AB, and BA. The first letter labels the atom that releases the electron that fills the hole, while the second letter refers to the atom from which the Auger electron is emitted. The probability that the process will go through one or the other of the possible mechanisms will be determined by the overlap of the wave function $\Psi_A(1)$ with $\Phi_{A,B}(1)$ and $\Psi(2)$ with $\Phi'_{A,B}(2)$. The mechanism AA is the most direct. It determines the lifetime of the hole in core atomic levels in pure metals and monatomic gases. When the number of electrons that can fill the hole increases, the lifetime decreases.

The Auger process goes through the BB mechanism when the energetically highest electron of atom A takes part in the chemical binding. A typical example is that of the $2p$ states of sodium in NaOH. In metallic sodium, the $2p$ hole is filled by the $3s$ electrons. In the compound NaOH, the mechanism *AA* for the Auger processes is not possible, since the $3s$ electron of sodium is involved in the hydroxyl group. It is, therefore, to be expected that the $2p$ line of sodium in NaOH will be narrower. It has been found, however, that it is much broader than in pure sodium. This phenomenon can be directly related to the fact that a larger number of valence electrons in the hydroxyl group can fill the $2p$ hole states of sodium.

Citrin [506] has deviced a parameter determined by the density of valence electrons of the ligand and by the volume of the ligand inside a sphere around the central atom. This parameter has allowed him to explain the level widths observed for a complete group of compounds, namely, the oxides and fluorides of sodium, magnesium, and aluminum, for which the hypothesis of the existence of ionic spheres is valid.

The mechanism *AB* is the most interesting, since in this case both intra-atomic and interatomic effects occur. This is evident, for example, in the case of the widths of $2s$ lines in the compounds mentioned above, for which Coster-Kronig (intraatomic) transitions are very efficient. It can be assumed that changes of the ligand do not result in significant changes in the probability of Coster–Kronig transitions. However, the lifetime of the hole state in the $2s$ level changes drastically as a function of the environment. Interatomic processes of the *AB* type are characteristic for copper, Cu_2O, and CuO crystals. In the compounds V_2O_3, VO_2, and V_2O_5, the Auger processes also go through the *AB* mechanism, but the intraatomic process dominates (we have considered here the $2p$ states).

In conclusion, the lifetime of hole states is determined by contributions from intra- and interatomic processes, and in some cases, the interatomic processes are dominant.

The shape of the photoelectron lines of core electrons has been studied theoretically by Doniach and Sunjič [507], who investigated the influence of the interaction between the created hole and valence electrons in metals on the shape of core-electron photoelectron lines. They made use of the results of Nozières and De Dominici [508] regarding the influence of multielectron effects on the shape of X-ray emission spectra. Doniach and Sunjič [507] showed that the intensity of the photoelectron current can be expressed as

$$I(\varepsilon) = \frac{\Gamma(1-\alpha)}{(\varepsilon^2+\gamma^2)^{(1-\alpha)/2}} \cos\left[{}^1/_2\pi\alpha + \theta(\varepsilon)\right], \tag{133}$$

where Γ is the Γ function, γ is the natural width of the hole state, and ϵ is the energy measured with respect to the Fermi level. For $\alpha < 1$, formula (133) predicts an asymmetric distribution for the intensity of photoelectron lines of core electrons. A larger tail is observed on the high-binding-energy side of the line. Formula (133) can be used for a rough determination of the magnitude of α. The precision of this determination will, of course, depend on how precisely the spectrometer resolution function is known.

The theoretical conclusions of Doniach and Sunjič [507] were obtained on the basis of a study of simple metals, for which it can be assumed that the electrons in the valence bands behave as nearly free electrons. It is, however, difficult to test their result experimentally because the simple metals have a high reactivity. This reactivity results in the formation of oxide or hydroxide layers on their surface, which also cause a broadening on the high-energy side of the photoelectron lines of the core electrons. Identification of these effects is extremely difficult. For this reason Hüfner and Wertheim [509] studied some less reactive metals, namely, silver, rhodium, palladium, iridium, platinum, and gold, although the theoretical analysis of the asymmetry of photoelectron lines is in these cases more difficult. As core lines, the $3d$ and $4f$ lines were chosen. For all of the metals investigated, except gold, an asymmetry of the photoelectron lines was observed. However, attempts to describe the asymmetry directly by using the functions entering formula (133) have failed as a result of the instrumental broadening, which in gold and silver is greater than the natural width of the hole state. The function of instrumental resolution was determined from the smearing of the Fermi edge in silver and copper. It was found that this function has a Gaussian shape with a half-width of 0.55 ± 0.05 eV. Using this function, the natural shape of the line was determined by choosing parameters to best fit formula (133). The parameters obtained are given in Table 53. Figure 187 shows the

TABLE 53. Binding Energy and Parameters of the Core Photoelectron Lines of Several Metals

Metal	Level	Energy	2γ	α
Rh	$3d_{3/2}$	311.9	0.80	0.20
	$3d_{5/2}$	307.2	0.60	0.10
Pd	$3d_{3/2}$	340.6	0.66	0.25
	$3d_{5/2}$	335.2	0.74	0.11
Ag	$3d_{3/2}$	374.3	0.40	0.07
	$3d_{5/2}$	368.2	0.38	0.03
Ir	$4f_{5/2}$	63.7	0.46	0.13
	$4f_{7/2}$	60.7	0.40	0.12
Pt	$4f_{5/2}$	74.7	0.53	0.19
	$4f_{7/2}$	71.3	0.50	0.19
Au	$4f_{5/2}$	87.7	0.43	0.00
	$4f_{7/2}$	84.0	0.40	0.02

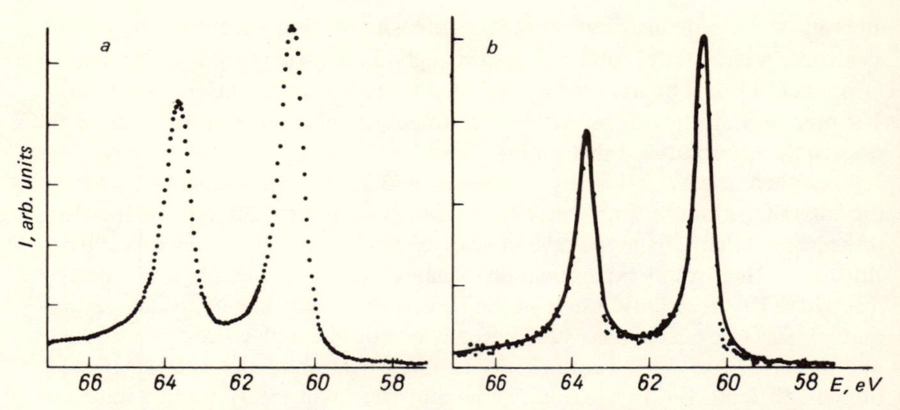

Figure 187. 4f photoelectron spectra of iridium (a) before and (b) after accounting for experimental broadening.

photoelectron spectra of $4f$ electrons in iridium before and after their theoretical processing.

The parameter α can be expressed as follows:

$$\alpha = 2 \sum (2l + 1) \left(\frac{\delta_l}{\pi} \right)^2,$$

where δ_l is the phase shift at the Fermi energy of the partial wave with angular momentum l. It also determines the details of the structure of the X-ray emission spectra in the vicinity of the Fermi level:

$$J \sim \frac{1}{\zeta_0} \frac{1}{\varepsilon^{1-\alpha}},$$

where ζ_0 is a parameter having a value close to the width of the valence band. Therefore, in future studies, it will be interesting to compare the magnitudes of α obtained by X-ray emission spectroscopy and X-ray photoelectron spectroscopy.

Shevchik [510] has studied the influence of the density of electronic states in alloys on the shape of the core-level photoelectron lines. The investigated alloys were prepared from metals with a low density of electron states at the Fermi level combined with metals with high density of states. The photoelectron spectra were measured with an HP-5950A electron spectrometer. The $K\alpha_{1,2}$ line of aluminum was used as source of photons. It was observed that the asymmetry of the photoelectron lines of the $4f$ core electrons of platinum and of the $2p$ electrons of nickel disappears when these metals are dissolved in cadmium. In contrast, asymmetry of the $4f$ lines of platinum is preserved when platinum is dissolved in gold. When gold and cadmium are dissolved in platinum, their core lines remain symmetric, as in the pure metals. If the atoms of platinum and nickel are dissolved in cadmium, their local density of states has a maximum not situated at the Fermi level, which is not the case in the pure metals. Therefore, the number of electron–hole pairs that can be excited when a hole is created in a core level is strongly lowered, and consequently the asymmetry of the core levels in this case disappears. The $4d$ levels of cadmium are energetically rather far away from the valence states of platinum and nickel, and therefore they do not interact strongly with each other. The d levels of nickel and platinum form virtual bound states in the s,p valence band of cadmium. The asymmetry of the core lines of platinum in gold does not disappear, although the density of states at the Fermi level is significantly lower than in pure platinum. Since the d level of gold is energetically situated close to the d level of platinum, it follows that the d states of platinum do not move out of the Fermi level, and form virtual states in the s,p band of gold.

When cadmium and gold are dissolved in platinum and nickel, in spite of the fact that the total density of states at the Fermi level is high, the local density of states of cadmium and gold remains low. In this case, the hole states of the dissolved atoms do not cause any asymmetry of the electron lines. As Table 53 shows, for pure metals, the asymmetry is higher for those metals that have the highest density of electron states at the Fermi level.

In alloys, the asymmetry of core states is determined by the local density of electron states of the components.

References

1. K. Siegbahn, C. Nordling, A. Falman, R. Nordberg, K. Hamrin, J. Hedman, G. Johansson, T. Bergmark, S. Karlsson, J. Lindberg, and B. Lindberg, *ESCA, Atomic, Molecular and Solid State Structure Studied by Means of Electron Spectroscopy*, Nova Acta Socientatis Scientarum Upsaliensis, Ser. IV, Vol. 20, Uppsala, 1967.
2. K. Siegbahn, C. Nordling, G. Johansson, J. Hedman, P. F. Hedén, K. Hamrin, U. Gelius, T. Bergmark, L. O. Werme, R. Manne and Y. Baer, *ESCA Applied to Free Molecules*, North-Holland, Amsterdam, 1971.
3. V. I. Nefedov, *Primenenie Rentgeno-Elektronnoi Spektroskopii v Khimii*, Izd. VINITI Moskva, 1973.
4. V. V. Nemoshkalenko, *Rentgenovskaya Emissionnaya Spektroskopiya Metallov i Splavov*, Naukova Dumka, Kiev, 1972.
5. V. V. Nemoshkalenko and V. G. Aleshin, *Teoreticheskie Osnovy Rentgenovskoi Emissionnoi Spektroskopii*, Naukova Dumka, Kiev, 1974.
6. U. Gelius, E. Basilier, S. Svensson, T. Bergmark, and K. Siegbahn, *J. Electron Spectrosc. Relat. Phenom.* **2**(5), 405 (1974).
7. B. Wannberg, U. Gelius, and K. Siegbahn, Design Principles in Electron Spectroscopy, Uppsala University, UUIP-818, April 1973.
8. U. Gelius, H. Fellner–Feldegg, B. Wannberg, A. G. Nilsson, E. Basilier, and K. Siegbahn, New Developments in ESCA Instrumentation, Uppsala University UUIP-855, April 1974.
9. C. A. Lucchesi and J. E. Lester, *J. Chem. Educ.* **50**, 205 (1973).
10. C. A. Lucchesi and J. E. Lester, *J. Chem. Educ. A* **50**, 269 (1973).
11. G. Johansson, J. Hedman, A. Berntsson, M. Klasson, and R. Nilsson, *J. Electron Spectrosc. Relat. Phenom.* **2**(4) 295 (1973).
12. D. J. Khatowich, J. Hudis, M. L. Perlman, and R. C. Ragaini, *J. Appl. Phys.* **42**(12), 4883 (1971).
13. W. Bremser and F. Linneman, *Chem. Ztg.* **95**, 1011 (1971).
14. M. F. Ebel and H. Ebel, *J. Electron Spectrosc. Relat. Phenom.* **3**(3), 169 (1974).
15. J. Hedman, Y. Baer, A. Berndtsson, M. Klasson, G. Leonhardt, R. Nilsson, and C. Nordling, *J. Electron Spectrosc. Relat. Phenom.* **1**(1) 101. (1972).
16. L. Ley, R. A. Pollack, F. R. McFeely, S. P. Kowalczyk, and D. A. Shirley, *Phys. Rev. B* **9**(2), 600 (1974).
17. L. F. Wagner, and W. E. Spicer, *Phys. Rev. Lett.* 1972, **28**(21), 1381 (1972).
18. D. E. Eastman, and W. D. Grobman, *Phys. Rev. Lett.* **28**(21), 1378 (1972).
19. C. J. Veseley, R. L. Hengehold, and D. W. Langer, *Phys. Rev. B* **5**, 2296 (1972).
20. D. Betteridge, J. C. Carver, and D. M. Hercules, *J. Electron Spectrosc. Relat. Phenom.* **2**(4), 327 (1973).

21. D. R. Urch and M. Webber, *J. Electron Spectrosc. Relat. Phenom.* **5**, 791 (1974).
22. I. Lindau and W. E. Spicer, *J. Electron Spectrosc. Relat. Phenom.* **3**(5), 409 (1974).
23. R. G. Steinhardt, J. Hudis, and M. L. Perlman, *Electron Spectroscopy*, D. A. Shirley, ed., North Holland, Amsterdam-London, 1972, p. 557.
24. T. A. Carlson and G. E. McGuire, *J. Electron Spectrosc. Relat. Phenom.* **1**(2), 161 (1973).
25. M. Klasson, J. Hedman, A. Berndtsson, R. Nilsson, C. Nordling, and P. Melnik, *Phys. Scr.* **5**(1–2) 93 (1972).
26. S. Tolansky, *Surface Microtopography*, Longmans, London, 1960.
27. M. McCargo, J. A. Davies, and F. Brown, *Can. J. Phys.* **41**, 1231 (1963).
28. M. Klasson, A. Berndtsson, J. Hedman, R. Nilsson, R. Nyholm, C. Nordling, *J. Electron Spectrosc. Relat. Phenom.* **3**(6), 427 (1974).
29. W. F. Krolikowski and W. E. Spicer, *Phys. Rev.* **185** (3), 882 (1969).
30. W. F. Krolikowski and W. E. Spicer, *Phys. Rev. B* **1**(2) 478 (1970).
31. H. Kanter, *Phys. Rev. B* **1**(2), 522 (1970).
32. P. W. Palmberg and T. N. Rhodin, *J. Appl. Phys.* **39**(5), 2425 (1968).
33. Y. Baer, P. F. Hedén, J. Hedman, M. Klasson, and C. Nordling, *Solid State Commun.* **8**(18), 1479 (1970).
34. T. Huen and F. Wooten, *Solid State Commun.* **9**(18), 871 (1971).
35. T. F. Gesell and E. T. Arakawa, *Phys. Rev. Lett.* **26**(7), 377 (1971).
36. H. Kanter, *Phys. Rev. B* **1**(5), 2357 (1970).
37. R. G. Steinhardt, J. Hudis, and M. L. Perlman, *Phys. Rev. B* **5**(3), 1016 (1972).
38. M. L. Tarng and G. K. Wehner, *J. Appl. Phys.* **44**(4), 1534 (1973).
39. H. Thomas, *Z. Phys.* **147**, 395 (1957).
40. N. V. Smith and G. B. Fischer, *Phys. Rev. B* **3**(11), 3662 (1971).
41. G. Brodén, *Phys. Kondens. Materie* **15**(1), 171 (1972).
42. D. E. Eastman, *Solid State Commun.* **8**(1), 41 (1970).
43. J. W. T. Ridgway and D. Hanemann, *Surface Sci.* **24**(2), 451 (1971).
44. L. F. Wagner and W. E. Spicer, *Phys. Rev. B* **9**(4), 1512 (1974).
45. G. W. Gobeli and F. G. Allen, *Phys. Rev.* **127**(1), 141 (1962).
46. J. L. Shay and W. E. Spicer, *Phys. Rev.* **169**(3), 650 (1968).
47. W. E. Spicer, *J. Phys. Chem. Solids* **22**, 365 (1961).
48. M. P. Seah, *Surface Sci.* **32**(3), 703 (1972).
49. C. J. Todd and H. G. Heckingbottom, *Phys. Lett. A* **42**(6), 455 (1973).
50. F. L. Battye, J. G. Jenkin, J. Liesegang, and R. G. G. Leckey, *Phys. Rev. B* **9**(7), 2887 (1974).
51. C. S. Fadley, Core and Valence Electronic States Studies with X-Ray Photoelectron Spectroscopy, Lawrence Rad. Lab. Rept. UCRL-19535, Ph.D. Thesis, 1970.
52. Y. Baer, P. F. Hedén, J. Hedman, M. Klasson, C. Nordling, and K. Siegbahn, *Phys. Scr.* **1**(1), 55 (1970).
53. S. Hüfner, R. L. Cohen, G. K. Wertheim, *Phys. Scr.* **5**(1–2), 91 (1972).
54. V. V. Nemoshkalenko and V. G. Aleshin, *Izv. Akad. Nauk SSSR, Ser. Fiz.* **38**(3), 462 (1974).
55. M. K. Bahl, *J. Phys. F* **4**(3), 497 (1974).
56. Y. Baer and G. Busch, *Phys. Rev. Lett.* **30**(7), 280 (1973).
57. J. C. Fuggle, L. M. Watson, D. J. Fabian, and P. R. Norris, *Solid State Commun.* **13**(4), 507 (1973).
58. S. Hüfner, G. K. Wertheim, and D. N. E. Buchanan, *Chem. Phys. Lett.* **24**(4), 527 (1974).
59. Y. S. Knodeyev, H. Siegbahn, K. Hamrin, K. Siegbahn, *ESCA Applied to High Tem-*

perature Molecular Beams of Bismuth and Lead, Uppsala University UUIP-802, December 1972.
60. H. Siegbahn and K. Siegbahn, *J. Electron Spectrosc. Relat. Phenom.* **2**(4), 319 (1973).
61. T. Koopmans, *Physica* **1**, 104. (1933).
62. I. Lindgren, *Arkiv Physik* **31** (4), 59 (1965).
63. E. C. Snow, J. M. Canfield, and J. T. Waber, *Phys. Rev. A* **135**(4), 969 (1964).
64. A. Rosén and I. Lindgren, *Phys. Rev.* **176**(1), 114 (1968).
65. W. Meyer, *Int. J. Quantum Chem.* **5**, 341 (1971).
66. U. Gelius, *Phys. Scr.* **9**(3), 133 (1974).
67. U. Gelius, G. Johansson, H. Siegbahn, C. J. Allan, D. A. Allison, and K. Siegbahn, *J. Electron Spectrosc.* **1** (3), 285 (1972).
68. L. C. Snyder, *J. Chem. Phys.* **55**, 95 (1971).
69. L. Hedin and A. Johansson, *J. Phys. B* **2**(12), 1336 (1969).
70. D. A. Shirley, *Chem. Phys. Lett.* **16** (2), 220 (1972).
71. J. C. Slater, *Quantum Theory of Atomic Structure*, McGraw-Hill, New York, 1960, Vol. 2, p. 287.
72. G. Verhaegen, J. J. Berger, J. P. Desclaux, and C. M. Moser, *Chem. Phys. Lett.* **9**, 479 (1971).
73. C. M. Moser, R. K. Nesbet, and G. Verhaegen, *Chem. Phys. Lett.* **12**, 230 (1971).
74. D. A. Shirley, ESCA, Preprint LBL-610, Lawrence Berkeley Laboratory, Berkeley, 1972, p. 12.
75. R. E. Watson, *Phys. Rev.* **118**(4), 1036 (1960).
76. J. A. Pople, D. P. Santry, and G. A. Segal, *J. Chem. Phys.* **43** (10), S136 (1965).
77. S. B. M. Hagström and C. S. Fadley, *X-Ray Spectroscopy*, L. V. Azaroff, ed., McGraw-Hill, New York, 1974, p. 379.
78. P. A. Cox and A. F. Orchard, *Chem. Phys. Lett.* **7**, 273 (1970).
79. M. O. Krause, M. L. Vestal, W. H. Johnston, and T. A. Carlson, *Phys. Rev. A* **133** (2), 385 (1964).
80. M. O. Krause and T. A. Carlson, *Phys. Rev.* **149**(1), 52 (1966).
81. M. O. Krause and T. A. Carlson, *Phys. Rev.* **158**(1), 18 (1967).
82. M. O. Krause, T. A. Carlson, and R. D. Dismukes, *Phys. Rev.* **170**(1), 37 (1968).
83. M. O. Krause, *Phys. Rev.* **177**(1), 151 (1969).
84. T. A. Carlson and M. O. Krause, *Phys. Rev. A* **137** (6), 1655 (1965).
85. T. A. Carlson and M. O. Krause, *Phys. Rev. A* **140**(4), 1057 (1965).
86. A. S. Davydov, *Kvantovaya Mekhanika*, Nauka, Moskva, 1973.
87. T. Åberg, *Phys. Rev.* **156** (1), 35 (1967).
88. C. S. Fadley, in *Electron Emission Spectroscopy*, W. Dekeyser, ed., D. Reidel, Dordrecht, Holland, 1973, p. 151.
89. R. Manne and T. Åberg, *Chem. Phys. Lett.* **7**(2), 282 (1970).
90. H. A. Bethe and E. E. Salpeter, *Handbuch der Physik*, Vol. 35, Springer Verlag, Berlin-Göttingen-Heidelberg, 1955.
91. U. Fano and J. W. Cooper, *Rev. Mod. Phys.* **40**, 441 (1968).
92. J. W. Cooper, *Phys. Rev.* **128**(2), 681 (1962).
93. S. T. Manson and J. W. Cooper, *Phys. Rev.* **165**(1), 126 (1968).
94. J. W. Cooper and S. T. Manson, *Phys. Rev.* **177**(1), 157 (1969).
95. J. Cooper and R. N. Zare, *J. Chem. Phys.* **48**(2), 942 (1968).
96. J. Cooper and R. N. Zare, in *Lectures in Theoretical Physics Vol. 11C*, S. Geltman, K. Mahanthappa, W. Brittin, eds., Gordon and Breach, New York, 1969.
97. A. J. Bearden, *J. Appl. Phys.* **37** (4), 1681 (1966).

98. G. Rakavy and A. Ron, *Phys. Rev.* **159**(1), 50 (1967).
99. H. Brysk and C. D. Zerby, *Phys. Rev.* **171**(2), 292 (1968).
100. D. J. Kennedy and S. T. Manson, *Phys. Rev. A* **5**(1), 227 (1972).
101. J. C. Tully, R. S. Berry, and B. J. Dalton, *Phys. Rev.* **176**(1), 95 (1968).
102. F. A. Grimm, in *Electron Spectroscopy*, D. A. Shirley, ed., North-Holland, Amsterdam, London, 1972, p. 199.
103. U. Gelius, in *Electron Spectroscopy*, D. A. Shirley, ed., North-Holland, Amsterdam-London, 1972, p. 311.
104. V. I. Nefedov, N. P. Sergushin, I. M. Band, and M. B. Trzhaskovskaya, *J. Electron Spectrosc. Relat. Phenom.* **2**(5), 383 (1973).
105. C. S. Fadley, *Chem. Phys. Lett.* **25**(2), 225. (1974).
106. T. Åberg, *Ann. Acad. Scient. Fenn., Ser. A, Physica*, **6**, 308 (1969).
107. M. Ya. Amusia, *Atomic Physics II*, Plenum Press, New York, 1971, p. 219.
108. G. Wendin, *J. Phys. B* **5**(1), 110 (1972).
109. G. Wendin, *J. Phys. B* **6**(1), 42. (1973).
110. S. Lundqvist and G. Wendin, *J. Electron Spectrosc. Relat. Phenom.* **5**(513) (1974).
111. M. Ya. Amusia, N. A. Cherepkov, and S. I. Sheftel, *Phys. Lett. A* **24**, 541 (1967).
112. L. Hedin, B. I. Lundqvist, and S. Lundqvist, *Electronic Density of States*, H. Bennett, ed., National Bureau of Standards Special Publication 323, Washington, 1971, p. 233.
113. J. C. Slater, *Quantum Theory of Molecules and Solids*, McGraw-Hill, New York, 1965.
114. J. Callaway, *Energy Band Theory*, Academic Press, New York, 1964.
115. U. Harrison, *Pseudopotentials in the Theory of Metals*, W. A. Benjamin, New York, Amsterdam, 1966.
116. G. C. Fletcher, *The Electron Band Theory of Solids*, North-Holland, Amsterdam-London, 1971.
117. T. L. Loucks, *Augmented Plane Wave Method: A Guide to Performing Electronic Structure Calculations*, Benjamin, New York, 1967.
118. J. Ziman, in *Solid State Physics*, H. Ehrenreich, F. Seitz, D. Turnbull, eds., Vol. 26, Academic Press, New York, London, 1971, p. 1.
119. A. A. Levin, *Vvedenie v Kvantovuyu Khimiyu Tverdogo Tela. Khimicheskaya Sviaz i Struktura Energeticheskikh Zon v Tetraedricheskikh Poluprovodnikakh*, Khimiya, Moskva, 1974.
120. V. Heine, M. Cohen, and D. Weir, in *Solid State Physics*, H. Ehrenreich, F. Seitz, D. Turnbull, eds., Vol. 24, Academic Press, New York, London, 1970.
121. *Methods in Computer Physics*, B. Adler, S. Fernbach, M. Rotenberg, eds. Academic Press, New York-London, 1968.
122. *Computational Methods in Band Theory*, P. M. Marcus, J. F. Janak, A. R. Williams, eds., Plenum Press, New York-London, 1971.
123. *Computational Solid State Physics*, F. Herman, N. W. Dalton, T. R. Koehler, eds., Plenum Press, New York-London, 1972.
124. T. O. Woodruff, *Solid State Physics*, F. Seitz, D. Turnbull, eds. Academic Press, New York-London, 1957, Vol. 4, p. 367.
125. J. O. Dimmock, *Solid State Physics*, F. Seitz, D. Turnbull, eds. Academic Press, New York-London, 1971, Vol. 26, p. 103.
126. F. Bloch, *Z. Physik* **52**, 555 (1928).
127. J. C. Slater, *Phys. Rev.* **51**, 846 (1937).
128. C. Herring, *Phys. Rev.* **57**, 1169 (1940).
129. E. E. Lafon and C. C. Lin, *Phys. Rev.* **152**(2), 579 (1966).
130. E. E. Lafon, Roy C. Chaney, and C. C. Lin, in *Computational Methods in Band Theory*, P. M. Marcus, J. F. Janak, A. R. Williams, eds., Plenum Press, New York-London, 1971, p. 284.

131. R. N. Euwema, D. J. Stukel, and T. C. Collins, in *Computational Methods in Band Theory*, P. M. Marcus, J. F. Janak, A. R. Williams, eds., Plenum Press, New York-London, 1971, p. 82.
132. F. Herman, R. L. Kortum, C. D. Kuglin, and R. A. Short, *Quantum Theory of Atoms, Molecules and the Solid State*, P. O. Löwdin, ed., Academic Press, New York, 1966, p. 381.
133. F. Herman, R. L. Kortum, C. D. Kuglin, J. P. van Dyke, and F. Skilman, *Methods Comput. Phys.* **8,** 193 (1968).
134. J. L. Shay, W. E. Spicer, and F. Herman, *Phys. Rev. Lett.* **18,** 649 (1967).
135. D. C. Slater, and P. J. de Ciceo, *Mass. Inst. Technol. Sol. St. and Mol. Theory Group Quart. Progr. Rep.* **50,** 147 (1963).
136. J. Phillips and L. Kleinman, *Phys. Rev.* **116** (1), 287 (1959).
137. I. V. Abarenkov and V. Heine, *Phil. Mag.* **12**(117), 529 (1965).
138. M. L. Cohen and T. K. Bergstresser, *Phys. Rev.* **141** (2), 789 (1966).
139. K. H. Johnson, and F. C. Smith, Jr., *Phys. Rev. B* **5** (3), 831 (1972).
140. R. E. Watson, *Phys. Rev.* **111** (4), 1108 (1958).
141. J. Korringa, *Physica* **13,** 392 (1947).
142. J. M. Ziman, *Proc. Phys. Soc.* **86** (2), 337 (1965).
143. K. H. Johnson, *Phys. Rev.* **150**(2), 429 (1966).
144. P. P. Ewald, *Ann. Phys.* **64,** 253 (1921).
145. J. C. Slater and G. Koster, *Phys. Rev.* **94** (6), 1498 (1954).
146. L. Hodges, H. Ehrenreich, and N. D. Lang, *Phys. Rev.* **152**(2), 505 (1966).
147. V. G. Aleshin and V. P. Smirnov, *Fiz. Tverdogo Tela*, **11**(7), 1920 (1969).
148. V. G. Aleshin and V. V. Nemoshkalenko, Zonnaya Struktura i Rentgenovskie Emissionnye Spektry Kristallov, Preprint IMF-AN SSSR, 1970.
149. G. Dresselhaus and M. S. Dresselhaus, *Phys. Rev.* **160**(3), 649 (1967).
150. J. C. Slater, *Phys. Rev.* **81** (3), 385 (1951).
151. W. Kohn and L. J. Sham, *Phys. Rev. A* **140**(4), 1133 (1965).
152. J. C. Slater and K. H. Johnson, *Phys. Rev. B* **5** (3), 844 (1972).
153. J. C. Slater and J. H. Wood, *Intern. J. Quantum Chem. IVS*, 3 (1971).
154. I. Lindgren, *Phys. Lett.* **19,** 382 (1965).
155. F. Herman, I. B. Ortenburger, and J. P. van Dyke, *Intern. J. Quantum Chem. IIIS,* 827 (1970).
156. D. A. Liberman, *Phys. Rev.* **171** (1), 1 (1968).
157. I. B. Orthenburger and F. Herman, *Computational Methods in Band Theory*, P. M. Marcus, J. F. Janak, A. R. Williams, eds., Plenum Press, New York-London, 1971, p. 469.
158. P. Soven, *Phys. Rev.* **156** (3), 809 (1967).
159. P. Soven, *Phys. Rev.* **178**(3), 1136 (1969).
160. V. G. Aleshin and V. P. Smirnov, *Fiz. Tverdogo Tela* **11** (7), 2010 (1969).
161. J. Topol, K. Unger, G. Leonhardt, and E. Heiss, *Phys. Stat. Sol.* (*b*) **61** (2), 485 (1974).
162. L. van Hove, *Phys. Rev.* **89**(6), 1189 (1953).
163. J. Philips, in *Solid State Physics*, F. Seitz, D. Turnbull, eds., Academic Press, New York, London, Vol. 18, 1966.
164. N. V. Smith, *Phys. Rev. B* **3**(6), 1862 (1971).
165. D. Brust, *Phys. Rev. A* **139**(2), 489 (1965).
166. R. L. Park, I. E. Houston, and D. G. Schreiner, *Rev. Sci. Inst.* **41,** 1810 (1970).
167. M. Barber, P. Swift, D. Cunningham, and M. J. Frazer, *Chem. Commun.* **1970,** 1338 (1972).
168. I. Adams, J. M. Thomas, G. M. Barcroft, K. D. Butler, and M. Barber, *Chem. Commun.* **1972,** 751 (1972).

169. R. E. Block, *J. Magn. Resonance* **5**, 155 (1971).
170. D. T. Clark and D. Kihcast, *J. Chem. Soc. (A)* **1971**, 3286 (1971).
171. B. J. Lindberg, *J. Electron Spectrosc. Relat. Phenom.* **5**, 149 (1974).
172. S. P. Kowalczyk, L. Ley, F. R. McFeely, R. A. Pollak, and D. A. Shirley, *Phys. Rev. B* **8**(8), 3583 (1973).
173. P. H. Citrin, *Phys. Rev. B* **8**(13), 5545 (1973).
174. R. S. Crisp, and S. E. Williams, *Phil. Mag.* **6**, 625 (1961).
175. D. Weare, *Proc. Phys. Soc.* **92**(4), 956 (1967).
176. F. S. Ham, *Phys. Rev.* **128**(6), 2524 (1962).
177. J. B. Swan, *Phys. Rev. A* **135**(5), 1467 (1964).
178. C. Kunz, *Phys. Lett.* **15**(4), 312 (1965).
179. R. H. Ritchie, *Phys. Rev.* **106**(5), 874 (1957).
180. E. A. Stern and R. A. Ferrel, *Phys. Rev.* **120**(1), 130 (1960).
181. C. J. van Oirschot, M. van den Brink, and W. M. H. Sachtler, *Surf. Sci.* **29**(1), 189 (1973).
182. G. A. Rooke, *J. Phys. C* **1**(3), 767 (1968).
183. R. K. Dimond, *Soft X-Ray Band Spectra and the Electronic Structure of Metals and Materials,* D. J. Fabian, ed., Academic Press, London, 1968, p. 45.
184. V. V. Nemoshkalenko and V. V. Gorskii, *Ukr. Fiz. Zhur.* **12**(6), 818 (1967).
185. L. Smrčka, *Czechosl. J. Phys. B* **21**(6), 683 (1971).
186. Y. Baer, P. F. Hedén, J. Hedman, M. Klasson, C. Nordling, and K. Siegbahn, *Phys. Scr.* **1**(1), 55 (1970).
187. C. S. Fadley and D. A. Shirley, *J. Res. NBS A* **74**, 543 (1970).
188. C. S. Fadley and D. A. Shirley, *Electronic Density of States*, H. Bennett, ed., NBS Spec. Publ. 323, Washington, D.C., 1971, 163.
189. S. B. M. Hagström, *Band Structure Spectroscopy of Metals and Alloys*, D. J. Fabian and L. M. Watson, ed., Academic Press, London-New York, 1973, p. 73.
190. S. B. M. Hagström, *Electron Spectroscopy*, D. A. Shirley, ed., North-Holland, Amsterdam-London, 1972, p. 515.
191. S. Hüfner and G. K. Wertheim, *Phys. Lett. A* **44**(1), 47 (1973).
192. S. Hüfner, G. K. Wertheim, N. V. Smith, and M. M. Traum, *Solid State Commun.* **11**(2), 323 (1972).
193. S. Hüfner, G. K. Wertheim, and J. H. Wernick, *Phys. Rev B* **8**(10), 4511 (1973).
194. D. A. Shirley, *Phys. Rev. B* **5**(12), 4709 (1972).
195. M. G. Ramchandani, *J. Phys. C* **3**(1), S1 (1970).
196. N. V. Smith and M. M. Traum, *Electron Spectroscopy*, D. A. Shirley, ed., North-Holland, Amsterdam-London, 1972, p. 541.
197. J. W. D. Connolly and K. H. Johnson, *MIT Solid State and Molecular Theory Group Report*, 1970, Vol. 72, p. 19.
198. S. Kupratakuln, *J. Phys. C* **2**(2), S109 (1970).
199. N. E. Christensen and B. O. Seraphin, *Phys. Rev. B* **4**, 3321 (1971).
200. C. B. Sommers and H. Amar, *Phys. Rev.* **188**(3), 1117 (1969).
201. D. E. Eastman and W. D. Grobman, *Phys. Rev. Lett.* **28**(20), 1327 (1972).
202. J. L. Freeouf, M. Erbudak, and D. E. Eastman, *Solid State Commun.* **13**(7), 771 (1973).
203. A. J. McAlister, M. L. Williams, J. R. Cuthill, and R. C. Dobbya, *Solid State Commun.* **9**(20), 1775 (1971).
204. V. V. Nemoshkalenko, V. G. Aleshin, Yu. N. Kucherenko, and L. M. Sheludchenko, *Solid State Commun.* **15**(11/12), 1745 (1974).
205. V. V. Nemoshkalenko, V. G. Aleshin, Yu. N. Kucherenko, and L. M. Sheludchenko, *J. Electron Spectrosc. Relat. Phenom.* **6**(2), 145 (1974).

206. G. M. Stocks, R. W. Williams, and J. S. Faulkner, *Phys. Rev. B* **4** (12), 4390 (1971).
207. G. M. Stocks, R. W. Williams, and J. S. Faulkner, *J. Phys. F* **3** (9), 1688 (1973).
208. G. A. Burdick, *Phys. Rev.* **129** (1), 138 (1963).
209. M. Chodorow, *Phys. Rev.* **55**, 675 (1939).
210. J. O. Dimmock, *Solid State Physics*, H. Ehrenreich, F. Seitz, D. Turnbull, eds., Vol. 26, Academic Press, New York-London, 1971.
211. L. F. Mattheiss, *Phys. Rev.* **133** (1), 184 (1964).
211a. G. Apai *et al.*, *Phys. Rev. B* **15** (2), 584 (1977).
211b. P. S. Wehner *et al.*, *Phys. Rev B* **14** (6), 2411 (1976).
212. V. V. Nemoshkalenko, A. I. Senkevich, M. A. Mindlina, and V. G. Aleshin, *Phys. Stat. Sol.* (*b*) **56** (2), 771 (1973).
213. R. C. Dobbyn, M. L. Williams, J. R. Cuthill, and A. J. McAlister, *Phys. Rev. B* **2** (16), 1563 (1970).
214. R. J. Liefeld, *Soft X-Ray Band Spectra*, D. J. Fabian, ed., Academic Press, New York, 1968, p. 133.
214a. P. S. Wehner *et al.*, *Phys. Rev. B* **14** (6), 2411 (1976).
215. S. Hüfner and G. K. Wertheim, *Phys. Lett. A* **47** (5), 349 (1974).
216. E. J. Zornberg, *Phys. Rev. B* **1** (1), 244 (1970).
217. K. C. Wong, E. P. Wohlfarth, D. M. Hum, *Phys. Lett. A* **29** (8), 452 (1969).
218. J. Yamashita, S. Asano, S. Wakoh, *J. Appl. Phys.* **39** (2), 1274 (1968).
219. S. Hüfner, G. K. Wertheim, and D. N. E. Buchanan, *Solid State Commun.* **14** (11), 1173 (1974).
220. N. E. Christensen, *Phys. Stat. Sol.* (*b*) **55** (1), 117 (1973).
221. S. P. Kowalczyk, L. Ley, R. A. Pollak, and D. A. Shirley, *Phys. Lett. A* **41** (5), 455 (1972).
222. R. A. Pollak, S. P. Kowalczyk, L. Ley, and D. A. Shirley, *Phys. Rev. Lett. A* **29** (5), 274 (1972).
223. T. A. Carlson, C. C. Lu, T. C. Tucker, C. W. Nestor, Jr., and F. B. Malik, Oak Ridge National Laboratory Report ORNL-4614, 1970.
224. L. Ley, R. A. Pollak, S. Kowalczyk, and D. A. Shirley, *Phys. Lett. A* **41** (5), 429 (1972).
225. P. Soven, *Phys. Rev. A* **6**, 1706 (1965).
226. T. L. Loucks, *Phys. Rev. Lett.* **14** (26), 1072 (1965).
227. L. G. Ferriera, *J. Phys. Chem. Sol.* **28** (10), 1891 (1967).
228. C. C. Lu *et al.*, *Atomic Data* **3**, 1 (1971).
229. S. B. M. Hagström, P. O. Hedén, and H. Lofgren, *Solid State Commun.* **8** (15), 1245 (1970).
230. P. O. Hedén, H. Löfgren, and S. B. M. Hagström, *Phys. Rev. Lett.* **26** (8), 432 (1971).
231. S. B. M. Hagström, *Electron Spectroscopy*, D. A. Shirley, ed., North-Holland, Amterdam-London, 1972, p. 515.
232. G. Brodén, S. B. M. Hagström, P. O. Hedén, and C. Norris, *Electronic Density of States*, H. Bennett, ed., NBS Spec. Publ. 323, Washington, D.C. 1971, 217.
233. Y. Baer and G. Busch, *J. Electron Spectrosc. Relat. Phenom.* **5**, 611 (1974).
234. P. A. Cox, Y. Baer, and C. K. Jorgensen, *Chem. Phys. Lett.* **22**, 443 (1973).
235. P. A. Cox, S. Evans, and A. F. Orchard, *Chem. Phys. Lett.* **13**, 386 (1972).
236. F. R. McFeely, S. P. Kowalczyk, L. Ley, and D. A. Shirley, *Phys Lett. A* **45**, 227 (1973).
237. S. C. Keeton and T. L. Loucks, *Phys. Rev.* **168** (3), 672 (1968).
238. J. C. Fuggle, A. F. Burr, L. M. Watson, D. J. Fabian, and W. Land, *J. Phys. F* **4** (2), 235 (1974).

239. V. V. Nemoshkalenko and V. G. Aleshin, in *Metallofizika*, Naukova Dumka, Kiev, 1975, p. 60.
240. L. Schwartz, F. Brouers, A. V. Vedyaev, and H. Ehrenreich, *Phys. Rev. B* **4** (10), 3383 (1971).
241. F. Brouers and A. V. Vedyaev, *Phys. Rev. B* **5** (2), 348 (1972).
242. U. Harrison, *Teoria Tverdogo Tela*, Mir, Moskva, 1972.
243. V. V. Nemoshkalenko, M. G. Chudinov, V. G. Aleshin, Yu. N. Kucherenko, and L. M. Sheludchenko, in *Metallofizika*, Naukova Dumka, Kiev, 1975, p. 60.
244. V. V. Nemoshkalenko, V. G. Aleshin, Yu. N. Kucherenko, and L. M. Scheludchenko, *Dokl. Akad. Nauk SSSR* **219** (2), 329 (1974).
245. V. V. Nemoshkalenko, V. G. Aleshin, A. I. Senkevich, Yu. N. Kucherenko, L. M. Scheludchenko, and M. T. Panchenko, *Phys. Stat. Sol.* **65** (1), K105 (1974).
246. S. Kirkpatrick, B. Velicky, and H. Ehrernreich, *Phys. Rev. B* **1** (8), 3250 (1970).
247. V. V. Nemoshkalenko, V. G. Aleshin, and M. G. Chudinov, *Ukr. Fiz. Zh.* **19** (6), 983 (1974).
248. J. Clift, C. Curry, and B. J. Thompson, *Phil. Mag.* **8**, 593 (1963).
249. S. Hüfner, G. K. Wertheim, R. L. Cohen, and J. H. Wernick, *Phys. Rev. Lett.* **28** (8), 488 (1972).
250. J. C. Love, F. E. Obenshaim, and F. Czjzek, *Phys. Rev. B* **3** (3), 2827 (1971).
251. J. Hedman, M. Klasson, R. Nilsson, C. Nordling, M. F. Sorokina, O. I. Kljushnikov, S. A. Nemnonov, V. A. Trapeznikov, and V. G. Zyryanov, *Phys. Ser.* **4** (1), 1 (1971).
252. F. M. Mueller, A. J. Freeman, J. O. Dimmock, and A. M. Furdynna, *Phys. Rev. B* **1** (12), 4617 (1970).
253. S. Hüfner, G. K. Wertheim, and J. H. Wernick, *Solid State Commun.* **11** (1), 259 (1972).
254. V. V. Nemoshkalenko, V. G. Aleshin, and A. I. Senkevich, in *Metallofizika*, Naukova Dumka, Kiev, 1974, pp. 22, 52.
255. V. V. Nemoshkalenko, M. G. Chudinov, V. G. Aleshin, Yu. N. Kucherenko, and L. M. Sheludchenko, *Solid State Commun.* **16** (6), 755 (1975).
256. V. V. Nemoshkalenko, V. G. Aleshin, and A. I. Senkevich, *Solid State Commun.* **13** (8), 1069 (1973).
257. V. V. Nemoshkalenko, M. G. Chudnikov, V. G. Aleshin, Yu. N. Kucherenko, and A. I. Senkevich, *Dokl. Akad. Nauk SSSR*, **7**, 641 (1975).
258. I. Ya. Dekhtyar and V. V. Nemoshkalenko, *Elektronnaya Struktura i Elektronnye Svoistva Perekhodnykh Metallov i ikh Splavov*., Naukova Dumka, Kiev, 1971.
259. V. V. Nemoshkalenko, A. I. Senkevich, V. G. Aleshin, and V. V. Dorskü, *Phys. Stat. Sol.* (*b*) **58** (2), K 125 (1973).
260. K. Levin and H. Ehrenreich, *Phys. Rev. B* **3** (12), 4172 (1971).
261. S. Hüfner, J. H. Wernick, and K. W. West, *Solid State Commun.* **10** (11), 1013 (1972).
262. D. J. Fabian, J. C. Fuggle, P. R. Norris, and L. M. Watson, *Proc. Int. Symp. for Synchrotron Radiation Users*, G. V. Marr, ed., Publ. DNPL/R 26, Washington, 1973.
263. A. C. Switendick and A. Narath, *Phys. Rev. Lett.* **22** (26), 1423 (1969).
264. H. Jones, *Proc. Roy. Soc. A* **144**, 255 (1934).
265. P. W. Anderson, *Phys. Rev.* **124** (1), 41 (1961).
266. P. O. Nilsson, *Phys. Kondens. Materie,* **11**, 1 (1970).
267. E. A. Stern, *Phys. Rev.* **144** (2), 545 (1966).
268. J. H. O. Varley, *Phil. Mag.* **45**, 887 (1954).
269. *Elektronnaya Struktura Perekhodnykh Metallov u Khimia ikh Splavov*, Sbornik Statei pod Red. Umanskogo Ya. S., Surisa R. A., Metallurgia, Moskva, 1966.
270. S. A. Nemnonov, *Fiz. Met. Metalov.* **24** (6), 1016 (1967).
271. S. Kirkpatrick, B. Velicky', N. D. Lang, and H. Ehrenreich, *J. Appl. Phys.* **40** (4), 1283 (1969).

272. N. D. Lang and H. Ehrenreich, *Phys. Rev.* **168**(2), 605 (1968).
273. D. H. Seib and W. E. Spicer, *Phys. Rev. Lett.* **20**(25), 1441 (1968).
274. D. H. Seib and W. E. Spicer, *Phys. Rev. B* **2**(6), 1676 (1970).
275. S. Kirkpatrick, B. Velicky´, and H. Ehrenreich, *Phys. Rev. B* **1**(8), 3520 (1970).
276. G. M. Stocks, R. W. Williams, and J. S. Faulkner, *Phys. Rev. Lett.* **26**(5), 253 (1971).
277. V. V. Nemoshkalenko, V. G. Aleshin, M. T. Panchenko, and V. Ya. Nagornyi, *Dokl. Akad. Nauk SSSR* **210**(3), 581 (1973).
278. V. V. Nemoshkalenko, V. G. Aleshin, M. T. Panchenko, and A. I. Senkevich, *Dokl. Akad. Nauk SSSR* **214**(3), 543 (1974).
279. V. G. Aleshin, Preprint *IMF AN USSR* 73, 10, Kiev, 1973.
280. V. V. Nemoshkalenko, V. G. Aleshin, and M. T. Panchenko, *Int. Symp. X-Ray Spectra and Electronic Structure of Matter*, München, Sept. 1972, A. Faestler, G. Weich, eds., München, 1973, p. 115.
281. V. V. Nemoshkalenko, V. G. Aleshin, M. T. Panchenko, and A. I. Senkevich, *Fiz. Tv. Tela* **15**(11), 3465 (1973).
282. V. V. Nemoshkalenko and V. G. Aleshin, *Fiz. Tv. Tela* **12**(1), 59 (1970).
283. R. A. Pollak, L. Ley, S. P. Kowalczyk, D. A. Shirley, J. D. Joanopoulos, D. J. Chadi, and M. L. Cohen, *Phys. Rev. Lett.* **29**(16), 1103 (1972).
284. J. Chelikowski, D. J. Chadi, and M. L. Cohen, *Phys. Rev. B* **8**(6), 2786 (1973).
285. T. Cora, R. Staley, J. D. Rimstidt, and J. Sharma, *Phys. Rev. B* **5**(6), 2309 (1972).
286. R. G. Cavell, S. P. Kowalczyk, L. Ley, R. A. Pollak, B. Mills, D. A. Shirley, and W. Perry, *Phys. Rev. B* 7(12), 5313 (1973).
287. F. R. McFeely, S. P. Kowalczyk, L. Ley, R. G. Cavell, R. A. Pollak, and D. A. Shirley, *Phys. Rev. B* 9(12), 5268 (1974).
288. G. S. Painter, D. E. Ellis, and A. R. Lubinsky, *Phys. Rev. B* **4**(10), 3610 (1971).
289. G. Wiech and E. Zöpf, *Electronic Density of States*, Ed. H. Bennett, NBS Spec. Publ. 323, Washington, D.C. 1971, p. 335.
290. M. A. Blokhin and V. P. Satchenko, *Izv. Akad. Nauk SSSR, Ser. Fiz.* **24,** 397 (1960).
291. V. V. Nemoshkalenko, V. G. Aleshin, and M. T. Panchenko, *Phys. Fenn.* **9**(S1), 424 (1974).
292. F. Herman, R. L. Kortum, C. D. Kuglin, and R. A. Short, *J. Phys. Soc. Jpn.*, Suppl. **21,** 7 (1966).
293. R. Keown, *Phys. Rev.* **150**(2), 568 (1966).
294. W. Van Haeringen and H. G. Junginger, *Solid State Commun.* **7**(16), 1135 (1969).
295. G. S. Painter and D. E. Ellis, *Phys. Rev. B* **1**(12), 4747 (1970).
296. E. W. Kortela and R. Manne, *J. Phys. C* **7**(9), 1749 (1974).
297. I. B. Borovski, V. I. Matiskin, and V. I. Nefedov, *J. Phys., Paris*, **10**(C4), 207 (1971).
298. Ch. Beyreuther and G. Wiech, *Phys. Fenn.* **9**(S1), 176 (1974).
299. O. Brümmer, G. Dräger, W. A. Fomichev, and A. S. Shulakov, in *Int. Symp. X-Ray Spectra and Electronic Structure of Matter*, A. Faessler and G. Wiech, eds., München, Section Physik der Universität, Vol. 1, 1973, p. 78.
300. J. Müller, K. Feser, G. Weich, and A. Faessler, *Phys. Lett.* **A44**(4), 263 (1973).
301. D. Weaire and M. F. Thorpe, *Phys. Rev. B* **4**(8), 2508 (1971).
302. L. B. Leder and J. A. Suddeth, *J. Appl. Phys.* **31**(8), 1442 (1960).
303. W. Y. Liang and S. L. Cundy, *Phil. Mag.* **19,** 1031 (1969).
304. N. R. Whetten, *Appl. Phys. Lett.* **8,** 135 (1966).
305. E. A. Taft and H. R. Philipp, *Phys. Rev. A* **138**(1), 197 (1965).
306. T. Noda and N. Inagaki, *Bull. Chem. Soc. Japan* **37,** 1534 (1964).
307. J. Kakinoki, *Acta Crystallogr.* **18,** 518 (1963).
308. K. Länger, Thesis, Univ. München, 1968.
309. E. O. Kane, *Phys. Rev.* **146**(2), 558 (1966).

310. L. Ley, S. P. Kowalczyk, R. A. Pollak, and D. A. Shirley, *Phys. Rev. Lett.* **29**(22), 1088 (1972).
311. R. A. Pollak, L. Ley, S. Kowalczyk, D. A. Shirley, J. D. Joanopoulos, D. J. Chadi, and M. L. Cohen, *Phys. Rev. Lett.* **29**(16), 1103 (1972).
312. W. E. Spicer and R. C. Eden, *Proc. Ninth Int. Conf. on the Physics of Semiconductors*, Vol. 1, Nauka, Leningrad, 1968, p. 61.
313. D. E. Eastman, *Phys. Rev. Lett.* **29**(22), 1508 (1972).
314. J. R. Chelikowsky and M. L. Cohen, *Phys. Rev. B* **10**(12), 5095 (1974).
315. F. Herman, R. L. Kortum, and C. D. Kuglin, *Methods in Computational Physics*, B. Adeer, S. Fernbach, and M. Rotenberg, eds., Academic Press, New York-London, 1968, p. 193.
316. K. C. Pandy and J. C. Phillips, *Phys. Rev. B* 9(4), 1552 (1974).
317. D. E. Eastman, W. D. Grobman, J. L. Freeouf, and M. Erbudak, *Phys. Rev. B* 9(8), 3473 (1974).
318. D. E. Eastman and W. D. Grobman, *Phys. Rev. Lett.* **28**(21), 1378 (1972).
319. W. E. Spicer, *Phys. Rev.* **154**(1), 385 (1967).
320. D. E. Eastman, J. L. Freeouf, and M. Erbudak, *Congrès du Centenaire de la Société Française de Physique*, Vittel, France, 1973.
321. I. B. Ortenburger and W. E. Rudge, IBM Research Lab. Report RJ-1041, 1972.
322. D. J. Stukel, T. C. Collins, and R. N. Euwema, *Electronic Density of States*, H. Bennett, ed., NBS Spec. Publ. 323, Washington, 1971, p. 93.
323. G. Wiech, *Z. Phys.* **216**, 472 (1968).
324. G. Wiech in *Rentgenovskie Spektry i Elektronnaya Struktura Veshchestva*, Vol. 1, Izd. IMF AN USSR, Kiev, 1969.
325. F. Herman, R. L. Kortum, I. B. Ortenburger, and J. P. van Dyke, Aerospace Research Lab. (ARL) Report 69-0080, 1969.
326. D. J. Stukel, R. N. Euwema, T. C. Collins, F. Herman, and R. L. Kortum *Phys. Rev.* **179**(3), 740 (1969).
327. P. Eckelt, *Phys. Stat. Sol.* **23**(1), 307 (1967).
328. U. Rössler and M. Lietz, *Phys. Stat. Sol.* **17**(2), 597 (1966).
329. W. D. Grobman, D. E. Eastman, and M. L. Cohen, *Phys. Lett. A* **43**(1), 49 (1973).
330. J. C. Phillips, *Rev. Mod. Phys.* **42**, 317 (1970).
331. W. A. Harrison, *Phys. Rev. B* 8(10), 4487 (1973).
332. L. Ley, S. P. Kowalczyk, F. R. McFeely, and D. A. Shirley, *Phys. Rev. B* **10**(12), 4881 (1974).
333. R. T. Poole, P. C. Kemeny, J. Liesegang, J. G. Senkin, and R. G. G. Leckey, *J. Phys. F* **3**, S46 (1973).
334. C. E. Moore, *Atomic Energy Levels*, NBS Circular No 467, US GRO, Washington, D. C., 1958.
335. C. J. Ballhausen, *Introduction to Ligand Field Theory*, McGraw-Hill, New York, 1962.
336. N. J. Shevchik, J. Tejeda, M. Cardona, *Phys. Rev.* 9(6), 2627 (1974).
337. C. J. Veseley and D. W. Langer, *Phys. Rev. B* **4**, 451 (1971).
338. F. R. McFeely, S. P. Kowalczyk, L. Ley, R. A. Pollak, and D. A. Shirley, *Phys. Rev. B* 7(12), 5228 (1973).
339. R. A. Pollak, S. Kowalczyk, L. Ley, and D. A. Shirley, *Phys. Rev. Lett.* **29**(5), 274 (1972).
340. J. D. Joanopoulos and M. L. Cohen, *Phys. Rev. B* 7(6), 2644 (1973).
341. R. A. Pollak, *Electronic Density of States of Solids from X-Ray Photoemission Spectroscopy*, Ph.D. Thesis, Lawrence. Berkeley Laboratory, Univ. of California, Berkeley, 1972, p. 51.

342. L. Ley, R. A. Pollak, S. P. Kowalczyk, F. R. McFeely, and D. A. Shirley, *Phys. Rev. B* **8**(2), 641 (1973).
343. S. Golin, *Phys. Rev. A* **140**(9), 993 (1965).
344. L. M. Falicov and P. J. Lin, *Phys. Rev.* **141**(2), 562 (1966).
345. L. G. Ferreira, *J. Chem. Phys. Sol.* **28**(10), 1891 (1967).
346. S. Golin, *Phys. Rev.* **166**(3), 643 (1968).
347. S. Mase, *J. Phys. Soc. Jpn.* **13**(5), 434 (1958).
348. S. Mase, *J. Phys. Soc. Jpn.* **14**(5), 584 (1959).
349. C. C. Lu, T. A. Carlson, F. B. Malik, T. C. Tucker, and C. W. Nestor, Jr., *Atomic Data*, **3**, 1 (1971).
350. V. V. Nemoshkalenko, A. I. Senkevich, and V. G. Aleshin, *Dokl. Akad. Nauk SSSR* **206**(3), 593 (1972).
351. V. V. Nemoshkalenko, A. I. Senkevich, and V. G. Aleshin, *Band Structure Spectroscopy of Metals and Alloys*, D. J. Fabian and L. M. Watson, eds., Academic Press, London-New York, 1973, p. 107.
352. P. H. Citrin and T. D. Thomas, *J. Chem. Phys.* **57**(10), 4446 (1972).
353. S. P. Kowalczyk, F. R. McFeely, L. Ley, R. A. Pollak, and D. A. Shirley, *Phys. Rev. B* **9**(8), 3573 (1974).
354. R. A. Pollak, Ph.D. Thesis, Univ. California, Berkeley, 1972, p. 220.
355. C. Y. Fong and M. L. Cohen, *Phys. Rev.* **185**(3), 1168 (1969).
356. L. P. Howland, *Phys. Rev.* **109**(6), 1927 (1958).
357. P. D. Cicco, *Phys. Rev.* **153**(3), 931 (1967).
358. A. B. Kunz, *Phys. Rev.* **175**(3), 1147 (1968).
359. A. B. Kunz, *Phys. Stat. Sol.* **29**(1), 115 (1968).
360. A. B. Kunz, *Phys. Rev.* **180**(3), 934 (1969).
361. P. E. Best, *Proc. Phys. Soc. London* **79**(1), 133 (1962).
362. M. Creuzberg, *Z. Phys.* **196**, 433 (1966).
363. F. C. Brown, C. Gähwiller, and H. Fujita, *Phys. Rev. B* **2**(6), 2126 (1970).
364. T. Sagawa, Y. Igushi, M. Sasanuma, T. Nasu, S. Yamaguchi, M. Nakamura, A. Ejiri, T. Masuoka, T. Sasaki, and T. Oshio, *J. Phys. Soc. Jpn.* **12**(12), 2587 (1966).
365. S. Nakai and T. Sagawa, *J. Phys. Soc. Jpn.* **26**(6), 1427 (1969).
366. Y. Igushi, T. Sagawa, S. Sato, M. Watanabe, H. Yamashita, A. Ejiri, M. Sasanuma, S. Nakai, M. Nakamura, S. Yamaguchi, T. Nakai, and T. Oshio, *Solid State Commun.* **6**(8), 575 (1968).
367. C. Sugiura and S. Kiyono, *Technol. Report, Tohoku Univ.* **35**(1), 61 (1970).
368. R. Haensel, C. Kunz, and B. Sonntag, *Phys. Rev. Lett.* **20**(6), 262 (1968).
369. R. Haensel, C. Kunz, T. Sasaki, and B. Sonntag, *Phys. Rev. B* **9**(8), 3601 (1974).
370. W. Gudat, C. Kunz, and H. Peterson, *Phys. Rev. Lett.* **32**(24), 1370 (1974).
371. S. T. Pantelides and F. C. Brown, *Phys. Rev. Lett.* **33**(5), 298 (1974).
372. W. P. Menzel, C. C. Lin, D. F. Fouguet, E. A. Lafon, and R. C. Chaney, *Phys. Rev. Lett.* **30**(26), 1313 (1973).
373. A. B. Kunz, D. J. Mickish, and T. C. Collins, *Phys. Rev. Lett.*, **31**(12), 756 (1973).
374. S. T. Pantelides, *Phys. Rev. B* **11**(6), 2391 (1975).
375. W. Gudat, C. Kunz, *Phys. Rev. Lett.* **29**(3), 169 (1972).
376. D. M. Roessler and W. C. Walker, *J. Phys. Chem. Solids* **28**(8), 1507 (1967).
377. C. E. Moore, *Atomic Energy Levels as Derived from Analysis of Optical Spectra*, USA NBS Circular No. 467, US GPO, Washington, D.C., 1949, p. 1.
378. F. C. Brown, C. Gähwiller, A. B. Kunz, and N. O. Lipari, *Phys. Rev. Lett.* **25**(14), 927 (1970).
379. A. B. Kunz, *J. Phys. C: Proc. Phys. Soc. London*, **3**, 1542 (1970).

380. A. B. Kunz, *Phys. Rev. B* **2**(12), 5015 (1970).
381. A. B. Kunz and N. O. Lipari, *J. Phys. Chem. Solids* **32**(6), 1141 (1971).
382. S. T. Pantelides, *Solid State Commun.* **16**(1), 95 (1975).
383. W. H. Strehlow and E. L. Cook, *J. Phys. Chem. Ref. Data* **2**, 163 (1973).
384. T. R. Robinson, *Proc. Phys. Soc. B* **65**, 910 (1952).
385. F. C. Jahoda, *Phys. Rev.* **107**(5), 1261 (1957).
386. H. R. Philipp and E. A. Taft, *Phys. Rev.* **113**(4), 1002 (1959).
387. H. R. Philipp and E. A. Taft, *Phys. Rev.* **120**(1), 37 (1960).
388. D. M. Roessler, *Brit. J. Appl. Phys.* **16**, 1119 (1965).
389. G. Stephan, Y. L. Calver, J. C. Lemomer, and S. Rabin, *J. Phys. Chem. Solids* **30**(3), 601 (1969).
390. G. W. Rubloff, *Phys. Rev. B* **5**(2), 662 (1972).
391. T. Tomiki and T. Miyata, *J. Phys. Soc. Jpn.* **27**(3), 658 (1969).
392. W. Hayes, A. B. Kunz, and E. E. Koch, *J. Phys. C* **4**, 1200 (1971).
393. J. Frandon, B. Lahaye, and F. Pradal, *Phys. Stat. Sol.* **53**(2), 565 (1973).
394. V. V. Nemoshkalenko, K. K. Sidorin, and V. G. Aleshin, Preprint IMF AN USSR 74 I, K, 1974.
395. N. V. Starostin and V. A. Ganin, *Fiz. Tv. Tela* **15**(11), 3403 (1973).
396. N. V. Starostin and M. P. Shepilov, *Fiz. Tv. Tela* **15**(12), 3709 (1973).
397. N. V. Starostin and V. A. Ganin, *Fiz. Tv. Tela* **16**(2), 572 (1974).
398. B. Hoenerlage and H. Wiesner, *Z. Phys.* **242**(5), 406 (1971).
399. J. A. Bearden and A. F. Burr, *Rev. Mod. Phys.* **39**, 128 (1967).
400. T. M. Zimkina and A. S. Vinogradov, *J. Phys., Paris, Colloque C* **32**(4), 278 (1971).
401. T. Åberg, *Phys. Lett. A* **26**, 515 (1968).
402. V. V. Nemoshkalenko, V. G. Aleshin, I. A. Brytov, K. K. Sidorin, and Yu. N. Romashchenko, *Uzv. Akad. Nauk SSSR, Ser. Fiz.* **38**(3), 626 (1974).
403. K. K. Sidorin, V. V. Nemoshkalenko, V. G. Aleshin, I. A. Brytov, Yu. N. Romashchenko, and A. I. Senkevich, in *Metallofizika*, Naukova Dumka, Kiev, 1974, pp. 48, 60.
404. N. F. Mott, *Proc. Phys. Soc. A, London*, **62**, 416 (1949).
405. D. Adler, *Solid State Physics*, H. Ehrenreich, F. Seitz, and D. Turnbull, eds., Vol. 21, Academic Press, New York-London, 1968.
406. G. K. Wertheim and S. Hüfner, *Phys. Rev. Lett.* **28**(16), 1028 (1972).
407. I. A. Brytov, N. I. Komyak, and A. P. Lukirski, in *Rentegenovskie Spektry i Elektronnaya Struktura Veshchestva*, Vol. 1, Izd. IMF AN USSR, Kiev, 1969, p. 284.
408. E. A. Bakulin, A. A. Balobanova, and V. A. Vasiliev, *Fiz. Tv. Tela* **13**(2), 653 (1971).
409. A. C. Switendick, *Mass. Inst. Technol. Sol. St. and Mol. Theory Group Quart. Progr. Report*, July 15, **49**, 41 (1963).
410. L. F. Mattheiss, *Phys. Rev. B* **5**(1), 290 (1972).
411. A. S. Koster, *J. Phys. Chem. Solids* **32**(12), 2685 (1971).
412. J. Liefeld, in *Soft X-Ray Spectra and the Electronic Structure of Metals and Alloys*, D. J. Fabian, ed., Academic Press, London-New York, 1968, 133.
413. V. G. Zyryanov and C. A. Nemnonov, *Fiz. Met. Metallov.* **31**, 515 (1971).
414. S. Hüfner and G. K. Wertheim, *Phys. Rev. B* **8**(10), 4857 (1973).
415. J. P. Dahl and A. C. Switendick, *J. Phys. Chem. Solids* **27**(6), 931 (1966).
416. G. K. Wertheim, L. F. Mattheiss, and M. Campagna, *Phys. Rev. Lett.* **32**(18), 997 (1974).
417. L. F. Mattheiss, *Phys. Rev.* **181**(3), 987 (1969).
418. L. F. Mattheiss, *Phys. Rev. B* **2**(10), 3918 (1970).
419. L. F. Mattheiss, *Phys. Rev. B* **6**(12), 4718 (1972).
420. S. Hüfner and G. K. Wertheim, *Phys. Lett. A* **44**(2), 133 (1973).

421. J. T. Sparks and T. Komoto, *Rev. Mod. Phys.* **40,** 752 (1968).
422. T. M. Tyler and J. L. Fry, *Phys. Rev. B* **1**(12), 4604 (1970).
423. R. M. White and N. F. Mott, *Phil. Mag.* **24**(190), 845 (1971).
424. A. Oshawa, H. Yamamoto, and H. Watanabe, *J. Phys. Soc. Jpn.* **37**(2), 568 (1974).
425. H. S. Jarrett, W. H. Cloud, R. J. Bouchard, S. R. Butler, C. G. Frederick, and J. L. Gilson, *Phys. Rev. Lett.* **21**(9), 617 (1968).
426. S. Ogawa, S. Waki, and D. Teranishi, *Int. J. Magnetism* **5,** 349 (1974).
427. E. K. Li, K. H. Johnson, D. E. Eastman, and J. L. Freeouf, *Phys. Rev. Lett.* **32**(9), 470 (1974).
428. S. Kono, T. Ishii, T. Sagawa, and T. Kobayashi, *Phys. Rev. Lett.* **28**(21), 1385 (1972).
429. S. Kono, T. Ishii, T. Sagawa, and T. Kobayashi, *Phys. Rev. B* 8(2), 795 (1973).
430 Y. Sakisaka, T. Ishii, and T. Sagawa, *J. Phys. Soc. Jpn.* **36**(5), 1372 (1974).
431. F. Herman and B. S. McClure, *Bull. Amer. Phys. Soc.* **5,** 48 (1960).
432. T. Novakov, *Phys. Rev. B* 3(8), 2693 (1971).
433. D. C. Frost, A. Ishitani, and C. A. McDowell, *Mol. Phys.* **24,** 861 (1972).
434. P. A. Jansson, *J. Opt. Soc. Am.* **60,** 184 (1970).
435. A. Goldmann, J. Tejeda, N. J. Shevchik, and M. Cardona, *Phys. Rev. B* **10**(10), 4388 (1974).
436. V. V. Nemoshkalenko, T. B. Shashkina, V. G. Aleshin, A. I. Senkevich, *J. Phys. Chem. Solids* **36**(1), 37 (1975).
437. B. Aronsson, T. Lundström, and S. Rundqvist, *Borides, Silicides and Phosphides*, Wiley, New York; Methuen, London, 1965.
438. G. V. Samsonov, Yu. M. Goryachev, B. A., Kovenskaya and R. Ya. Telnikov, *Izv. Vuzov. Fizika*, **6,** 37 (1972).
439. M. C. Cadeville, Ph.D. Thesis, Strasbourg, 1965.
440. L. Ramqvist, K. Hamrin, G. Johansson, A. Fahlman, and C. Nordling, *J. Phys. Chem. Solids* **30**(7), 1835 (1969).
441. L. Ramqvist, R. Ekstig, E. Källne, E. Noreland, and R. Manne, *J. Phys. Chem. Solids* **32**(1), 149 (1971).
442. T. B. Shashkina, *Phys. Stat. Sol.* **44**(2), 571 (1971).
443. E. T. Bezruk and L. Ya. Markovskii, *Izv. Akad. Nauk SSSR, Ser. Neorg. Khim.* **4,** 447 (1965).
444. C. S. Fadley and D. A. Shirley, *Phys. Rev. A* **2,** 1109 (1970).
445. V. V. Nemoshkalenko, V. G. Aleshin, K. P. Tsomaya, and A. I. Senkevich, *Ukr. Fiz. Zh.* **11,** 1039 (1975).
446. D. N. Hendrickson, J. M. Hollander, and W. L. Jolly, *Inorg. Chem.* **9,** 612 (1970).
447. W. Bremser and F. Linneman, *Chem. Ztg.* **96,** 36 (1972).
448. C. J. Vesely and D. W. Langer, *Phys. Rev. B* **6,** 3770 (1972).
449. W. Gudat, E. E. Koch, P. Y. Yu, M. Cardona, and C. M. Penchina, *Phys. Stat. Sol.* **52**(16), 12, 505 (1972).
450. I. N. Shabanova, N. P. Sergushin, K. M. Kolobova, V. A. Trapeznikov, and V. I. Nefedov, *Fiz. Met. Metallov*, **34,** 1187 (1972).
451. C. J. Vesely and D. W. Langer, *Phys. Rev. B* **4**(1), 36 (1972).
452. J. C. Carver and G. K. Schweitzer, *J. Chem. Phys.* **57,** 973 (1972).
453. R. Kuentzler, *C. R. Acad. Sci., Paris*, **270,** 197 (1970).
454. L. Ramqvist, K. Hamrin, G. Johansson, U. Gelius, and C. Nordling, *J. Phys. Chem. Solids* **31**(12), 2669 (1970).
455. H. Ihara, E. Kumashiro, A. Iton, and K. Maeda, *Jpn. J. Appl. Phys.* **12**(9), 1462 (1973).
456. P. Auger, *J. Phys. Radium* **6,** 205 (1925).
457. M. A. Listengarten, *Izv. Akad. Nauk SSSR, Ser. Fiz.* **24**(9), 1050 (1960).

458. K. D. Sevier, *Low Energy Electron Spectroscopy*, Wiley Interscience, New York, 1972.
459. P. W. Palmberg, *Electron Spectroscopy*, D. A. Shirley, ed., North-Holland Publishing Company, Amsterdam-London, 1972, p. 835.
460. C. C. Chang, *Surf. Sci.* **25,** 53 (1971).
461. M. F. Chung, L. H. Jenkins, *Surf. Sci.* **22**(3), 479 (1970).
462. J. P. Coad and J. C. Rivière, *Z. Phys.* **244**(1), 19 (1971).
463. R. C. Oswald and R. E. Weber, Technical Report AFAL-TR-70-12 (1970).
464. H. E. Bishop and J. C. Rivière, *J. Appl. Phys.* **40**(4), 1740 (1969).
465. C. R. Brundle, *J. Electron Spectrosc. Relat. Phenom.* **5**(6), 291 (1974).
466. K. S. Kim, W. E. Baitinger, J. W. Amy, and N. Winograd, *J. Electron Spectrosc.* **5,** 351 (1974).
467. C. R. Brundle, *J. Vac. Sci. Tech.* **11,** 212 (1974).
468. S. T. Manson, *J. Electron Spectrosc. Relat. Phenom.* **1**(5), 413 (1972).
469. W. J. Carter, G. K. Schweitzer, and T. A. Carlson, *J. Electron Spectrosc. Relat. Phenom.* **5,** 827 (1974).
470. J. H. Seofield, Lawrence Livermore Lab. Report UCRL-51326, 1973.
471. C. D. Wagner, *Anal. Chem.* **44,** 1050 (1972).
472. C. R. Brundle and M. W. Roberts, *Proc. Roy. Soc. London, Ser. A.*, **331,** 383 (1972).
473. W. A. Fraser, J. V. Florino, W. N. Deglass, and W. D. Robertson, *Surf. Sci.* **36**(3), 661 (1973).
474. T. E. Madey, J. T. Yates, and N. E. Erickson, *Chem. Phys. Lett.* **19,** 487 (1973).
475. C. S. Fadley and S. Å. L. Bergström, *Phys. Lett. A* **35**(5), 375 (1971).
476. B. Henke, *Phys. Rev. A* **6**(1), 94 (1972).
477. K. Siegbahn, U. Gelius, H. Siegbahn, and E. Olsen, *Phys. Lett. A* **32**(4), 221 (1970).
478. C. S. Fadley and S. Å. L. Bergström, *Electron Spectroscopy*, D. A. Shirley, ed., North-Holland Publishing Company, Amsterdam-London, 1972, p. 233.
479. C. S. Fadley, R. J. Baird, W. Siekhaus, T. Novakov, and S. Å. L. Bergström, *J. Electron Spectrosc. Relat. Phenom.* **4**(1), 93 (1974).
480. C. S. Fadley, *J. Electron Spectrosc. Relat. Phenom.* **5,** 725 (1974).
481. S. S. Brinen, *J. Electron Spectrosc. Relat. Phenom.* **5,** 377 (1974).
482. C. S. Fadley, D. A. Shirley, A. J. Freeman, P. S. Bagus, and J. V. Mallow, *Phys. Rev. Lett.* **23**(24), 1397 (1969).
483. C. S. Fadley and D. A. Shirley, *Phys. Rev. A* **2**(4), 1109 (1970).
484. S. Hüfner and G. K. Wertheim, *Phys. Rev. B* **7**(6), 2333 (1973).
485. J. C. Carver, T. A. Carlson, L. C. Cain, and G. K. Schweitzer, *Electron Spectroscopy*, D. A. Shirley, ed., North-Holland Publishing Company, Amsterdam-London 1972, p. 803.
486. S. P. Kowalczyk, L. Ley, R. A. Pollak, F. R. McFeely, and D. A. Shirley, *Phys. Rev. B* **7**(8), 4009 (1973).
487. J. C. Carver, G. K. Schweitzer, and T. A. Carlson, *J. Chem. Phys.* **57**(2), 973 (1972).
488. F. R. McFeely, S. P. Kowalczyk, L. Ley, and D. A. Shirley, *Solid State Commun.* **15**(6), 1051 (1974).
489. G. K. Wertheim, R. L. Cohen, A. Rosencwaig, and H. J. Guggenheim, *Electron Spectroscopy*, D. A. Shirley, ed., North-Holland Publishing Company, Amsterdam-London, p. 813.
490. R. L. Cohen, G. K. Wertheim, A. Rosencwaig, and H. J. Guggenheim, *Phys. Rev. B* **5**(3), 1037 (1972).
491. F. R. McFeely, S. P. Kowalczyk, L. Ley, and D. A. Shirley, *Phys. Lett. A* **49**(4), 301 (1974).
492. F. R. McFeely, S. P. Kowalczyk, L. Ley, and D. A. Shirley, *Phys. Lett. A* **45,** 227 (1973).

493. A. Rosencwaig, G. K. Wertheim, and H. J. Guggenheim, *Phys. Rev. Lett.* **27**(8), 479 (1971).
494. B. Wallbank, I. G. Main, and C. E. Johnson, *J. Electron Spectrosc. Relat. Phenom.* **5**, 259 (1974).
495. T. A. Carlson, J. C. Carver, L. J. Saethre, F. G. Santibánez, and G. A. Vernon, *J. Electron Spectrosc. Relat. Phenom.* **5**, 247 (1974).
496. T. Novakov, *Phys. Rev. B* **3**(8), 2693 (1971).
497. L. Yin, I. Adler, T. Tsang, L. J. Matienso, and S. O. Grim, *Chem. Phys. Lett.* **24**(1), 81 (1974).
498. R. P. Gupta and S. K. Sen, *Phys. Rev. B* **10**(1), 71 (1974).
499. T. A. Carlson and C. W. Nestor, Jr., *Phys. Rev. A* **8**(6), 2887 (1973).
500. V. F. Demekhin, V. V. Nemoshkalenko, V. G. Aleshin, Yu. I. Bairachnyi, and V. L. Sukhorukov, in *Metallofizika*, Naukova Dumka, Kiev, 1975, pp. 27, 60.
501. S. Larson and J. W. D. Connolly, *Chem. Phys. Lett.* **20**, 323 (1973).
502. V. P. Sachenko, *Mnogoelektronnye Protsessy i Rentgenovskie Spektry*, Avtoref. Dokt. Dis., Rostov-na-Donu, 1974.
503. W. Bambynen, B. Crasemann, R. W. Fink, H. U. Freund, H. Mark, C. D. Swift, R. E. Price, and P. V. Rao, *Rev. Mod. Phys.* **44**, 716 (1972).
504. R. M. Friedman, J. Hudis, and M. L. Perlman, *Phys. Rev. Lett.* **29**(11), 692 (1972).
505. R. W. Shaw, Jr. and T. D. Thomas, *Phys. Rev. Lett.* **29**, 689 (1972).
506. P. H. Citrin, *Phys. Rev. Lett.* **31**(19), 1164 (1973).
507. S. Doniach and M. Sunjič, *J. Phys. C* **3**, 285 (1970).
508. P. Nozières and C. T. De Dominicis, *Phys. Rev.* **178**(4), 1097 (1969).
509. S. Hüfner and G. K. Wertheim, *Phys. Rev. B* **11**(2), 678 (1975).
510. N. J. Shevchik, *Phys. Rev. Lett.* **33**(22), 1336 (1974).

Index